CLASSICAL MECHANICS

Lecture Notes

CLASSICAL MECHANICS

Lecture Notes

Helmut Haberzettl

The George Washington University, Washington, D.C.

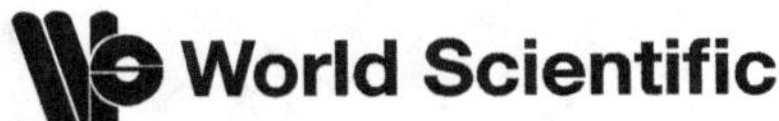 **World Scientific**

NEW JERSEY · LONDON · SINGAPORE · BEIJING · SHANGHAI · HONG KONG · TAIPEI · CHENNAI · TOKYO

Published by

World Scientific Publishing Co. Pte. Ltd.

5 Toh Tuck Link, Singapore 596224

USA office: 27 Warren Street, Suite 401-402, Hackensack, NJ 07601

UK office: 57 Shelton Street, Covent Garden, London WC2H 9HE

Library of Congress Control Number: 2021938270

British Library Cataloguing-in-Publication Data
A catalogue record for this book is available from the British Library.

CLASSICAL MECHANICS
Lecture Notes

ISBN 978-981-123-827-7 (hardcover)
ISBN 978-981-123-849-9 (paperback)
ISBN 978-981-123-828-4 (ebook for institutions)
ISBN 978-981-123-829-1 (ebook for individuals)

For any available supplementary material, please visit
https://www.worldscientific.com/worldscibooks/10.1142/12314#t=suppl

Desk Editors: George Vasu/Nur Syarfeena Binte Mohd Fauzi

Typeset by Stallion Press
Email: enquiries@stallionpress.com

Preface

The present notes are based on several graduate lecture courses on *Advanced Mechanics* I have given in our Ph.D. program at The George Washington University for a good number of years. They are *lecture notes* — that is, they provide a personal selection of topics which, I believe, gives a reasonably good introductory overview of the field *and* which can be covered in the 28, or so, 100-minute lectures of a single semester. Needless to say, they do not cover classical mechanics in its entirety. In view of the time limitations, a number of topics can only be treated at an elementary level and some do not appear at all. Moreover, since the target audience are first-year graduate students with primary interests ranging from experimental to theoretical physics, the present *Lecture Notes* provide a traditional picture of mechanics, without any reference to more advanced treatments in terms of differential geometry. Such treatments should be left to a special-topics course for mathematically inclined theory students.

Textbooks on classical mechanics usually provide far more material than can be covered in a one-semester lecture course. The resulting discrepancy between the limited selection presented to the students and the wealth of information found in textbooks often is confusing — sometimes even intimidating — to students. The present notes provide a self-contained, streamlined version of the topics actually covered in my course, to guide the students in their studies, and to free them of the need to take extensive notes during my lectures. Also, they are intended as a road map through the materials available in other textbooks for further studies. By making my lecture notes available now in the present book, I hope that this may also be useful to other students and instructors. However, I emphasize that the present lecture notes *do not* make the study of more comprehensive textbooks obsolete. Indeed, students are strongly urged not to rely on one single textbook in their course preparations. Being presented with the same facts from different points of view helps in gaining understanding, *and* it often prevents misunderstandings.

The present *Notes* have benefited a great deal from consulting a good number of existing textbooks while preparing, updating, and revising my lectures over the course of two decades or so. Over the years, however, I found myself coming back again and again to a few excellent textbooks that provide the backbone for much of the material found in the present *Notes*.

As a general, more comprehensive textbook, I recommend *Classical Mechanics* by Goldstein, Poole, and Safko [1], based on earlier editions by Goldstein [2]. The annotated, extensive bibliography of this book may be used as a guide for further studies. Additional highly recommended books — despite their somewhat dated appearances — are the classic texts by Landau and Lifshitz [3] and by Sommerfeld [4]. A mathematically advanced treatment is given by Arnold [5]. The text by Barger and Olsson [6] — albeit at an undergraduate level — is useful for its many practical examples. The section on deterministic chaos in Chapter 8 has benefited greatly from the very lucid presentation in Taylor's textbook [7]. For the study of more general mathematical questions, I recommend the book on mathematical physics by Hassani [8] that contains also a large number of interesting historical references. Another standard mathematical reference is the textbook by Arfken, Weber, and Harris [9]. In preparing the present notes I have also extensively used the textbooks by Fließbach [10], Budó [11], and Kuypers [12] available only in German. The latter, in particular, contains a large number of very useful, nontrivial problems with complete solutions. Last but not least, I have also drawn heavily on my own notes taken as a student sitting in on Professor Werner Sandhas's lecture on *Klassische Mechanik* at Mainz University [13].

I should add that it is *not* an oversight that the present *Notes* do not contain homework problems. Working through many and varied problems is indeed an indispensable learning step for students to achieve mastery of the subject matter, however, including sets of homework problems with textbooks nowadays is not useful because of the internet. Textbook homework sets usually become 'stale' very quickly because ready-made solutions for any particular textbook can be pulled from the net within a very short while. Instead, it is much more effective for instructors to find their own selections of problems suitable for their needs, allowing for greater flexibility and giving them better control over the level of difficulty of their courses.

Finally, I would like to thank all students and colleagues who have pointed out misprints, errors, and inconsistencies in previous iterations of the present *Notes*. I particularly would like to thank Professor Kanzo Nakayama from the University of Georgia for his many valuable and useful insights which have helped in improving the *Notes*. Needless to say, all errors that remain are entirely mine.

H. Haberzettl

January 2021

Suggested Reading

[1] H. Goldstein, C. P. Poole, and J. Safko, *Classical Mechanics*, 3rd edition (Addison-Wesley, San Francisco, CA, 2002).

[2] H. Goldstein, *Classical Mechanics*, 2nd edition (Addison-Wesley, Reading, MA, 1980).

[3] L. D. Landau and E. M. Lifshitz, *Course of Theoretical Physics, Vol. 1: Mechanics*, 3rd edition (Pergamon Press, Oxford, 1976).

[4] A. Sommerfeld, *Lecures on Theoretical Physics, Vol. 1: Mechanics* (Academic Press, New York, NY, 1952).

[5] V. I. Arnold, *Mathematical Methods of Classical Mechanics*, 2nd edition (Springer-Verlag, Berlin, 1989).

[6] V. D. Barger and M. G. Olsson, *Classical Mechanics — A Modern Perspective*, 2nd edition (McGraw-Hill, New York, NY, 1995).

[7] J. R. Taylor, *Classical Mechanics* (University Science Books, Sausalito, CA, 2005).

[8] S. Hassani, *Mathematical Physics — A Modern Introduction to Its Foundations*, 2nd edition (Springer-Verlag, New York, NY, 2013).

[9] G. B. Arfken, H. J. Weber, and F. E. Harris, *Mathematical Methods for Physicists: A Comprehensive Guide*, 7th edition (Academic Press, Waltham, MA, 2013).

Other References

[10] T. Fließbach, *Lehrbuch zur Theoretischen Physik I: Mechanik* (in German), 3rd edition (Spektrum Akademischer Verlag, Heidelberg, 1999).

[11] A. Budó, *Theoretische Mechanik* (in German), 6th edition (VEB Verlag der Wissenschaften, Berlin, 1971). Translated from the original Hungarian edition, *Mechanika Tankönyvkiadó* (Budapest, 1953).

[12] F. Kuypers, *Klassische Mechanik* (in German), 5th edition (Wiley-VCH, Weinheim, 1997).

[13] H. Haberzettl, *Vorlesungsmitschrift* of lecture class *Klassische Mechanik*, W. Sandhas, Johannes Gutenberg-Universität Mainz, Winter Semester 1971/72.

[14] *Isaac Newton's 'Philosophiae Naturalis Principia Mathematica'*, Facsimile of third edition (1726), Vol. I and II (in Latin), edited by I. Bernard Cohen (Harvard University Press, Cambridge, MA, 1972). The first edition was published 1687 in London.

Newer editions may be available for some of the textbooks listed here.

Contents

Preface v

1. Newtonian Mechanics   1

  1.1 Basic Kinematics 1

      1.1.1 Coordinate systems 2

      1.1.2 Trajectory of a point mass 6

  1.2 The Fundamental Laws of Mechanics 10

      1.2.1 Newton's axioms 11

      1.2.2 Newton's second law for variable mass 14

      1.2.3 Applying Newton's laws 16

  1.3 One-dimensional Motion 17

      1.3.1 Energy conservation 17

      1.3.2 Simple harmonic oscillation 19

      1.3.3 Harmonic oscillation with damping 21

      1.3.4 Forced oscillation 23

      1.3.5 Variable mass: Vertical launch of a rocket 24

      1.3.6 Green's function method 26

  1.4 Conservation Laws 27

      1.4.1 Momentum conservation 27

      1.4.2 Angular momentum conservation 27

      1.4.3 Energy conservation 29

  1.5 Systems of Point Particles 34

      1.5.1 Total momentum 35

      1.5.2 Total angular momentum 36

      1.5.3 Total energy 38

      1.5.4 The ten integrals of motion 41

  1.6 Inertial Frames of Reference 42

      1.6.1 Galilei transformations 42

      1.6.2 Closed systems 51

  1.7 Accelerated Frames of Reference 52

      1.7.1 Linearly accelerated frame 52

      1.7.2 Rotating frame 53

1.8	The Central-Force Problem	59
	1.8.1 The two-body problem	59
	1.8.2 The Kepler problem	65
	1.8.3 Newton's cannonball	73
	1.8.4 The restricted three-body problem	77
	1.8.5 Scattering in a central-force field	80

2. Lagrangian Mechanics   **94**

2.1	Lagrange Equations of the First Kind	94
	2.1.1 Constraints	95
	2.1.2 Forces of constraint	96
	2.1.3 Equations of motion with constraints	100
	2.1.4 Virtual work: The d'Alembert Principle	102
	2.1.5 Energy conservation	103
	2.1.6 Applications	104
2.2	Lagrange Equations of the Second Kind	108
	2.2.1 Generalized coordinates	108
	2.2.2 Lagrange function and equations of motion	111
	2.2.3 Conservation laws	135
	2.2.4 Summary of Lagrange equations	138

3. Variational Principles   **141**

3.1	Basics of Variational Calculus	141
	3.1.1 Variation without constraints	143
	3.1.2 Variation with constraints	148
3.2	Hamilton's Principle	157
	3.2.1 Gauge transformations	159
	3.2.2 Equivalence of Hamilton and d'Alembert	161
	3.2.3 Equivalence to Lagrange equations with differential constraints	162
3.3	Noether's Theorem	163
	3.3.1 Generalized Noether Theorem	166

4. Hamiltonian Mechanics   **169**

4.1	Hamilton's Equations	169
	4.1.1 Conservation laws	173
	4.1.2 Hamilton's principle revisited	173
	4.1.3 Poisson brackets	174
	4.1.4 Legendre transformations	176

4.2   Canonical Transformations 176
     4.2.1   Infinitesimal canonical transformations 181
4.3   Liouville's Theorem 182
     4.3.1   Hamiltonian systems 182
     4.3.2   Dissipative systems 188
4.4   Hamilton–Jacobi Equations 189
     4.4.1   Time-independent Hamilton function 191
     4.4.2   Separation of variables 191
     4.4.3   Examples 193

5. Mechanics of Rigid Bodies   **196**

5.1   Kinematics 196
     5.1.1   Rotations: orthogonal transformations 197
     5.1.2   Angular velocities 207
5.2   Inertia Tensor 210
     5.2.1   Kinetic energy 210
     5.2.2   Similarity transformation 212
     5.2.3   Parallel-axis theorem 214
     5.2.4   Angular momentum 215
     5.2.5   Principal-axis transformation 216
5.3   Euler's Equations 219
     5.3.1   Torque-free rotation 221
     5.3.2   The torque-free symmetric top 223
5.4   The Heavy Top 226
     5.4.1   The heavy symmetric top 227
     5.4.2   Earth as a rotating top 230

6. Small Oscillations   **237**

6.1   Systems with One Degree of Freedom 237
6.2   Systems with Many Degrees of Freedom 239
     6.2.1   Normal coordinates 244
     6.2.2   External forces 246
6.3   Applications 247
     6.3.1   Two-dimensional motion 247
     6.3.2   Coupled pendula 249
     6.3.3   Vibrational modes of a triatomic molecule 251

7. Continuum Mechanics   254

  7.1   Vibrating String   254
         7.1.1   Solving the one-dimensional wave equation   258
         7.1.2   Generalizations   260
  7.2   Beam in an External Force Field   261
         7.2.1   Static sagging in the gravitational field   264
  7.3   Fluid Dynamics   266
         7.3.1   Continuity equation   267
         7.3.2   Euler's equation of fluid dynamics   268
         7.3.3   Equation of state   270
         7.3.4   Examples   271
         7.3.5   Sound waves   277

8. Beyond Classical Mechanics   279

  8.1   Chaotic Systems   279
         8.1.1   The road to chaos   280
         8.1.2   Lyapunov exponent   286
         8.1.3   Phase-space trajectories and Poincaré maps   289
         8.1.4   The logistic map   293
         8.1.5   Nonlinear systems are normal   297
  8.2   Special Relativity   297
         8.2.1   The Einstein Principle of Relativity   298
         8.2.2   Four-dimensional spacetime   302
         8.2.3   Lorentz transformations defined   306
         8.2.4   Homogeneous Lorentz group $O(1,3)$ 311
         8.2.5   Constructing the Lorentz transformation   313
         8.2.6   Rotations and boosts   317
         8.2.7   Kinematics   322
         8.2.8   Relativistic point-particle mechanics   335

Appendices   344

  A   Coordinates, Vector Operations, etc.   344
         A.1   Cartesian coordinates x, y, z   344
         A.2   Cylindrical coordinates ρ, ϕ, z   345
         A.3   Spherical coordinates r, φ, θ   346
         A.4   Arbitrary curvilinear coordinates q_1, q_2, q_3   347

A.5	Vector-operator identities	349
A.6	Integral theorems	349
B	Dirac δ Distribution	350
C	Green's Function Method: An Example	354
Index		359

Chapter 1

Newtonian Mechanics

Classical Mechanics describes the rules and laws governing the nonrelativistic motion of material bodies, where motion is defined as the change of location as a function of time. This motion is subject to forces that are presumed to be known. The mere description of the motion of a body is called *kinematics*. The study of the underlying laws that govern a mechanical motion is called *dynamics*. A large part of mechanics deals with point masses and systems of point masses, where the bodies are idealized as point-like objects without spatial extension, with only one intrinsic property, called the mass m.

1.1 Basic Kinematics

It is assumed that time and length measurements are well defined operations that, in principle, can be refined to any degree of accuracy. In classical mechanics, time is a mere parameter that is taken to be the same everywhere. In other words, it is presumed that we can define the state of the universe 'at any given time' like we would take a snap-shot photograph of a group of people in a room. We know now that this is not true. This is only an acceptable approximation of physical reality for velocities small compared to the speed of light and for physical phenomena that take place in a region of space small enough so that a light beam traversing it takes a negligible time compared to the typical times associated with the physical phenomena we are interested in. For typical mechanical phenomena taking place on Earth and in the surrounding space, this is an excellent approximation. For example, even though light takes about 8.3 min to travel from Earth to the Sun, which may seem a long time, it takes Earth one year to go around the Sun, at a speed of about 30 km/s. Compared to the speed of light, $c \approx 300\,000$ km/s, this is four orders of magnitude smaller.

We note in this context that the speed of light in vacuum is actually *defined* to be given *exactly* by $c = 299\,792\,458$ m/s. With this fixed value for c, the length unit of a meter follows from the definition of the second as $9\,192\,631\,770$ periods associated with the radiative transition between two given states of a cesium-133 atom. In other words, the fundamental units of length and time are defined by *one* measurement *and* the axiom of special relativity that the speed of light c has the same constant value in all frames of reference.

1.1.1 *Coordinate systems*

In mechanics, the three-dimensional space in which the motion takes place is assumed to be *flat*; such a space is called a *Euclidean* space. Any point in the three-dimensional Euclidean space can be described by its location relative to a common, fixed reference point, called the *origin O*. Each point can thus be characterized uniquely by its distance r from that origin and by the direction in space that one must follow to reach this point from the origin. This can be described concisely by a *vector* $\mathbf{r}$, which can be thought of as an arrow whose base is at the origin and whose tip is at the point of interest (see Fig. 1.1). The length, or magnitude, of that arrow corresponds to the point's distance from the origin; one writes this as $r = |\mathbf{r}|$.

One should note, however, that the introduction of an origin is a matter of convenience but not of necessity. Physical space *per se* does *not* possess a unique origin. Physical space is a so-called *affine space*[1] where the difference between two points can be expressed as a vector, a vector can be added to any point to obtain another point, but points cannot be added (or subtracted) from one another.

Hence, with respect to a different (fixed) reference point O', located a constant distance $\mathbf{d}$ from the first origin O, the location $\mathbf{r}$ is given by a simple translation,

$$\mathbf{r}' = \mathbf{r} - \mathbf{d} \tag{1.1}$$

(see Fig. 1.1). Note that $\mathbf{r}$ and $\mathbf{r}'$ describe the same point in space. The physics happening at that point should not depend on which point is chosen as the reference point, i.e., space is presumed to be *homogeneous* and all basic laws of physics must be *translationally invariant*.

This is of course notwithstanding the fact that for a given problem, it may be very advantageous to choose one particular reference point over

---

[1]See the mathematical literature for the details of affine vector spaces.

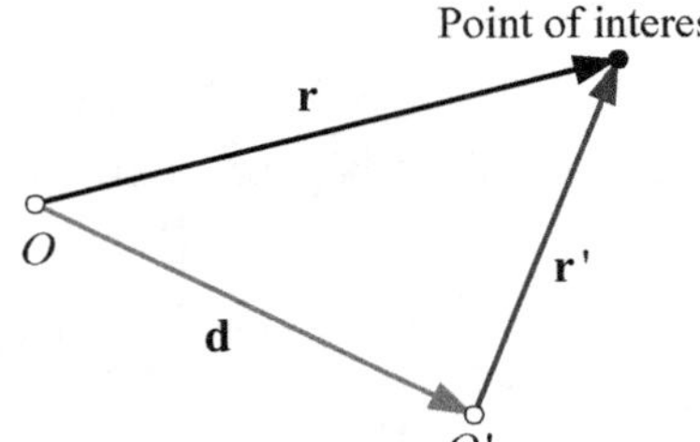

Fig. 1.1   Point in space as seen from origin O (vector $\mathbf{r}$) and as seen from origin O' (vector $\mathbf{r}'$).

any other. For example, for a problem exhibiting spherical symmetry about one particular point, one would naturally choose that point as the reference point since this immediately reduces the problem to one where only distances from the center of symmetry play any role.

For practical purposes, it is often convenient to consider a representation of the abstract vector $\mathbf{r}$ in a so-called *Cartesian* coordinate system of the three-dimensional coordinate space $\mathbb{R}^3$ defined by three (dimensionless) unit vectors $\mathbf{e}_1$, $\mathbf{e}_2$, and $\mathbf{e}_3$ that provide three orthogonal axes for the physical space satisfying[2]

$$\mathbf{e}_i \cdot \mathbf{e}_j = \delta_{ij}, \tag{1.2}$$

$$\mathbf{e}_i \times \mathbf{e}_j = \sum_{k=1}^{3} \varepsilon_{ijk}\mathbf{e}_k, \tag{1.3}$$

where the last equation signifies a conventional right-handed coordinate system. The Kronecker delta is given by

$$\delta_{ij} = \begin{cases} 1, & \text{if } i = j, \\ 0, & \text{if } i \neq j, \end{cases} \tag{1.4}$$

and for the epsilon symbol one has

$$\varepsilon_{ijk} = \begin{cases} +1, & ijk \text{ cyclic permutation of 123,} \\ -1, & ijk \text{ anti-cyclic permutation of 123,} \\ 0, & \text{else.} \end{cases} \tag{1.5}$$

With the help of the given unit vectors, any vector $\mathbf{r}$ can then be represented in terms of three numbers x_1, x_2, and x_3 (carrying the dimension of a length) according to

$$\mathbf{r} = x_1\mathbf{e}_1 + x_2\mathbf{e}_2 + x_3\mathbf{e}_3 = \sum_{i=1}^{3} x_i\mathbf{e}_i. \tag{1.6}$$

---

[2]It is assumed here that the notions of scalar product ("·") and vector product (" $\times$ ") for a Euclidean vector space are known concepts. However, one may actually take Eqs. (1.2) through (1.5) as the definitions of these products for $\mathbb{R}^3$ given the expansion (1.6).

where

$$x_i = \mathbf{r} \cdot \mathbf{e}_i \tag{1.7}$$

is the i-th *component* of the vector $\mathbf{r}$ in the Cartesian coordinate system. In other words,

$$\mathbf{r} = \sum_{i=1}^{3} (\mathbf{r} \cdot \mathbf{e}_i)\mathbf{e}_i = \mathbf{r} \cdot \sum_{i=1}^{3} \mathbf{e}_i\mathbf{e}_i = \mathbf{r} \cdot \mathbf{1}, \tag{1.8}$$

where

$$\mathbf{1} = \sum_{i=1}^{3} \mathbf{e}_i\mathbf{e}_i \tag{1.9}$$

is a dyadic representation of the unit operator, whose left- or right-handed action on a vector is defined by the scalar product, as in (1.8) [see also Eq. (1.20)]. The unit operator (1.9) is independent of the particular choice $\{\mathbf{e}_1, \mathbf{e}_2, \mathbf{e}_3\}$ of the coordinate system.[3]

For a given Cartesian coordinate system for $\mathbb{R}^3$, it is possible, therefore, to represent the vector $\mathbf{r}$ by a three-dimensional column vector,

$$\mathbf{r} \longrightarrow \vec{r} = \begin{pmatrix} x_1 \\ x_2 \\ x_3 \end{pmatrix}, \tag{1.10}$$

since there is a one-to-one correspondence between the three numbers $\{x_1, x_2, x_3\}$ and the vector $\mathbf{r}$ according to (1.6). In this three-dimensional representation, the length r of the vector $\mathbf{r}$ is given by

$$\mathbf{r} \cdot \mathbf{r} = r^2 = \vec{r}^{\mathrm{T}}\vec{r} = \begin{pmatrix} x_1 \\ x_2 \\ x_3 \end{pmatrix}^{\mathrm{T}} \begin{pmatrix} x_1 \\ x_2 \\ x_3 \end{pmatrix} = x_1^2 + x_2^2 + x_3^2, \tag{1.11}$$

where T stands for 'transposed',

$$\vec{r}^{\mathrm{T}} = \begin{pmatrix} x_1 \\ x_2 \\ x_3 \end{pmatrix}^{\mathrm{T}} = (x_1, x_2, x_3), \tag{1.12}$$

which corresponds to the row-vector representation of $\mathbf{r}$.

---

[3] A *dyad*, or *dyadic product*, $\mathbf{D} = \mathbf{a}\mathbf{b}$ (sometimes written as $\mathbf{D} = \mathbf{a} \otimes \mathbf{b}$) is an ordered product of two vectors $\mathbf{a}$ and $\mathbf{b}$ defined by its operation on another vector $\mathbf{c}$, where the left- and right-handed scalar multiplications, $\mathbf{D} \cdot \mathbf{c} = \mathbf{a}(\mathbf{b} \cdot \mathbf{c})$ and $\mathbf{c} \cdot \mathbf{D} = (\mathbf{c} \cdot \mathbf{a})\mathbf{b}$, produce vectors that are aligned with $\mathbf{a}$ or $\mathbf{b}$, respectively. Expanding the vectors $\mathbf{a}$ and $\mathbf{b}$ by the unit vectors $\mathbf{e}_i$ of some Cartesian frame, we have $\mathbf{D} = (\sum_i a_i\mathbf{e}_i)(\sum_j b_j\mathbf{e}_j) = \sum_{i,j} \mathbf{e}_i D_{ij}\mathbf{e}_j$, where the $D_{ij} = a_i b_j$ are the elements of the matrix representation $D = \{D_{ij}\}$ of the dyad operator $\mathbf{D}$. More generally, any linear matrix-type operator D on $\mathbb{R}^3$ can be expanded in terms of dyadic products of unit vectors, $\mathsf{D} = \sum_{i,j} \mathbf{e}_i D_{ij}\mathbf{e}_j$, where D_{ij} does not factorize. For the unit dyad $D_{ij} \to \delta_{ij}$. (Sometimes the notation $\overleftrightarrow{D} \equiv \mathsf{D}$ is employed to indicate that the dyad acts like a vector to the left and to the right.)

Moreover, abstract linear transformations A of the form

$$\mathbf{r}' = \mathsf{A}\mathbf{r} \tag{1.13}$$

that map the vector $\mathbf{r}$ into another vector $\mathbf{r}'$ may be expanded in terms of dyadic products of unit vectors,

$$\mathsf{A} = \sum_{i,j=1}^{3} \mathbf{e}_i\, a_{ij}\, \mathbf{e}_j. \tag{1.14}$$

The mapping (1.13) then may be written as

$$\mathsf{A}\mathbf{r} = \mathsf{A}\cdot\mathbf{r} = \sum_{i,j=1}^{3} \mathbf{e}_i a_{ij}(\mathbf{e}_j\cdot\mathbf{r}) = \sum_{i,j=1}^{3} \mathbf{e}_i a_{ij} x_j = \sum_{i=1}^{3}\mathbf{e}_i \underbrace{\sum_{j=1}^{3} a_{ij}x_j}_{=x_i'}, \tag{1.15}$$

where

$$x_i' = \sum_{j=1}^{3} a_{ij}x_j, \quad \text{for} \quad i = 1,2,3, \tag{1.16}$$

determines the Cartesian components of $\mathbf{r}'$. The 3×3 matrix representation of A,

$$\mathsf{A} \quad \longrightarrow \quad A = \begin{pmatrix} a_{11} & a_{12} & a_{13} \\ a_{21} & a_{22} & a_{23} \\ a_{31} & a_{32} & a_{33} \end{pmatrix}, \tag{1.17}$$

thus leads to the matrix relation

$$\vec{r}' = A\vec{r} \quad \longrightarrow \quad \begin{pmatrix} x_1' \\ x_2' \\ x_3' \end{pmatrix} = \begin{pmatrix} a_{11} & a_{12} & a_{13} \\ a_{21} & a_{22} & a_{23} \\ a_{31} & a_{32} & a_{33} \end{pmatrix} \begin{pmatrix} x_1 \\ x_2 \\ x_3 \end{pmatrix} \tag{1.18}$$

representing the abstract mapping (1.13).

The identity matrix is denoted by

$$\mathbb{1} = \begin{pmatrix} 1 & 0 & 0 \\ 0 & 1 & 0 \\ 0 & 0 & 1 \end{pmatrix}, \tag{1.19}$$

i.e., its elements are $\mathbb{1}_{ij} = \delta_{ij}$. They are related to the unit operator via

$$\mathbf{e}_i \cdot \mathbf{1} \cdot \mathbf{e}_j = \sum_{k=1}^{3}(\mathbf{e}_i\cdot\mathbf{e}_k)(\mathbf{e}_k\cdot\mathbf{e}_j) = \sum_{k=1}^{3}\delta_{ik}\delta_{kj} = \delta_{ij} = \mathbb{1}_{ij}. \tag{1.20}$$

In other words, the identity matrix $\mathbb{1}$ is the Cartesian representation of the (abstract) unit operator $\mathbf{1}$ the same way that $\vec{r}$ is a representation of $\mathbf{r}$.

The Cartesian coordinate system is just one of many possible ways of representing a vector. In the following other coordinate systems will be considered when appropriate.

While, strictly speaking, one should always distinguish between an abstract vector and its representation, we may not always do so in the following if what is meant is obvious from the context. Specifically, while a letter with an overarrow, like $\vec{r}$, will always refer to a representation of the abstract vector denoted by a bold-face letter, like $\mathbf{r}$, the latter may, on occasion, also be used for its representation.

1.1.2   *Trajectory of a point mass*

The motion of a point mass is described by a vector function $\mathbf{r}(t)$ for the body's location at any given time t. In a Cartesian coordinate system this *trajectory* can be represented as

$$\mathbf{r}(t) = x_1(t)\,\mathbf{e}_1 + x_2(t)\,\mathbf{e}_2 + x_3(t)\,\mathbf{e}_3 = \sum_{i=1}^{3} x_i(t)\,\mathbf{e}_i, \tag{1.21}$$

where the components x_i are functions of time. Other representations are possible. For example, instead of (x_1, x_2, x_3) we may choose *cylindrical* coordinates (ρ, ϕ, z) defined by

$$\left. \begin{array}{l} \rho = \sqrt{x_1^2 + x_2^2}, \\[4pt] \phi = \arctan \frac{x_2}{x_1}, \\[4pt] z = x_3, \end{array} \right\} \quad \Longleftrightarrow \quad \left\{ \begin{array}{l} x_1 = \rho \cos\phi, \\[4pt] x_2 = \rho \sin\phi, \\[4pt] x_3 = z, \end{array} \right. \tag{1.22}$$

or *spherical* coordinates (r, φ, θ) given by

$$\left. \begin{array}{l} r = \sqrt{x_1^2 + x_2^2 + x_3^2}, \\[4pt] \varphi = \arctan \frac{x_2}{x_1}, \\[4pt] \theta = \arctan \frac{\sqrt{x_1^2 + x_2^2}}{x_3}, \end{array} \right\} \quad \Longleftrightarrow \quad \left\{ \begin{array}{l} x_1 = r \sin\theta \, \cos\varphi, \\[4pt] x_2 = r \sin\theta \, \sin\varphi, \\[4pt] x_3 = r \cos\theta. \end{array} \right. \tag{1.23}$$

These coordinates are shown in Fig. 1.2. Moreover, the time dependence of the coordinate variables may be replaced by a parametric representation where $t = t(\lambda)$ is a function of some parameter λ.

As mentioned, in principle, we may use any set of coordinates to describe the particle trajectories. However, we emphasize that the Cartesian coordinates (and their variants obtained by trivial curvilinear or non-orthogonal

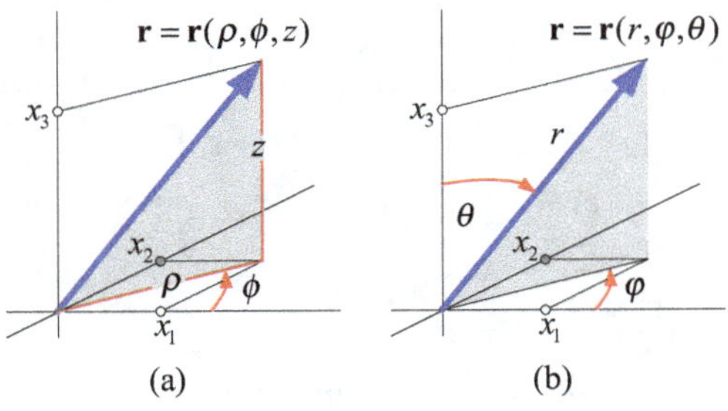

Fig. 1.2 (a) Cylindrical coordinates and (b) spherical coordinates.

distortions) are distinguished by the fact that every continuous trajectory *always* corresponds to continuously changing coordinates. By contrast, for cylindrical coordinates, for example, a continuous motion through the origin corresponds to a discontinuous jump of ϕ by π. This exemplifies that the use of non-Cartesian coordinates may lead to spurious unphysical singularities associated with such discontinuities whenever derivatives of the coordinates are involved.

Appendix A provides a listing of various coordinates, their relationships, and how to express various operations (gradient, divergence, curl, etc.) in these coordinates.

1.1.2.1 *Velocity and acceleration*

The *velocity* $\mathbf{v}$ of a point mass, given by

$$\mathbf{v}(t) = \frac{d\mathbf{r}}{dt} = \lim_{\Delta t \to 0} \frac{\mathbf{r}(t + \Delta t) - \mathbf{r}(t)}{\Delta t}, \tag{1.24}$$

describes the rate at which a body changes its location, with Cartesian components

$$\mathbf{v}(t) = v_1(t)\, \mathbf{e}_1 + v_2(t)\, \mathbf{e}_2 + v_3(t)\, \mathbf{e}_3. \tag{1.25}$$

The corresponding magnitude is

$$|\mathbf{v}| = v = \sqrt{v_1^2 + v_2^2 + v_3^2}. \tag{1.26}$$

A dot is used to denote the time derivative, i.e.,

$$\dot{\mathbf{r}}(t) = \frac{d\mathbf{r}}{dt} = \mathbf{v}, \qquad \text{or} \qquad \dot{x}_i = \frac{dx_i}{dt} = v_i. \tag{1.27}$$

The infinitesimal path-length increment ds along the trajectory associated with the time increment $dt \ (> 0)$ is related to the magnitude v by

$$ds = |d\mathbf{r}| = \frac{|d\mathbf{r}|}{dt} dt = \left| \frac{d\mathbf{r}}{dt} \right| dt = v\, dt \quad \Rightarrow \quad v = \frac{ds}{dt}. \tag{1.28}$$

The *acceleration* $\mathbf{a}$ is defined by

$$\mathbf{a}(t) = \frac{d\mathbf{v}}{dt} = \lim_{\Delta t \to 0} \frac{\mathbf{v}(t + \Delta t) - \mathbf{v}(t)}{\Delta t}, \tag{1.29}$$

with Cartesian components,

$$\mathbf{a}(t) = a_1(t)\,\mathbf{e}_1 + a_2(t)\,\mathbf{e}_2 + a_3(t)\,\mathbf{e}_3, \tag{1.30}$$

related to the components of $\mathbf{v}$ and $\mathbf{r}$ via

$$\dot{v}_i = \frac{dv_i}{dt} = \frac{d^2 x_i}{dt^2} = \ddot{x}_i = a_i, \tag{1.31}$$

where the double-dot on x_i signifies the second-order time derivative.

When performing the time derivatives here, it is important to note that the Cartesian unit vectors $\mathbf{e}_i$ are fixed in space and therefore time-independent. This is in contrast to curvilinear coordinates where the unit vectors depend on the position.

This is easily illustrated for the two-dimensional motion of a particle. In the xy-plane, the corresponding infinitesimal position element $d\mathbf{r}$ is given by[4]

$$d\mathbf{r} = dx\,\mathbf{e}_x + dy\,\mathbf{e}_y$$

$$= \left(\frac{\partial x}{\partial r} dr + \frac{\partial x}{\partial \varphi} d\varphi \right) \mathbf{e}_x + \left(\frac{\partial y}{\partial r} dr + \frac{\partial y}{\partial \varphi} d\varphi \right) \mathbf{e}_y$$

$$= dr\,\mathbf{e}_r + r\,d\varphi\,\mathbf{e}_\varphi, \tag{1.32}$$

which provides

$$\mathbf{e}_r = \frac{\partial x}{\partial r}\mathbf{e}_x + \frac{\partial y}{\partial r}\mathbf{e}_y = \cos\varphi\,\mathbf{e}_x + \sin\varphi\,\mathbf{e}_y, \tag{1.33a}$$

$$\mathbf{e}_\varphi = \frac{1}{r}\left(\frac{\partial x}{\partial \varphi}\mathbf{e}_x + \frac{\partial y}{\partial \varphi}\mathbf{e}_y \right) = -\sin\varphi\,\mathbf{e}_x + \cos\varphi\,\mathbf{e}_y, \tag{1.33b}$$

and therefore

$$\dot{\mathbf{e}}_r = \dot{\varphi}\,\mathbf{e}_\varphi, \quad \dot{\mathbf{e}}_\varphi = -\dot{\varphi}\,\mathbf{e}_r. \tag{1.34}$$

For the (planar) trajectory

$$\mathbf{r} = x(t)\,\mathbf{e}_x + y(t)\,\mathbf{e}_y = r(t)\,\mathbf{e}_r(t) \tag{1.35}$$

this means [cf. Eq. (1.32)]

$$\mathbf{v} = \frac{d\mathbf{r}}{dt} = \dot{r}\,\mathbf{e}_r + r\,\dot{\varphi}\,\mathbf{e}_\varphi, \tag{1.36}$$

and

$$\mathbf{a} = \ddot{\mathbf{r}} = \frac{d\mathbf{v}}{dt} = \left(\ddot{r} - r\dot{\varphi}^2 \right)\mathbf{e}_r + \left(r\ddot{\varphi} + 2\dot{r}\dot{\varphi} \right)\mathbf{e}_\varphi. \tag{1.37}$$

In these planar coordinates, the magnitude of the velocity is given by

$$\mathbf{v}^2 = v^2 = \dot{r}^2 + r^2\dot{\varphi}^2. \tag{1.38}$$

---

[4]The notation x, y, z is used interchangeably with x_1, x_2, x_3 for the Cartesian components.

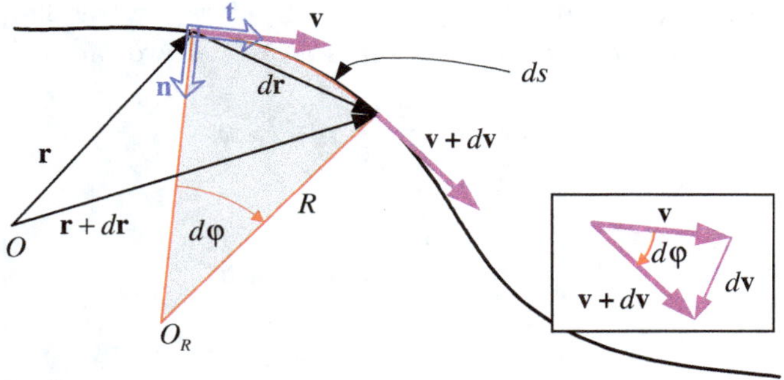

Fig. 1.3   Incremental change of trajectory from $\mathbf{r} \to \mathbf{r} + d\mathbf{r}$ with associated velocity change $\mathbf{v} \to \mathbf{v} + d\mathbf{v}$. The local curvature of the path is measured by the radius R of the inscribed circle around O_R. The corresponding line segment along the trajectory is $ds = R\,d\varphi$ where $d\varphi$ is the opening angle of the circle sector (shaded area) whose periphery is ds.

Tangential and centripetal acceleration

Introducing

$$\mathbf{t} = \frac{\mathbf{v}}{v},\tag{1.39}$$

this unit vector points along the tangent of the trajectory at the particle's location (see Fig. 1.3). In view of

$$v\mathbf{t} = \mathbf{v} = \frac{d\mathbf{r}}{dt} = \frac{d\mathbf{r}}{ds}\frac{ds}{dt} = v\frac{d\mathbf{r}}{ds},\tag{1.40}$$

it can also be defined by

$$\mathbf{t} = \frac{d\mathbf{r}}{ds}.\tag{1.41}$$

The acceleration is then expressed as

$$\mathbf{a} = \frac{d\mathbf{v}}{dt} = \dot{v}\mathbf{t} + v\dot{\mathbf{t}},\tag{1.42}$$

with $\dot{\mathbf{t}} \perp \mathbf{t}$, as seen immediately from the time derivative of the normalization,

$$\mathbf{t}\cdot\mathbf{t} = 1 \quad\xrightarrow{\text{apply } d/dt}\quad 2\dot{\mathbf{t}}\cdot\mathbf{t} = 0.\tag{1.43}$$

Consulting Fig. 1.3, we see that as $\mathbf{v}$ changes to $\mathbf{v} + d\mathbf{v}$, the tangent unit vector changes from $\mathbf{t}$ to $\mathbf{t} + d\mathbf{t}$, with $|\mathbf{t}| = |\mathbf{t} + d\mathbf{t}| = 1$ and $|d\mathbf{t}| = d\varphi$. With the normal unit vector $\mathbf{n}$,

$$d\mathbf{t} = |d\mathbf{t}|\,\mathbf{n} = d\varphi\,\mathbf{n},\tag{1.44}$$

which is perpendicular to the trajectory and points toward the point O_R from which the *local curvature radius* R is measured, we can now write

$$\dot{\mathbf{t}} = \frac{d\varphi}{dt}\,\mathbf{n} = \frac{1}{R}\frac{ds}{dt}\,\mathbf{n} = \frac{v}{R}\,\mathbf{n}, \tag{1.45}$$

and therefore

$$\mathbf{a} = \dot{v}\mathbf{t} + \frac{v^2}{R}\mathbf{n} = a_t\mathbf{t} + a_n\mathbf{n}, \tag{1.46}$$

where

$$a_t = \dot{v} \qquad \text{and} \qquad a_n = \frac{v^2}{R} \tag{1.47}$$

are the *tangential*, or *orbital*, and the *normal*, or *centripetal*, accelerations, respectively.

1.2   The Fundamental Laws of Mechanics

Physical theories assume certain unproven, fundamental laws which are formulated as a result of observations. These laws should be as simple as possible and they should allow one to describe a class of phenomena as large as possible.[5] Laws of this kind are referred to as *laws of nature*. In physics, the description of phenomena in terms of such laws — once well established — is synonymous with their explanation. A law of nature implies infinitely many predictions. Experiments that show (within acceptable experimental errors) these predictions to be correct provide corroboration of the general validity of the law. However, in view of the necessarily finite number of observations, a law of nature *cannot* be proved in this manner. Confidence in the general validity of a law is generated by an ever increasing number of independent observations that either can be explained by the law or follow directly from predictions based on the law.[6] However, a single — *repeatable* — observation in contradiction to such a prediction will immediately *falsify* the law — or, at the very least, limit the range of applicability of the law.

---

[5]The guiding principles of simplicity on the one hand and broad applicability on the other for the formulation of physical laws are sometimes referred to as *Occam's razor*, attributed to William of Ockham (1285–1349). In fact, however, this principle had been invoked already by Aristotle (384–322 BCE) and it is perhaps even older than that.

[6]This process is sometimes referred to as *verification* of the law. This is perhaps an unfortunate choice of words since a finite number of observations can never absolutely ascertain that the law is indeed true.

1.2.1 *Newton's axioms*

In Mechanics, the three axioms formulated by Isaac Newton (1643-1727) in his monumental work *Philosophiae Naturalis Principia Mathematica* (London, 1687; see also [14]) may be taken as forming a set of laws of nature governing mechanical phenomena. As we now know, their general validity is restricted to phenomena at velocities small compared to the speed of light.

The *first axiom* deals with frames of reference.

$$
\boxed{\begin{array}{ll} \textbf{\textit{Lex}} \\ \textbf{\textit{Prima}} \end{array} \quad \begin{array}{l} \text{There exist frames of reference in which the} \\ \text{free motion (i.e., motion not subject to} \\ \text{forces) is described by} \quad \dot{\mathbf{r}}(t) = \mathbf{v} = const. \end{array}} \tag{1.48}
$$

Such frames are called *inertial* frames of reference. Inertial frames are distinguished from others because the laws of physics become particularly simple when formulated in such frames. An example of a non-inertial frame would be a rotating frame (in which a force-free straight-line motion appears curved). Force-free motion in compliance with the first axiom is said to be *uniform*.

Non-uniform motion in an inertial frame of reference arises from a force $\mathbf{F}$, as stated by the *second axiom*,

$$
\boxed{\begin{array}{ll} \textbf{\textit{Lex}} \\ \textbf{\textit{Secunda}} \end{array} \quad \mathbf{F} = \frac{d\mathbf{p}}{dt} \quad \text{(inertial frame)}} \tag{1.49}
$$

where, generically, the *momentum* $\mathbf{p}$ is the product of the mass m and its velocity,

$$
\mathbf{p} = m\mathbf{v}. \tag{1.50}
$$

It is crucial here that $\mathbf{F}$ is the total *net* force (including dissipative forces) acting on the body. In particular, this means that the system needs to be *closed* such that all forces on all mass contributions of the system are accounted for for all times. The way forces are to be added is given by the *superposition of forces*, which serves as an amendment to the second axiom:

$$
\boxed{\begin{array}{ll} \textbf{\textit{Superposition}} \\ \textbf{\textit{Principle}} \end{array} \quad \begin{array}{l} \text{If several forces } \mathbf{F}_i \text{ act on a body,} \\ \text{the total force } \mathbf{F} \text{ is the vector sum of} \\ \text{all individual forces:} \\ \qquad \mathbf{F} = \sum_i \mathbf{F}_i. \end{array}} \tag{1.51}
$$

The superposition principle is often also called *Newton's fourth axiom*.

The requirement of a closed system for the second axiom can, in principle, always be satisfied by making the system large enough to include all dynamically relevant components. However, this is not the most economical way of treating a system with variable mass where the main system of interest gains or loses mass. (An example for mass loss is a rocket burning off fuel.) Accounting for the incremental mass contributions dm that provide the gains or losses in terms of the closed-system scenario of the full second axiom (1.49) would be too cumbersome (or even impossible) to describe in purely mechanical terms. Instead, one may write the second axiom in the form

$$\mathbf{F} = \frac{\Delta \mathbf{p}}{dt}, \tag{1.52}$$

where $\mathbf{F}$ is the total net force on the main system alone and

$$\Delta \mathbf{p} = \mathbf{p}_f - \mathbf{p}_i \tag{1.53}$$

is the difference between final and initial momenta, $\mathbf{p}_f$ and $\mathbf{p}_i$, respectively, across an infinitesimal time increment dt. Clearly, in this description $\Delta \mathbf{p} \neq d\mathbf{p} \equiv d(m\mathbf{v})$ because the main system is not closed and 'leaks' momentum via incremental mass changes dm. More details of this procedure are discussed in Sec. 1.2.2.

If the mass does not change during the motion, the second axiom may be expressed simply as

$$\mathbf{F} = m\ddot{\mathbf{r}} \quad \text{(constant mass)}. \tag{1.54}$$

This form of the second axiom introduces the notion of an *inertial mass* m that may be thought of as an intrinsic property of the body in question describing how much resistance the body offers to changing its state of motion (i.e., for a given — fixed — force, a larger mass is subject to a smaller acceleration, and vice versa).

The notion of an inertial mass is to be juxtaposed to that of a *gravitational* mass that is proportional to the gravitational force on the body. Experimentally, inertial and gravitational masses are found to be the same to better than 1 part in 10^{12}. We will therefore not distinguish here between inertial and gravitational masses. Indeed, the equivalence of inertial and gravitational masses is one of the central tenets of the theory of general relativity.

Note that the form (1.54) defines *both* mass and force: If length and time can be measured, one can measure the acceleration. Therefore, subjecting an arbitrarily chosen unit mass m_0 to a given, fixed force of magnitude

F_0 (even if its value is not known), we can measure the acceleration a_0 and write it as $a_0 = F_0/m_0$. Subjecting an unknown mass m to the *same* force F_0, its measured acceleration can be written as $a = F_0/m$ and thus $m = m_0 a_0/a$ which determines m in terms of the unit mass m_0. Then, with the body's mass defined in this manner, Eq. (1.54) allows one to determine any force acting on the body by simply measuring the resulting acceleration.

The most important aspect about the second law is that it provides a *dynamical* formulation of mechanical phenomena: The second law is a second-order differential equation for the trajectory of a body, and its solution allows one to make non-trivial predictions for these trajectories that are amenable to experimental scrutiny. For example, we may immediately predict that for a constant force $\mathbf{F}$ (and hence constant acceleration $\mathbf{a} = \mathbf{F}/m$) the velocity changes according to $\mathbf{v}(t) = \mathbf{v}_0 + \mathbf{a}t$ (where $\mathbf{v}_0$ is the initial velocity). This has indeed been found to be true in experiments *provided the velocities are small compared to the speed of light*. However, the implication that if we only let $\mathbf{F}$ act long enough, the velocity will grow to arbitrarily large values, is *not* true. We know now that velocities cannot grow without bounds and we observe these limitations for large velocities approaching the speed of light. Experimentally, the second axiom is thus found to have a restricted range of validity which is usually referred to as *classical mechanics*, or *nonrelativistic mechanics*. (Some of the pertinent issues regarding this restriction are explored in Sec. 8.2 on special relativity.)

Newton's **third axiom** reads:

<table>
<tr><td rowspan="5">Lex
Tertia</td><td>For every (active) force exerted on a body by</td><td rowspan="5">(1.55)</td></tr>
<tr><td>its environment, the body exerts a (reactive)</td></tr>
<tr><td>force of equal magnitude, acting in the opposite direction, on its environment:</td></tr>
<tr><td>$\mathbf{F}_{\text{action}} = -\mathbf{F}_{\text{reaction}}.$</td></tr>
</table>

For pairwise forces between two bodies 1 and 2 (see Fig. 1.4), this means in particular

$$\mathbf{F}_{12} = -\mathbf{F}_{21}, \qquad (1.56)$$

where $\mathbf{F}_{ij}$ denotes the force on body i due to body j. (Such forces are also referred to as "two-body forces", for obvious reasons.) The third law allows for both possibilities depicted in Fig. 1.4. However, in many instances the forces between two bodies lie on the line connecting the two bodies, as

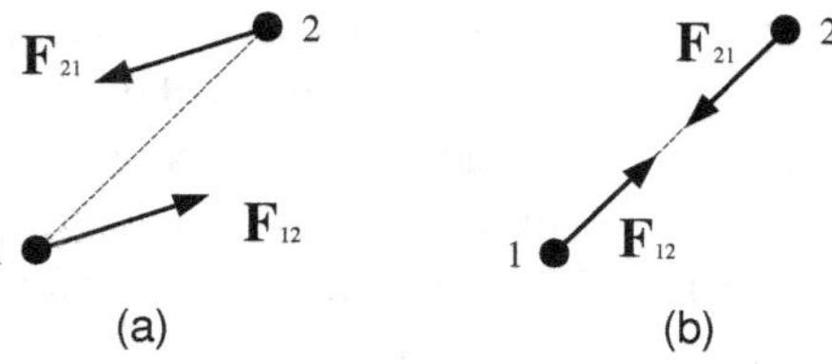

Fig. 1.4   Two situations of forces between two bodies compatible with the third axiom: The situation depicted in (a) is *not* compatible with the collinearity constraint, Eq. (1.57), while (b) is.

shown in Fig. 1.4(b). Such special cases satisfy the additional constraint:

$$\boxed{\begin{array}{ll} \textbf{\textit{Collinearity}} & \text{The force between two bodies lies on} \\ \textbf{\textit{Constraint:}} & \text{the line connecting the two bodies:} \\ & (\mathbf{r}_1 - \mathbf{r}_2) \times \mathbf{F}_{12} = 0. \end{array}} \qquad (1.57)$$

This additional constraint is satisfied by, for example, gravitational and Coulomb forces. Magnetic forces between moving charges, however, do not comply with it. Unless explicitly stated otherwise, we will assume here that the collinearity constraint holds true for the forces considered in the following.

1.2.2 *Newton's second law for variable mass*

If one considers a dynamical problem with variable mass, applying Newton's second law in the form (1.49) taking the momentum change simply as $d\mathbf{p} = d(m\mathbf{v})$ may produce paradoxical results. However, as alluded to above, one must realize that *for variable mass, the system is no longer closed* and one therefore must account also for the momentum associated with the mass increment dm. If at a given instant the body of mass m subject to force $\mathbf{F}$ has velocity $\mathbf{v}$ in an inertial frame and the infinitesimal mass increment dm has *relative* velocity $\mathbf{u}$ with respect to the body,[7] the initial and final momenta, $\mathbf{p}_i$ and $\mathbf{p}_f$, for cases of mass accretion ($dm > 0$) or mass ejection

---

[7]It is assumed here that the entire change dm can be associated with one single relative velocity $\mathbf{u}$. If this simplifying assumption cannot be applied, the present derivation must be modified accordingly by introducing appropriate velocity distributions. For example, if the mass change dm happens isotropically relative to the original body of mass m, the net momentum change is simply $d\mathbf{p} = m\,d\mathbf{v}$, albeit with a time-dependent mass, $m = m(t)$.

$(dm < 0)$ across an infinitesimal time step dt are

$$dm > 0: \quad \begin{cases} \mathbf{p}_i = m\mathbf{v} + dm(\mathbf{v} + \mathbf{u}), \\ \mathbf{p}_f = (m + dm)(\mathbf{v} + d\mathbf{v}), \end{cases} \tag{1.58}$$

$$dm < 0: \quad \begin{cases} \mathbf{p}_i = m\mathbf{v}, \\ \mathbf{p}_f = (m + dm)(\mathbf{v} + d\mathbf{v}) - dm(\mathbf{v} + \mathbf{u}), \end{cases} \tag{1.59}$$

respectively. Hence, the momentum change, $\Delta\mathbf{p} = \mathbf{p}_f - \mathbf{p}_i$, for both cases is

$$\Delta\mathbf{p} = d(m\mathbf{v}) - dm(\mathbf{v} + \mathbf{u})$$
$$= m\,d\mathbf{v} - dm\,\mathbf{u}, \tag{1.60}$$

producing the equation of motion

$$m\frac{d\mathbf{v}}{dt} - \frac{dm}{dt}\mathbf{u} = \mathbf{F}. \tag{1.61}$$

An application of this result to the vertical launch of a rocket is given in Sec. 1.3.5.

This derivation shows that the usual closed-system equality $\Delta\mathbf{p} = d\mathbf{p} \equiv d(m\mathbf{v})$ holds true for the variable-mass case only if $\mathbf{u} = -\mathbf{v}$, i.e., if the mass increment dm is or remains at rest in the inertial frame and thus has no momentum of its own. This happens when the body sweeps up stationary mass particles (dust, etc.) or when it loses mass due to a slow leak, etc.

Note that we may also interpret Eq. (1.61) as

$$m\frac{d\mathbf{v}}{dt} = \mathbf{F} + \mathbf{F}_{\text{recoil}}, \quad \text{where} \quad \mathbf{F}_{\text{recoil}} = \frac{dm}{dt}\mathbf{u}, \tag{1.62}$$

which produces an equivalent equation of motion similar in appearance to the constant-mass case, with the recoil force added to the original force. This form is particularly well suited to assess the effect of the variable mass on the motion. For a rocket, for example, where $dm < 0$ and the direction $\mathbf{u}$ of the ejected mass is opposite the direction of motion $\mathbf{v}$, the recoil force acts along $\mathbf{v}$ and thus propels the rocket forward.

(Parenthetically, we add here that it is sometimes claimed that Newton's second law should always be written in the form $m\ddot{\mathbf{r}} = \mathbf{F}$, with $\mathbf{F}$ subsuming the necessary additional recoil-type forces for variable-mass cases. While this results in an equivalent formulation for the equation of motion, as seen here, we point out that the forms of these additional forces follow from considering incremental momentum changes $\Delta\mathbf{p} = \mathbf{p}_f - \mathbf{p}_i$ in terms of the modified second law (1.52). There is no other *independent* avenue for their derivation because their forms depend on the dynamics of the problem. *Conceptually*, therefore, Newton's second law in its original form (1.49) provides the basic axiomatic formulation of the underlying dynamics; Eq. (1.62), by contrast, is a derived formulation.)

1.2.3   *Applying Newton's laws*

The forces of classical mechanics are only considered to be functions of position and velocity, and possibly directly of time itself, and thus can be written as

$$\mathbf{F} = \mathbf{F}\big(\mathbf{r}(t), \dot{\mathbf{r}}(t), t\big). \tag{1.63}$$

This excludes a dependence on the acceleration, or even higher derivatives, and it also excludes, for example, that the force would depend on the motion of the body at earlier times.[8]

It is crucial to realize that, for the time being, we presume that the force for a particular problem is *a priori* given, i.e., that it does not depend on the dynamics of the problem itself. This restriction of *Newtonian Mechanics* will be lifted when we consider the Lagrangian formulation of mechanics in Chapter 2. (To prevent possible misunderstanding: Of course, it is possible also in Newtonian mechanics to treat problems where at least part of the force depends on the dynamics of the problem. Typically, however, this requires splitting the problem into a part where the force is completely known and a part where it is not known *a priori*. The solution of the former then allows one to determine the explicit force function for the latter.[9])

The solution of a problem in Newtonian mechanics usually proceeds along the following lines.

- Formulation of the force law $\mathbf{F} = \mathbf{F}\big(\mathbf{r}(t), \dot{\mathbf{r}}(t), t\big)$.
- Solution of the second-order differential equation

$$m\ddot{\mathbf{r}}(t) = \mathbf{F}\big(\mathbf{r}(t), \dot{\mathbf{r}}(t), t\big). \tag{1.64}$$

  If the problem is sufficiently complicated, one may have to resort here to a numerical solution. In the case of a time-dependent mass, the modified procedure outlined in the preceding Sec. 1.2.2 applies (for a simple example, see Sec. 1.3.5).

---

[8]This really points to limitations stemming from the Newtonian notion of a force based on mechanical phenomena. In electromagnetism, by contrast, the Abraham-Lorentz radiation-damping force provides an example for a dependence on $d^3\mathbf{r}/dt^3$.

[9]As an example, consider the planar pendulum (see Fig. 2.1) where the unknown string tension can only be determined once the solution for the angular motion has been found since the latter depends only on the known force of gravity.

- Determination of the solution's integration constants. Usually they are given in terms of the initial conditions of position and velocity at a fixed time, say, $\mathbf{r}_0 = \mathbf{r}(0)$ and $\mathbf{v}_0 = \dot{\mathbf{r}}(0)$.

Finally, trivial as it may seem, one needs to study and *understand* the solution. Useful tools for this may be graphical representations or other computer-aided visualizations, like time-lapse simulations, etc. Furthermore, one should *always* investigate whether there are constants of motion (energy, momentum, etc.) and what their implications are for the physics of the problem at hand.

1.3 One-dimensional Motion

The restriction of the motion to one dimension, with equation of motion

$$m\ddot{x}(t) = F\big(x(t), \dot{x}(t), t\big), \tag{1.65}$$

describes many interesting situations that provide insights into the general principles of mechanics. Typical cases usually treated in less advanced undergraduate-level courses are:

$$\text{no forces:} \quad m\ddot{x} = 0 \; ; \tag{1.66}$$

$$\text{homogeneous gravitational field:} \quad m\ddot{x} = -mg \; ; \tag{1.67}$$

$$\text{gravitational field with friction:} \quad m\ddot{x} = -mg - \gamma\dot{x} \; ; \tag{1.68}$$

$$\text{simple harmonic motion:} \quad m\ddot{x} = -kx \; ; \tag{1.69}$$

$$\text{harmonic motion with damping:} \quad m\ddot{x} = -kx - 2m\lambda\dot{x} \; ; \tag{1.70}$$

$$\text{forced oscillation:} \quad \ddot{x} + \omega_0^2 x + 2\lambda\dot{x} = a_0 \cos(\omega t). \tag{1.71}$$

Even though students may already be familiar with most, if not all, of these problems, we shall briefly outline the last three examples to review some basic notions of mechanics. In addition, we consider the vertical launch of a rocket as an example of a problem with a time-dependent mass.

1.3.1 *Energy conservation*

As an first illustration, let us consider the solution for a force that only depends on the position, i.e.,

$$m\ddot{x} = F(x). \tag{1.72}$$

Introducing

$$V(x) = -\int dx \, F(x) \, + \, C, \tag{1.73}$$

where $V(x)$ is referred to as the *potential energy*, or simply as the *potential*,[10] the force equation can be rewritten as

$$\frac{m}{2}\frac{d}{dt}\dot{x}^2 = -\frac{d}{dt}V(x) \tag{1.74}$$

upon multiplying both sides by $\dot{x}(t)$. The constant C in (1.73) is chosen by convention; it is of no significance for Eq. (1.74). Integration yields

$$\frac{m}{2}\dot{x}^2 = E - V(x), \tag{1.75}$$

where $m\dot{x}^2/2$ is the *kinetic energy*. The integration constant E, called the *total energy*, is only determined up to the choice of the constant C in (1.73). Solving now for $\dot{x} = dx/dt$ yields[11]

$$dt = \frac{dx}{\sqrt{\frac{2}{m}\left[E - V(x)\right]}}, \tag{1.76}$$

or

$$\boxed{t - t_0 = \int_{x_0}^{x}\frac{dx'}{\sqrt{\frac{2}{m}\left[E - V(x')\right]}}.} \tag{1.77}$$

This integral determines $t = t(x)$ and therefore implicitly, by inversion, the desired function $x = x(t)$. The two integration constants x_0 and E are related by

$$E = \frac{1}{2}mv_0^2 + V(x_0) \tag{1.78}$$

to the initial conditions $x_0 = x(t_0)$ and $v_0 = \dot{x}(t_0)$.

In special cases, for certain potential functions $V(x)$, the integral in (1.77) can be solved explicitly. One of the standard examples of this kind is the harmonic oscillator which will be treated in the following section. In all cases, however, the qualitative behavior of the solution can be discussed by plotting the function $V(x)$ and its intersections with the horizontal line $E = const$, as in Fig. 1.5. According to (1.75), the kinetic energy $m\dot{x}^2/2$ is

---

[10]It should be emphasized that *potential energy* is the correct term. *Potential*, by contrast, is the potential energy — here — per mass unit. It is common, however, to use the somewhat sloppy abbreviated term because it is clear from the context what is meant.

[11]Note that by choosing the positive square root here, we assume that x increases with increasing t. This need not be the case, but we will ignore this question for now. We will come back to it on page 64 where we show that the solutions for positive and negative square roots are linked by orbital symmetries. Determining one solution, therefore, is sufficient.

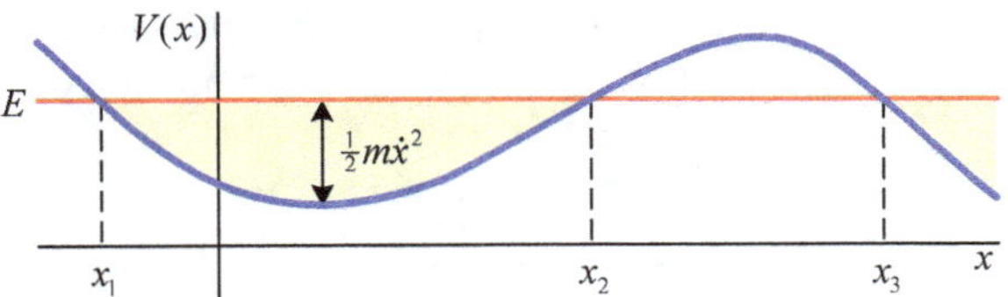

Fig. 1.5   Generic example showing a plot of the potential function $V(x)$ and the constant energy E. The physically accessible regions here (with positive kinetic energies in the shaded areas) are the intervals $x_1 \leq x \leq x_2$ and $x \geq x_3$.

determined by the vertical distance between E and $V(x)$ in this plot. Since in this example $E > 0$ and the kinetic energy must always be positive, this immediately shows that only the regions where the potential curve dips below the horizontal energy line correspond to allowed physical motion. One such region in the figure is given by $x_1 \leq x \leq x_2$. For any direction of motion (i.e., whether $\dot{x} > 0$ or $\dot{x} < 0$) the value of $\dot{x}^2$ is directly determined by the vertical distance. At the intersections x_1 and x_2 with $V(x)$, $\dot{x} \to 0$ and the direction of the motion reverses at these points. For this example, therefore, the trajectory is an oscillatory motion between the two *turning points* x_1 and x_2, where the time to go from x_1 to x_2 and back to x_1,

$$T = 2 \int_{x_1}^{x_2} \frac{dx'}{\sqrt{\frac{2}{m}\left[E - V(x')\right]}}, \qquad (1.79)$$

is the period T of oscillation. (The oscillatory motion here need *not* be sinusoidal.)

Note that in view of the energy conservation described by (1.75), the actual solution procedure involves here only a first-order differential equation. The existence of a conserved quantity — here, the energy — therefore simplifies the solution of the problem. In Sec. 3.3 on *Noether's Theorem*, we shall see that simplifications of this kind are always due to underlying symmetries.

1.3.2   *Simple harmonic oscillation*

One of the simplest, yet widely encountered, phenomena in physics concerns *simple harmonic motion*.[12] It is defined by *Hooke's law*,

$$F(x) = -kx, \qquad (1.80)$$

---

[12]See also Chapter 6, Small Oscillations, on page 237.

describing a linear restoring force opposing the displacement x from an equilibrium position at $x = 0$. The corresponding equation of motion (1.69) is written as

$$\boxed{\ddot{x} + \omega_0^2 x = 0}\,, \tag{1.81}$$

with $\omega_0^2 = k/m$. This equation describes an *harmonic oscillator*. The corresponding potential is given by

$$V(x) = -\int_{x_0}^{x} dx'\, F(x') = \int_{x_0}^{x} dx'\, kx' = \tfrac{1}{2}kx^2 - \tfrac{1}{2}kx_0^2, \tag{1.82}$$

where x_0 is an arbitrary reference point. We may choose $x_0 = 0$ without loss of generality and obtain

$$\boxed{V(x) = \tfrac{1}{2}kx^2 \qquad \text{(harmonic oscillator)}}\,. \tag{1.83}$$

According to (1.77), when choosing initial conditions $x(t_0) = 0$ we have

$$t - t_0 = \sqrt{\frac{m}{2}} \int_0^x \frac{dx'}{\sqrt{E - \tfrac{1}{2}kx'^2}}$$

$$= \sqrt{\frac{m}{2E}} \int_0^x \frac{dx'}{\sqrt{1 - \tfrac{k}{2E}x'^2}}$$

$$= \sqrt{\frac{m}{k}} \int_0^y \frac{dy'}{\sqrt{1 - y'^2}}$$

$$= \frac{1}{\omega_0} \arcsin\sqrt{\frac{k}{2E}}x, \tag{1.84}$$

where the substitution $y' = \sqrt{k/2E}\, x'$ was used. Inverting the arcsin produces the *harmonic solution*

$$x(t) = \sqrt{\frac{2E}{k}} \sin\left(\omega_0 t + \phi\right). \tag{1.85}$$

The constant

$$\omega_0 = \sqrt{\frac{k}{m}} \tag{1.86}$$

is thus seen to be the *angular frequency* of the harmonic oscillation. For the initial condition chosen here, the *phase angle* is $\phi = -\omega_0 t_0$. Other initial conditions will produce a different phase angle, but the general structure of the solution will not change. In general, the two required integration constants can always be expressed in terms of E and ϕ as

$$x(t_0) = \sqrt{\frac{2E}{k}}\,\sin\left(\omega_0 t_0 + \phi\right), \tag{1.87a}$$

$$\dot{x}(t_0) = \sqrt{\frac{2E}{m}}\,\cos\left(\omega_0 t_0 + \phi\right). \tag{1.87b}$$

1.3.3 *Harmonic oscillation with damping*

Modifying the force of the simple harmonic oscillator by a friction term $-2m\lambda\dot{x}$ $(\lambda > 0)$, i.e.,

$$F(x,\dot{x}) = -kx - 2m\lambda\dot{x}, \tag{1.88}$$

the differential equation now becomes

$$\ddot{x} + 2\lambda\dot{x} + \omega_0^2 x = 0. \tag{1.89}$$

The force F now depends also on $\dot{x}$, and not just x. In this case, therefore, we cannot use the integral (1.77).

Making the ansatz[13]

$$x(t) = c\,e^{\eta t}, \tag{1.90}$$

we find

$$\eta^2 + 2\lambda\eta + \omega_0^2 = 0, \tag{1.91}$$

with roots

$$\eta_{1,2} = -\lambda \pm \sqrt{\lambda^2 - \omega_0^2}. \tag{1.92}$$

For the time being, we will assume $\lambda^2 \neq \omega_0^2$. The general solution then is an arbitrary superposition of these two possibilities,

$$x(t) = e^{-\lambda t}\left[c_1\,e^{\sqrt{\lambda^2-\omega_0^2}\,t} + c_2\,e^{-\sqrt{\lambda^2-\omega_0^2}\,t}\right], \tag{1.93}$$

---

[13]The German noun *Ansatz* is commonly used in the mathematics or physics literature; in this context, it means something like 'trial approach' or 'educated guess'. For the present example, the 'educated guess' is to try an exponential function (since this always works for homogeneous linear differential equations), with an open parameter η that can then be determined by the resulting algebraic equation (1.91). Ideally, the outcome of trying out an ansatz tells one how to set up the general solution, as in Eq. (1.93) here. Note that the proper plural of *Ansatz* is *Ansätze* (alternatively spelled as *Ansaetze* if one cannot use an umlaut) — however, the usage of *ansatz* is so commonplace in mathematics and physics that one occasionally also sees the anglicized version *ansatzes* (not recommended).

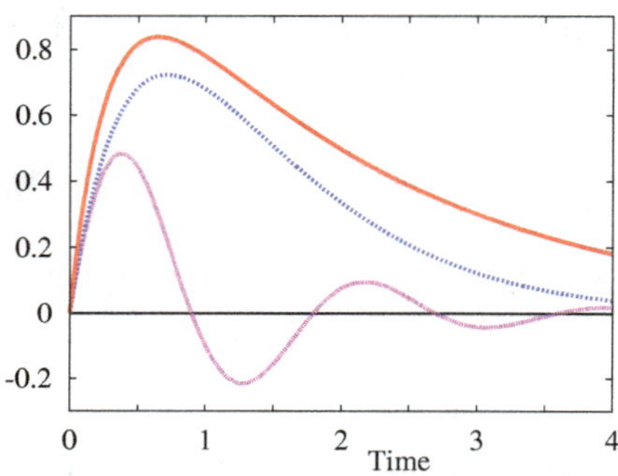

Fig. 1.6   Generic example solutions for the damped harmonic oscillator with initial conditions (1.94) for over-damping ($\lambda^2 > \omega_0^2$, top curve), critical damping ($\lambda^2 = \omega_0^2$, middle curve), and underdamping ($\lambda^2 < \omega_0^2$, bottom curve), in arbitrary units.

with coefficients c_1 and c_2 determined by initial conditions

$$x(t=0) = x_0, \qquad \dot{x}(t=0) = v_0. \tag{1.94}$$

Choosing $x_0 = 0$ for the following discussion produces

$$x(t) = \frac{v_0\, e^{-\lambda t} \left[e^{\sqrt{\lambda^2 - \omega_0^2}\, t} - e^{-\sqrt{\lambda^2 - \omega_0^2}\, t} \right]}{2\sqrt{\lambda^2 - \omega_0^2}}. \tag{1.95}$$

This particular choice for x_0 simply means that we start the clock when the mass goes through the origin, but does not otherwise prejudice the solution.

Underdamping: Considering now first the special case $\lambda^2 < \omega_0^2$, this yields

$$x(t) = e^{-\lambda t}\, \frac{v_0}{\Omega} \sin\left(\Omega t\right), \quad \text{where} \quad \Omega = \sqrt{\omega_0^2 - \lambda^2}. \tag{1.96}$$

This is the standard harmonic oscillator solution, with a frequency Ω instead of ω_0, but with an extra damping factor $e^{-\lambda t}$ that eventually will bring the amplitude down to zero (see Fig. 1.6, bottom curve).

Overdamping: The second special case, $\lambda^2 > \omega_0^2$, produces

$$x(t) = e^{-\lambda t}\, \frac{v_0}{\sqrt{\lambda^2 - \omega_0^2}} \sinh\left(\sqrt{\lambda^2 - \omega_0^2}\, t \right). \tag{1.97}$$

This solution does not show any oscillatory behavior at all: The initial rise of the hyperbolic sine is cut off by the exponential damping factor and the solution then goes to zero (see Fig. 1.6, top curve).

Critical damping: For the third case, $\lambda^2 = \omega_0^2$, we go back to the differential equation (1.89) and easily verify that both linearly independent expressions $\mathrm{e}^{-\lambda t}$ and $t\,\mathrm{e}^{-\lambda t}$ are solutions. Hence the general solution for this case is

$$x(t) = (c_1 + c_2 t)\,\mathrm{e}^{-\lambda t}. \tag{1.98}$$

With the present initial conditions, c_1 would vanish and $c_2 = v_0$ (see Fig. 1.6, middle curve). Of course, this solution can be obtained as well from both (1.96) and (1.97) by taking appropriate limits.

1.3.4 *Forced oscillation*

Next, we consider an harmonic damped oscillator driven by an external force, i.e.,

$$F(x, \dot{x}, t) = -kx - 2m\lambda\dot{x} + f(t), \tag{1.99}$$

where $f(t)$ is the external force term which we take of harmonic form,

$$f(t) = ma_0 \cos\omega t, \tag{1.100}$$

with an acceleration parameter a_0 and a driving frequency ω. The *inhomogeneous* linear second-order differential equation to be solved then reads

$$\ddot{x} + 2\lambda\dot{x} + \omega_0^2 x = a_0 \cos\omega t. \tag{1.101}$$

In solving this equation, we consider only $\omega_0^2 = k/m > \lambda^2$, which corresponds to the underdamped case of Eq. (1.96) of the previous homogeneous example.

The general solution of an inhomogeneous differential equation is obtained as the general solution of the homogeneous problem and any particular solution of the inhomogeneous equation. We know already the general solution of the homogeneous problem; it is given by Eq. (1.93).

To find a particular solution of the inhomogeneous equation, consider the following two differential equations,

$$\ddot{x} + 2\lambda\dot{x} + \omega_0^2 x = a_0 \cos\omega t, \tag{1.102a}$$

$$i\ddot{y} + 2\lambda i\dot{y} + \omega_0^2 iy = a_0 i \sin\omega t. \tag{1.102b}$$

Defining the complex function $z = x + iy$, they can be combined into

$$\ddot{z} + 2\lambda\dot{z} + \omega_0^2 z = a_0\,\mathrm{e}^{i\omega t}. \tag{1.103}$$

By construction, the solution we seek is the real part of the solution of this equation.

Inserting the ansatz

$$z(t) = A\,e^{i(\omega t - \varphi)} \tag{1.104}$$

into the differential equation produces[14]

$$A = \frac{a_0}{\sqrt{\left(\omega_0^2 - \omega^2\right)^2 + (2\lambda\omega)^2}} \tag{1.105}$$

and

$$\tan\varphi = \frac{2\lambda\omega}{\omega_0^2 - \omega^2}, \qquad \text{for} \quad 0 \le \varphi \le \pi. \tag{1.106}$$

With this phase angle, the particular solution x_{p} becomes

$$x_{\mathrm{p}}(t) = \frac{a_0 \cos(\omega t - \varphi)}{\sqrt{\left(\omega_0^2 - \omega^2\right)^2 + (2\lambda\omega)^2}} \tag{1.107}$$

and the general solution thus is [cf. (1.93)]

$$x(t) = e^{-\lambda t}\left[c_1\,e^{i\Omega t} + c_2\,e^{-i\Omega t}\right] + A\cos(\omega t - \varphi), \tag{1.108}$$

where $\Omega = \sqrt{\omega_0^2 - \lambda^2}$ is the frequency of the underdamped case (1.96) and the coefficients c_1 and c_2 are determined by initial conditions for this forced oscillator. In view of the damping term $e^{-\lambda t}$, the first term (called a *transient*) will eventually become negligibly small and for large times the behavior of the solution is entirely given by $x_{\mathrm{p}}(t)$.

The oscillatory behavior of the steady-state solution, therefore, is determined solely by the frequency ω of the external driving force, irrespective of the value of the intrinsic (or natural) frequency ω_0. For $\omega = \omega_0$, in particular, the corresponding phase angle is $\varphi = \pi/2$ and the amplitude A reaches its maximum. This phenomenon is called a *resonance*. Resonant behavior — an enhanced response to an external stimulus — is found for all types of physical systems driven by external frequencies equal to their natural frequencies.

1.3.5   Variable mass: Vertical launch of a rocket

As a simple example for the variable-mass case discussed in Sec. 1.2.2, we consider the vertical launch of a rocket. The rocket expels fuel as it rises from the launch pad making the total mass a function of time, $m = m(t)$.

---

[14] A is chosen here as having the same sign as a_0; any possible sign difference can be absorbed in the phase factor $e^{-i\varphi}$.

In the inertial system furnished by the stationary launch pad, the rocket rises with velocity v and its fuel is expelled with velocity

$$v' = v - c, \tag{1.109}$$

where $-c$ is the (constant) downward relative velocity of fuel leaving the rocket. The upward direction is taken as positive here. During the time increment dt, the change in total (vertical) momentum of the system consisting of the rocket and its expelled fuel is [see also (1.60)]

$$\Delta p = d(mv) + (-dm)v' = m\,dv + dm\,c, \tag{1.110}$$

where dv is the change in upward velocity and $-dm$ (> 0) is the amount of fuel mass expelled. Taking gravity as the only external force (neglecting air friction), Newton's second law then reads

$$\frac{\Delta p}{dt} = F \qquad \longrightarrow \qquad m\frac{dv}{dt} + \frac{dm}{dt}c = -mg. \tag{1.111}$$

Let us assume that fuel is consumed at a constant rate,

$$m(t) = M - \mu t, \qquad \text{for } t \le t_e, \tag{1.112}$$

where M is the initial total mass of the rocket at lift-off (at $t = 0$) and $\mu = -dm/dt = const > 0$, and that at $t = t_e$, the rocket burners are shut off and the mass does not change any longer, making $m_e = M - \mu t_e$ the final mass of the rocket. The equation of motion then becomes

$$\dot{v} = -g + \frac{\mu c}{M - \mu t}. \tag{1.113}$$

Integration yields

$$v(t) = -gt + c \ln \frac{M}{M - [\mu t]} \tag{1.114}$$

for the velocity and

$$h(t) = -\frac{1}{2}gt^2 + \frac{Mc}{\mu}\left(\frac{M - [\mu t]}{M}\ln\frac{M - [\mu t]}{M} + \frac{[\mu t]}{M}\right) \tag{1.115}$$

for the height above the launch pad. The expression $[\mu t]$ in square brackets here is defined as

$$[\mu t] = \begin{cases} \mu t, & \text{for} \quad t \le t_e, \\ \mu t_e, & \text{for} \quad t > t_e. \end{cases} \tag{1.116}$$

In other words, for times larger than t_e, the rocket behaves like a freely falling object tossed into the air with an initial upward velocity of $v_e = -gt_e + c \ln(M/m_e)$. Note that to be able to lift off at all, the rocket's fuel burn rate μ and ejection speed c must satisfy $c\mu > Mg$. The proof is left as an exercise.

1.3.6   *Green's function method*

The differential equations encountered here are all *linear* and can be written in the generic form

$$\mathcal{D}_t\, x(t) = f(t), \tag{1.117}$$

where $x(t)$ is the unknown function, $f(t)$ is the known inhomogeneity (which may be zero), and $\mathcal{D}_t$ is a linear combination of differential operators up to order n of the form

$$\mathcal{D}_t = a_0 + a_1 \frac{d}{dt} + \cdots + a_n \frac{d^n}{dt^n}, \tag{1.118}$$

with constant coefficients a_i.

Inhomogeneous linear differential equations, with $f(t) \neq 0$, can be solved by seeking a *Green's function* $g(t, t')$ that satisfies

$$\mathcal{D}_t\, g(t, t') = \delta(t - t'), \tag{1.119}$$

where $\delta(t - t')$ is the *Dirac delta distribution* (see Appendix B). In a manner of speaking, therefore, we may view the Green's function as the *inverse* of the differential operator $\mathcal{D}_t$. It is now easily seen that

$$x_{\mathrm{p}}(t) = \int_{-\infty}^{+\infty} g(t, t')\, f(t')\, dt' \tag{1.120}$$

is a particular solution of the differential equation (1.117), i.e.,

$$\mathcal{D}_t\, x_{\mathrm{p}}(t) = \mathcal{D}_t \int_{-\infty}^{+\infty} g(t, t') f(t')\, dt' = \int_{-\infty}^{+\infty} \left[\mathcal{D}_t\, g(t, t') \right] f(t')\, dt'$$

$$= \int_{-\infty}^{+\infty} \delta(t - t') f(t')\, dt'$$

$$= f(t). \tag{1.121}$$

The most general solution of Eq. (1.117) is then given by

$$x(t) = x_{\mathrm{h}}(t) + \int_{-\infty}^{+\infty} g(t, t')\, f(t')\, dt', \tag{1.122}$$

where $x_{\mathrm{h}}(t)$ is the general solution of the corresponding homogeneous differential equation.

What makes this *Green's function method* particularly useful is the fact that knowledge of the general solution $x_\mathrm{h}(t)$ of the homogeneous problem is sufficient to *construct* the Green's function $g(t, t')$ in a straightforward manner. This is illustrated in Appendix C for the example of the second-order differential equation for an externally driven harmonic oscillator. For more general treatments, see, for example, Refs. [8] or [9].

1.4 Conservation Laws

In general, conservation laws allow one to simplify the solution of a given problem. In the following the conservation of momentum, angular momentum, and the energy will be considered.

1.4.1 *Momentum conservation*

Newton's second axiom (for constant mass) immediately implies

$$\boxed{\mathbf{F} = 0 \quad \Rightarrow \quad \frac{d\mathbf{p}}{dt} = 0 \quad \Rightarrow \quad \mathbf{p} = const}, \tag{1.123}$$

i.e., the momentum is conserved if there is no force.

1.4.2 *Angular momentum conservation*

Defining the *angular momentum* by

$$\mathbf{L} = \mathbf{r} \times \mathbf{p}, \tag{1.124}$$

and the *torque* by

$$\mathbf{N} = \mathbf{r} \times \mathbf{F}, \tag{1.125}$$

one finds

$$\frac{d\mathbf{L}}{dt} = \dot{\mathbf{r}} \times \mathbf{p} + \mathbf{r} \times \dot{\mathbf{p}} = \underbrace{\dot{\mathbf{r}} \times m\dot{\mathbf{r}}}_{=0} + \mathbf{r} \times \mathbf{F} = \mathbf{r} \times \mathbf{F}, \tag{1.126}$$

i.e.,

$$\frac{d\mathbf{L}}{dt} = \mathbf{N}. \tag{1.127}$$

Thus, the rate of change of the angular momentum is equal to the torque. Therefore, if the torque vanishes, we have

$$\boxed{\mathbf{N} = 0 \quad \Rightarrow \quad \frac{d\mathbf{L}}{dt} = 0 \quad \Rightarrow \quad \mathbf{L} = const}, \tag{1.128}$$

thus providing a conserved angular momentum.

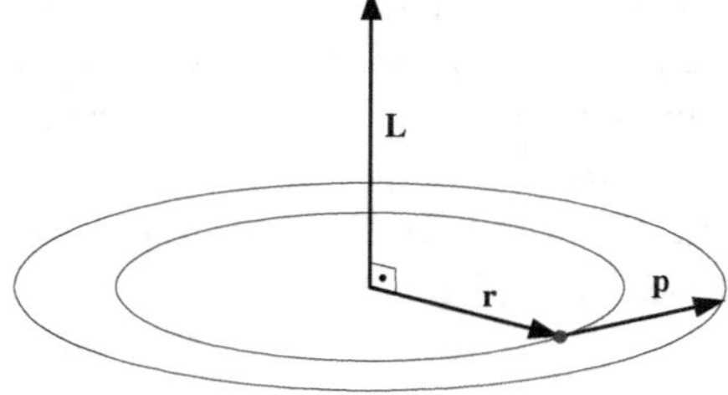

Fig. 1.7   Angular momentum $\mathbf{L} = \mathbf{r} \times \mathbf{p}$. If $\mathbf{L}$ is conserved, the motion takes place in the plane to which $\mathbf{L}$ is perpendicular.

Note that both angular momentum and torque are defined relative to the origin of the chosen inertial frame. If the frame of reference changes, $\mathbf{L}$ and $\mathbf{N}$ will change even if the change involves only a simple (fixed) translation.

Note also that a conserved angular momentum implies that the motion takes place only in the plane to which $\mathbf{L}$ is perpendicular (see Fig. 1.7). This plane is also called the *invariable plane* since its orientation in space stays fixed for given (conserved) $\mathbf{L}$.

1.4.2.1   *Central force*

If the force $\mathbf{F}$ is parallel (or anti-parallel) to the position vector $\mathbf{r}$, i.e.,

$$\mathbf{F} = f\frac{\mathbf{r}}{r} = f\mathbf{e}_r, \tag{1.129}$$

where $f = \mathbf{F} \cdot \mathbf{e}_r$ is a scalar function, then $\mathbf{F}$ is called a *central force*.

Evidently, the torque vanishes for central forces and therefore the angular momentum is conserved. Taking the xy-plane as the plane of motion and employing polar coordinates (r,φ,z), one has

$$\mathbf{L} = r\mathbf{e}_r \times m\left(\dot{r}\mathbf{e}_r + r\dot{\varphi}\mathbf{e}_\varphi\right) = mr^2\dot{\varphi}\,\mathbf{e}_z, \tag{1.130}$$

or

$$\mathbf{L} = \ell\,\mathbf{e}_z, \quad \text{with} \quad \ell = mr^2\dot{\varphi}, \tag{1.131}$$

where $\ell = const$, of course.

The equations of motion, according to the second axiom, read now [cf. Eq. (1.37)]

$$m\left(\ddot{r} - r\dot{\varphi}^2\right) = f, \tag{1.132}$$

$$m\left(r\ddot{\varphi} + 2\dot{r}\dot{\varphi}\right) = \frac{m}{r}\frac{d}{dt}(r^2\dot{\varphi}) = 0. \tag{1.133}$$

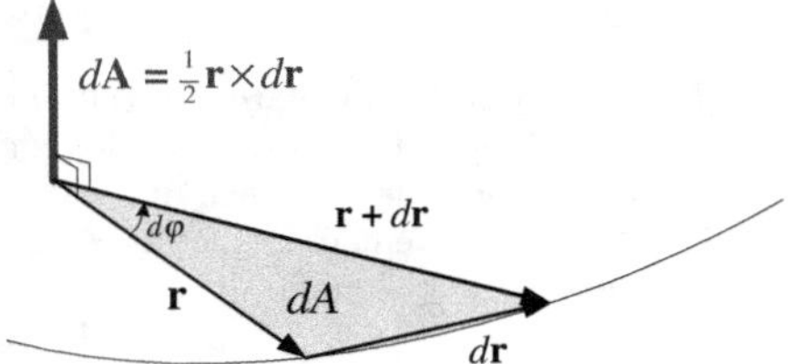

Fig. 1.8 Vector $d\mathbf{A}$ associated with the area dA swept out by the position vector when undergoing an infinitesimal change from $\mathbf{r}$ to $\mathbf{r} + d\mathbf{r}$ during an infinitesimal time interval dt.

The fact that there is no force term on the right-hand side of the second equation (which gives rise to angular momentum conservation), allows one to decouple the equations, i.e.,

$$m\ddot{r} = f + \frac{\ell^2}{mr^3} \equiv \hat{f}, \tag{1.134}$$

$$\dot{\varphi} = \frac{\ell}{mr^2}. \tag{1.135}$$

The first equation here appears now like a one-dimensional problem with an effective force $\hat{f}$, where the *centrifugal force term* $mr\dot{\varphi}^2 = \ell^2/mr^3$ was added, and the second equation has been reduced to a first-order differential equation, with ℓ being the corresponding integration constant. The full treatment of these equations of motion will be deferred until Sec. 1.8.

Area law

The area dA swept out by the position vector during an interval dt (see Fig. 1.8),

$$d\mathbf{A} = \frac{\mathbf{r} \times d\mathbf{r}}{2} = \frac{r\,\mathbf{e}_r \times r\,d\varphi\,\mathbf{e}_\varphi}{2} = \frac{r^2}{2}\,d\varphi\,\mathbf{e}_z = \frac{\mathbf{L}}{2m}\,dt, \tag{1.136}$$

is directly related to the angular momentum,

$$\boxed{\text{Area law:} \quad \frac{dA}{dt} = \frac{\ell}{2m} = const,} \tag{1.137}$$

thus providing the *area law*: *When angular momentum is conserved, the position vector sweeps out equal areas in equal times.* This is, therefore, a property of *all* central forces.

1.4.3 *Energy conservation*

In the one-dimensional example in Sec. 1.3, we have already encountered energy conservation in passing. The following provides a more formal and general treatment of this central topic of mechanics.

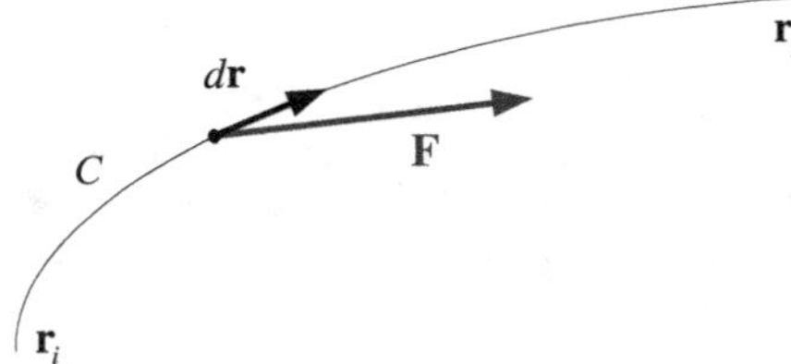

Fig. 1.9 A body moves from initial position $\mathbf{r}_i$ to the final position $\mathbf{r}_f$ along a curve C under the influence of a force $\mathbf{F}$. The path increment $d\mathbf{r}$ at each point is tangential to the curve C.

1.4.3.1 Work and power

If a body moves from $\mathbf{r}$ to $\mathbf{r} + d\mathbf{r}$ under the influence of a force $\mathbf{F}$, the force does *incremental work* dW on the body given by the scalar product

$$dW = \mathbf{F} \cdot d\mathbf{r}. \tag{1.138}$$

The *total work* W due to the force $\mathbf{F}$ along a curve C from an initial position $\mathbf{r}_i$ to a final position $\mathbf{r}_f$ (see Fig. 1.9) is given by

$$W = \int_C dW = \int_C \mathbf{F} \cdot d\mathbf{r}. \tag{1.139}$$

Obviously, in general this work will depend on the particular path C (even if the initial and final points of the path are kept fixed).

If the force $\mathbf{F}$ is composed of several contributions, $\mathbf{F} = \sum_i \mathbf{F}_i$, we may associate a corresponding individual work contribution W_i with each force $\mathbf{F}_i$ and the total work is $W = \sum_i W_i$.

Work dW done per time dt,

$$P = \frac{dW}{dt} = \mathbf{F} \cdot \dot{\mathbf{r}}, \tag{1.140}$$

is called *power*.

1.4.3.2 Kinetic and potential energies

From Newton's second axiom, one obtains

$$m\ddot{\mathbf{r}} \cdot \dot{\mathbf{r}} = \mathbf{F} \cdot \dot{\mathbf{r}} \tag{1.141}$$

or

$$\frac{d}{dt} \frac{m\dot{\mathbf{r}}^2}{2} = \mathbf{F} \cdot \dot{\mathbf{r}} = P, \tag{1.142}$$

which means that the energy transferred to (or from) the body changes its velocity. The entity

$$T = \frac{1}{2}m\dot{\mathbf{r}}^2, \tag{1.143}$$

therefore, describes an energy associated with the body's state of motion in the given frame; it is called the body's *kinetic energy*.

The total work can now be written as

$$W = \int_C \mathbf{F} \cdot d\mathbf{r} = \int_{t_i}^{t_f} \mathbf{F} \cdot \frac{d\mathbf{r}}{dt} dt = \int_{t_i}^{t_f} \frac{d\left(\frac{1}{2}m\dot{\mathbf{r}}^2\right)}{dt} dt, \qquad (1.144)$$

where t_i and t_f are the times associated with the initial and final points of the curve C. The total work is therefore given by the difference in kinetic energies at these times,

$$W_{fi} = \left(\tfrac{1}{2}m\dot{\mathbf{r}}^2\right)_{t_f} - \left(\tfrac{1}{2}m\dot{\mathbf{r}}^2\right)_{t_i} = T_f - T_i. \qquad (1.145)$$

This does *not* mean that this difference is independent of C, of course, since the initial and final times appearing in Eq. (1.144) depend on the parameterization $t = t[C]$ along C. In other words, in general, the kinetic energies at the initial and final points depend also on the history and not just on the locations.

Let us write the total force as

$$\mathbf{F} = \mathbf{F}_{\text{cons}} + \mathbf{F}_{\text{diss}}, \qquad (1.146)$$

where the *conservative force* $\mathbf{F}_{\text{cons}}$ is to subsume all contributions for which there exists a *scalar* function V such that

$$\mathbf{F}_{\text{cons}} \cdot \dot{\mathbf{r}} = -\frac{dV(\mathbf{r})}{dt}, \qquad (1.147)$$

where V depends on the position only. The *dissipative force* $\mathbf{F}_{\text{diss}}$ simply comprises all contributions that cannot be written in this way. The function $V(\mathbf{r})$ is the *potential energy* (or simply the *potential*[15]) associated with the conservative force $\mathbf{F}_{\text{cons}}$.

Equation (1.142) can now be rewritten as

$$\frac{d}{dt}\left[\frac{1}{2}m\dot{\mathbf{r}}^2 + V(\mathbf{r})\right] = \frac{d}{dt}\left[T + V(r)\right] = \mathbf{F}_{\text{diss}} \cdot \dot{\mathbf{r}}, \qquad (1.148)$$

and the entity in the square brackets on the left,

$$\boxed{E = T + V(r) = \frac{1}{2}m\dot{\mathbf{r}}^2 + V(\mathbf{r})}, \qquad (1.149)$$

is called the *total mechanical energy E* of the body. The presence of dissipative forces will transform this mechanical energy into other forms of energy.

---

[15] We often use here the terms potential and potential energy interchangeably, however, see footnote 10 on page 18.

If all dissipative forces vanish, $\mathbf{F}_{\text{diss}} = 0$, the total mechanical energy is conserved,

$$\boxed{\; \mathbf{F} \text{ conservative} \quad \Rightarrow \quad E = \frac{1}{2}m\dot{\mathbf{r}}^2 + V(\mathbf{r}) = const \;}, \qquad (1.150)$$

which is of course why the associated forces are called conservative.

Potential energy

To obtain a more direct relation between the potential and a given conservative force, rewrite Eq. (1.147) as

$$\mathbf{F}_{\text{cons}} \cdot \dot{\mathbf{r}} = -\frac{\partial V}{\partial x}\frac{dx}{dt} - \frac{\partial V}{\partial y}\frac{dy}{dt} - \frac{\partial V}{\partial z}\frac{dz}{dt} = -\dot{\mathbf{r}} \cdot \boldsymbol{\nabla}V, \qquad (1.151)$$

where $\boldsymbol{\nabla}$ denotes the gradient operator,[16]

$$\boldsymbol{\nabla} = \mathbf{e}_x \frac{\partial}{\partial x} + \mathbf{e}_y \frac{\partial}{\partial y} + \mathbf{e}_z \frac{\partial}{\partial z}. \qquad (1.152)$$

As suggested by (1.151), in the following we will consider only conservative forces that are functions of the position only and therefore may be written as

$$\boxed{\; \mathbf{F}(\mathbf{r}) = -\boldsymbol{\nabla}V(\mathbf{r}) \qquad \text{(conservative force)} \;}, \qquad (1.153)$$

where henceforth the index 'cons' will be dropped.[17]

A direct (necessary and sufficient) test whether a given force $\mathbf{F}(\mathbf{r})$ can be written in this way is whether the force is curl-free, i.e.,

$$\text{curl}\,\mathbf{F}(\mathbf{r}) = 0 \quad \Longleftrightarrow \quad \mathbf{F}(\mathbf{r}) = -\boldsymbol{\nabla}V(\mathbf{r}). \qquad (1.154)$$

The "$\Leftarrow$" direction of the proof of this equivalence is trivial since the curl of a gradient always vanishes. For the "$\Rightarrow$" direction, introduce two arbitrary independent paths C and C' from $\mathbf{r}_i$ to $\mathbf{r}_f$ and consider their integral

---

[16]The symbol $\boldsymbol{\nabla}$ is usually restricted to denote the gradient in Cartesian coordinates. In arbitrary coordinates, the notation "grad" is used instead.

[17]A more general form that includes also terms directly perpendicular to $\dot{\mathbf{r}}$ itself would be

$$\mathbf{F}_{\text{cons}} = -\boldsymbol{\nabla}V(\mathbf{r}) + \dot{\mathbf{r}} \times \mathbf{U}(\mathbf{r}, t),$$

where $\mathbf{U}$ may be any vector field (however, see also footnote 18). A well-known example for the second term is the Lorentz force acting on a particle with charge q in a magnetic field $\mathbf{B}(\mathbf{r}, t)$, i.e., $\mathbf{U} = q\mathbf{B}/c$. The Lorentz force conserves energy and hence is indeed conservative; it satisfies Eq. (1.147) with $V = const$.

difference. According to Stokes's theorem, this closed path is related to the integral of $\mathrm{curl}\,\mathbf{F}$ over the enclosed surface, and therefore

$$\int\limits_{\mathbf{r}_i,C}^{\mathbf{r}_f} \mathbf{F}\cdot dr' - \int\limits_{\mathbf{r}_i,C'}^{\mathbf{r}_f} \mathbf{F}\cdot dr' = \oint d\mathbf{A}\cdot\mathrm{curl}\,\mathbf{F} = 0. \tag{1.155}$$

This proves that the work integral

$$W_{fi} = \int\limits_{\mathbf{r}_i}^{\mathbf{r}_f} \mathbf{F}(\mathbf{r}')\cdot dr' \tag{1.156}$$

is *independent* of the path and thus the work W_{fi} done by the force in going from $\mathbf{r}_i$ to $\mathbf{r}_f$ is a function of these endpoints only.[18] Defining therefore the *potential* by

$$\boxed{V(\mathbf{r}) = -\int\limits_{\mathbf{r}_0}^{\mathbf{r}} \mathbf{F}(\mathbf{r}')\cdot dr'}\,, \tag{1.157}$$

where $\mathbf{r}_0$ is an arbitrary (fixed) reference point, the work can be written as

$$W_{fi} = V(\mathbf{r}_i) - V(\mathbf{r}_f), \tag{1.158}$$

which shows that the reference point $\mathbf{r}_0$ is of no physical consequence since the dependence on $\mathbf{r}_0$ drops out in this difference. Finally, Eq. (1.157) implies that the total differential[19] $dV = \boldsymbol{\nabla} V\cdot dr$ is given by $dV = -\mathbf{F}\cdot dr$ and uniqueness of the former representation therefore implies $\mathbf{F} = -\boldsymbol{\nabla} V$. This completes the proof of the equivalence (1.154). The equivalence diagram

$$\boxed{\begin{array}{ccc} \mathbf{F}\text{ is conservative} & \Longleftrightarrow & \displaystyle\int\limits_{i\to f} \mathbf{F}\cdot dr \ \ \text{path independent} \\[2mm] \Updownarrow & & \Updownarrow \\[2mm] \mathbf{F} = -\boldsymbol{\nabla} V(\mathbf{r}) & \Longleftrightarrow & \boldsymbol{\nabla}\times\mathbf{F} = 0 \end{array}} \tag{1.159}$$

summarizes these findings.

---

[18]Note that a force of the type $\dot{\mathbf{r}}\times\mathbf{U}$ considered in footnote 17 does not do any work — it is trivially independent of the path. Also, Eq. (1.155) is valid because of $d\mathbf{A}\cdot\mathrm{curl}(\dot{\mathbf{r}}\times\mathbf{U}) = 0$. For the curl to vanish by itself, it is required that $\mathbf{U}$ is without divergence, $\mathrm{div}\,\mathbf{U} = 0$. For the example of the Lorentz force this is true, of course.

[19]A *total* differential is also called an *exact* or a *perfect* differential. A differential $df = \sum_i h_i\, dx_i$ is exact if $\int df$ is independent of the path. This will be true if and only if $h_i = \partial f/\partial x_i$, which leads to the *integrability condition* $\partial h_i/\partial x_j = \partial h_j/\partial x_i$ as a necessary and sufficient test whether df is exact.

Note that Eq. (1.158) combined with (1.145),

$$W_{fi} = V_i - V_f = T_f - T_i, \tag{1.160}$$

provides an alternative proof of energy conservation since the arbitrariness of choosing the initial and final points implies

$$\boxed{T + V = E = const}, \tag{1.161}$$

independent of when this expression is evaluated.

1.5   Systems of Point Particles

The previous considerations are now to be generalized for systems of particles consisting of N point masses. Enumerating the particles by the index i, each particle is subject to Newton's second law,

$$m_i \ddot{\mathbf{r}}_i = \mathbf{F}_i \quad (i = 1, \ldots, N). \tag{1.162}$$

The total force acting on particle i can be decomposed into an *internal* force $\mathbf{F}_i^{\text{int}}$ due to the other particles in the system and an *external* force $\mathbf{F}_i^{\text{ext}}$ originating from causes outside of the system,

$$\mathbf{F}_i = \mathbf{F}_i^{\text{int}} + \mathbf{F}_i^{\text{ext}}. \tag{1.163}$$

To simplify the following considerations, we restrict the internal forces to be composed of two-body forces $\mathbf{F}_{ij}$ only that describe the action of particle j on particle i depending only on the properties (positions, velocities, etc.) of these two particles. We therefore have[20]

$$\mathbf{F}_i^{\text{int}} = \sum_{j=1}^{N} \mathbf{F}_{ij} \quad (i = 1, \ldots, N) \tag{1.164}$$

and

$$m_i \ddot{\mathbf{r}}_i = \sum_{j=1}^{N} \mathbf{F}_{ij} + \mathbf{F}_i^{\text{ext}} \quad (i = 1, \ldots, N). \tag{1.165}$$

---

[20]To simplify the notation when considering summations, etc., instead of explicitly excluding $j \neq i$, we define $\mathbf{F}_{ii} \equiv 0$.

1.5.1 *Total momentum*

The *center of mass* (CM) $\mathbf{R}$ of the particles is defined by

$$\mathbf{R} = \frac{1}{M} \sum_{i=1}^{N} m_i \mathbf{r}_i, \tag{1.166}$$

where

$$M = \sum_{i=1}^{N} m_i \tag{1.167}$$

is the *total mass* of the system. The entity

$$\mathbf{P} = M\dot{\mathbf{R}} = \sum_{i=1}^{N} m_i \dot{\mathbf{r}}_i = \sum_{i=1}^{N} \mathbf{p}_i, \tag{1.168}$$

the sum of all individual momenta, is called the *total momentum* of the system. It is like the momentum of a point particle located at the CM whose mass is the entire mass of the system.

Taking the second time derivative of Eq. (1.166) yields

$$M\ddot{\mathbf{R}} = \sum_{i=1}^{N} m_i \ddot{\mathbf{r}}_i = \sum_{i,j=1}^{N} \mathbf{F}_{ij} + \sum_{i=1}^{N} \mathbf{F}_i^{\text{ext}} = \sum_{i=1}^{N} \mathbf{F}_i^{\text{ext}} \tag{1.169}$$

where because of

$$\sum_{i,j=1}^{N} \mathbf{F}_{ij} \overset{(i\leftrightarrow j)}{=} \sum_{j,i=1}^{N} \mathbf{F}_{ji} = \sum_{i,j=1}^{N} \mathbf{F}_{ji} \overset{(1.55)}{=} - \sum_{i,j=1}^{N} \mathbf{F}_{ij} \tag{1.170}$$

the double sum over the internal two-body forces vanishes. We find, therefore, that dynamically the CM behaves like a point particle (whose mass is the entire mass of the system) subject to the sum total of all external forces acting on the system,

$$M\ddot{\mathbf{R}} = \dot{\mathbf{P}} = \mathbf{F}^{\text{ext}} \equiv \sum_{i=1}^{N} \mathbf{F}_i^{\text{ext}} \; ; \tag{1.171}$$

internal forces have no influence whatsoever on the motion of the CM.

If the sum of all external forces $\mathbf{F}^{\text{ext}}$ (but not necessarily individual contributions $\mathbf{F}_i^{\text{ext}}$) vanishes, this implies that

$$\boxed{\mathbf{F}^{\text{ext}} = 0 \quad \Rightarrow \quad \mathbf{P} = const}, \tag{1.172}$$

i.e., the total momentum of a system without any *net* external force is conserved, and the CM moves with uniform velocity $\dot{\mathbf{R}} = \mathbf{u} = const$. The CM therefore, can be written as

$$\mathbf{R}(t) = \mathbf{u}t + \mathbf{R}_0, \tag{1.173}$$

where $\mathbf{R}_0 = const$ is the position of the CM at $t = 0$. The particular inertial frame where the CM is at rest is called the *center of mass system* (CMS); it is the preferred frame to describe the motion within the system.

If all individual external forces vanish, $\mathbf{F}_i^{\text{ext}} = 0$, the system has no interaction with the outside world and it is referred to as a *closed system*. In practice, it is usually sufficient for the external interactions to be very small compared to the internal interactions to consider a system to be closed.

1.5.2 *Total angular momentum*

Defining the *total angular momentum* $\mathbf{L}$ by

$$\mathbf{L} = \sum_{i=1}^{N} \mathbf{L}_i = \sum_{i=1}^{N} \mathbf{r}_i \times \mathbf{p}_i, \tag{1.174}$$

where $\mathbf{L}_i$ is the angular momentum of particle i, the time derivative is given by

$$\frac{d\mathbf{L}}{dt} = \sum_{i=1}^{N} \mathbf{r}_i \times \mathbf{F}_i = \sum_{i=1}^{N} \mathbf{r}_i \times \mathbf{F}_i^{\text{ext}} + \sum_{i,j=1}^{N} \mathbf{r}_i \times \mathbf{F}_{ij}. \tag{1.175}$$

From Newton's 3rd law and the additional collinearity constraint (1.51) we find

$$\begin{aligned}
\sum_{i,j=1}^{N} \mathbf{r}_i \times \mathbf{F}_{ij} &= \frac{1}{2} \sum_{i,j=1}^{N} \mathbf{r}_i \times \mathbf{F}_{ij} + \frac{1}{2} \sum_{i,j=1}^{N} \mathbf{r}_i \times \mathbf{F}_{ij} \\
&= \frac{1}{2} \sum_{i,j=1}^{N} \mathbf{r}_i \times \mathbf{F}_{ij} + \frac{1}{2} \sum_{i,j=1}^{N} \mathbf{r}_j \times \mathbf{F}_{ji} \\
&= \frac{1}{2} \sum_{i,j=1}^{N} \mathbf{r}_i \times \mathbf{F}_{ij} - \frac{1}{2} \sum_{i,j=1}^{N} \mathbf{r}_j \times \mathbf{F}_{ij} \\
&= \frac{1}{2} \sum_{i,j=1}^{N} (\mathbf{r}_i - \mathbf{r}_j) \times \mathbf{F}_{ij} \\
&= 0. \tag{1.176}
\end{aligned}$$

In other words, internal forces do not produce a resulting torque and therefore cannot change the total angular momentum. We thus have

$$\frac{d\mathbf{L}}{dt} = \mathbf{N}^{\text{ext}} \equiv \sum_{i=1}^{N} \mathbf{N}_i^{\text{ext}} = \sum_{i=1}^{N} \mathbf{r}_i \times \mathbf{F}_i^{\text{ext}}, \qquad (1.177)$$

i.e., the change in total angular momentum is brought about by the *total torque* $\mathbf{N}^{\text{ext}}$ due to external forces.

If the total external torque vanishes, the total angular momentum is conserved,

$$\boxed{\mathbf{N}^{\text{ext}} = 0 \quad \Rightarrow \quad \mathbf{L} = const}. \qquad (1.178)$$

This is trivially true if the system is closed.

1.5.2.1   *Relationship to the center-of-mass system*

The total angular momentum is a quantity that depends on the reference point. It is a straightforward exercise to show that $\mathbf{L}$ with respect to an arbitrary reference point can be related to $\mathbf{L}_{\text{CM}}$ with respect to the center of mass $\mathbf{R}$ by

$$\mathbf{L} = \mathbf{L}_{\text{CM}} + M(\mathbf{R} \times \dot{\mathbf{R}}) = \mathbf{L}_{\text{CM}} + \mathbf{R} \times \mathbf{P}, \qquad (1.179)$$

i.e., the total angular momentum is the sum of the angular momentum within the center-of-mass system and the angular momentum of the center of mass with respect to the given inertial system. Using (1.171), Eq. (1.177) then becomes

$$\frac{d\mathbf{L}_{\text{CM}}}{dt} = \sum_{i=1}^{N} \underbrace{(\mathbf{r}_i - \mathbf{R})}_{=\mathbf{r}_i^{\text{CM}}} \times \mathbf{F}_i^{\text{ext}} \equiv \mathbf{N}_{\text{CM}}^{\text{ext}}, \qquad (1.180)$$

where the torque is now also to be evaluated in the CM. The time derivative is to be taken with respect to the inertial frame, as before.

These results mean that the center-of-mass motion (1.171) completely decouples from the equation of motion for the total angular momentum. This finding will be useful for many applications; in particular, for the treatment of the rigid-body motion in Sec. 5.3.

1.5.3 *Total energy*

Defining the *total kinetic energy*,

$$T = \sum_{i=1}^{N} \tfrac{1}{2} m_i \dot{\mathbf{r}}_i^2, \tag{1.181}$$

and taking the time derivative, one finds, similar to the one-particle case,

$$\frac{dT}{dt} = \sum_{i=1}^{N} \mathbf{F}_i \cdot \dot{\mathbf{r}}_i = \sum_{i=1}^{N} \left(\mathbf{F}_{i,\text{cons}} + \mathbf{F}_{i,\text{diss}} \right) \cdot \dot{\mathbf{r}}_i, \tag{1.182}$$

where the forces have been split into conservative and dissipative contributions,

$$\mathbf{F}_i = \mathbf{F}_{i,\text{cons}} + \mathbf{F}_{i,\text{diss}}. \tag{1.183}$$

The conservative forces presume the existence of a scalar function V depending on all particle coordinates such that

$$\sum_{i=1}^{N} \mathbf{F}_{i,\text{cons}} \cdot \dot{\mathbf{r}}_i = -\frac{dV(\mathbf{r}_1, \mathbf{r}_2, \ldots, \mathbf{r}_N)}{dt} = -\sum_{i=1}^{N} \boldsymbol{\nabla}_{\mathbf{r}_i} V \cdot \dot{\mathbf{r}}_i, \tag{1.184}$$

where

$$\boldsymbol{\nabla}_{\mathbf{r}_i} = \mathbf{e}_x \frac{\partial}{\partial x_i} + \mathbf{e}_y \frac{\partial}{\partial y_i} + \mathbf{e}_z \frac{\partial}{\partial z_i} \tag{1.185}$$

is the gradient with respect to the variable $\mathbf{r}_i = (x_i, y_i, z_i)$. As before, we restrict the considerations to conservative forces that are functions of the positions only and put

$$\mathbf{F}_{i,\text{cons}}(\mathbf{r}_1, \mathbf{r}_2, \ldots, \mathbf{r}_N) = -\boldsymbol{\nabla}_{\mathbf{r}_i} V(\mathbf{r}_1, \mathbf{r}_2, \ldots, \mathbf{r}_N). \tag{1.186}$$

We thus have

$$\frac{d}{dt}(T + V) = \sum_{i=1}^{N} \mathbf{F}_{i,\text{diss}} \cdot \dot{\mathbf{r}}_i. \tag{1.187}$$

If there are no dissipative forces then the energy is conserved,

$$\boxed{\text{Forces conservative} \quad \Rightarrow \quad T + V = E = const} \ . \tag{1.188}$$

Determination of the potentials

For given conservative forces, one can say more about the structure of the potential function V. Dropping the index 'cons', the conservative force contains external and internal contributions and we can introduce potentials for both of them, i.e.,

$$\mathbf{F}_i = \mathbf{F}_i^{\text{ext}}(\mathbf{r}_i) + \sum_{j=1}^{N} \mathbf{F}_{ij}(\mathbf{r}_i, \mathbf{r}_j)$$

$$= -\boldsymbol{\nabla}_{\mathbf{r}_i}\left[V^{\text{ext}}(\mathbf{r}_1, \mathbf{r}_2, \ldots, \mathbf{r}_N) + V^{\text{int}}(\mathbf{r}_1, \mathbf{r}_2, \ldots, \mathbf{r}_N)\right]. \tag{1.189}$$

For the conservative forces $\mathbf{F}_i^{\text{ext}}$, there exist potentials V_i, analogous to (1.157), with

$$\mathbf{F}_i^{\text{ext}}(\mathbf{r}_i) = -\boldsymbol{\nabla}V_i(\mathbf{r}_i) \equiv -\boldsymbol{\nabla}_{\mathbf{r}_i}V_i(\mathbf{r}_i). \tag{1.190}$$

Since

$$\boldsymbol{\nabla}_{\mathbf{r}_i}V_j(\mathbf{r}_j) = 0 \quad \text{for} \quad i \neq j, \tag{1.191}$$

we can put

$$V^{\text{ext}}(\mathbf{r}_1, \mathbf{r}_2, \ldots, \mathbf{r}_N) = \sum_{j=1}^{N} V_j(\mathbf{r}_j) \tag{1.192}$$

and thus have

$$\mathbf{F}_i^{\text{ext}}(\mathbf{r}_i) = -\boldsymbol{\nabla}_{\mathbf{r}_i}V^{\text{ext}}(\mathbf{r}_1, \mathbf{r}_2, \ldots, \mathbf{r}_N), \tag{1.193}$$

as desired.

For the conservative two-body forces $\mathbf{F}_{ij}$, there also exists a potential V_{ij},

$$\mathbf{F}_{ij}(\mathbf{r}_i, \mathbf{r}_j) = -\boldsymbol{\nabla}_{\mathbf{r}_i}V_{ij}(\mathbf{r}_i, \mathbf{r}_j) = -\boldsymbol{\nabla}_{\mathbf{r}_i-\mathbf{r}_j}V_{ij}(\mathbf{r}_i, \mathbf{r}_j), \tag{1.194}$$

where the last equality follows from the fact that $\mathbf{r}_j$ in this equation is a mere parameter and the derivative does not change if $\mathbf{r}_i - \mathbf{r}_j$ is substituted for $\mathbf{r}_i$. Newton's 3rd law then provides

$$\begin{aligned}
0 = \mathbf{F}_{ij} + \mathbf{F}_{ji} &= -\boldsymbol{\nabla}_{\mathbf{r}_i}V_{ij}(\mathbf{r}_i, \mathbf{r}_j) - \boldsymbol{\nabla}_{\mathbf{r}_j}V_{ji}(\mathbf{r}_j, \mathbf{r}_i) \\
&= -\boldsymbol{\nabla}_{\mathbf{r}_i-\mathbf{r}_j}V_{ij}(\mathbf{r}_i, \mathbf{r}_j) - \boldsymbol{\nabla}_{\mathbf{r}_j-\mathbf{r}_i}V_{ji}(\mathbf{r}_j, \mathbf{r}_i) \\
&= -\boldsymbol{\nabla}_{\mathbf{r}_i-\mathbf{r}_j}V_{ij}(\mathbf{r}_i, \mathbf{r}_j) + \boldsymbol{\nabla}_{\mathbf{r}_i-\mathbf{r}_j}V_{ji}(\mathbf{r}_j, \mathbf{r}_i) \\
&= -\boldsymbol{\nabla}_{\mathbf{r}_i-\mathbf{r}_j}\left[V_{ij}(\mathbf{r}_i, \mathbf{r}_j) - V_{ji}(\mathbf{r}_j, \mathbf{r}_i)\right]
\end{aligned} \tag{1.195}$$

and therefore

$$V_{ij}(\mathbf{r}_i, \mathbf{r}_j) = V_{ji}(\mathbf{r}_j, \mathbf{r}_i). \tag{1.196}$$

In view of the collinearity constraint (1.51), the two-body forces considered here have the structure [see Fig. 1.4(b)]

$$\mathbf{F}_{ij} = -f_{ij}\frac{\mathbf{r}_i - \mathbf{r}_j}{|\mathbf{r}_i - \mathbf{r}_j|}, \tag{1.197}$$

where f_{ij} is some scalar function, with $f_{ij} = f_{ji}$, that can depend only on the magnitude of the distance between the two particles,

$$r_{ij} = |\mathbf{r}_i - \mathbf{r}_j|. \tag{1.198}$$

Using then the fact that the gradient of a function that depends only on r is given by

$$\boldsymbol{\nabla} f(r) = \frac{df(r)}{dr}\frac{\mathbf{r}}{r}, \tag{1.199}$$

this means that the two-body potentials must have the structure

$$V_{ij}(\mathbf{r}_i, \mathbf{r}_j) = V_{ij}(r_{ij}), \tag{1.200}$$

and

$$\mathbf{F}_{ij} = -\boldsymbol{\nabla}_{\mathbf{r}_i - \mathbf{r}_j} V_{ij}(\mathbf{r}_i, \mathbf{r}_j) = -\frac{dV_{ij}(r_{ij})}{dr_{ij}}\frac{\mathbf{r}_i - \mathbf{r}_j}{|\mathbf{r}_i - \mathbf{r}_j|}. \tag{1.201}$$

Writing therefore ($V_{ii} \equiv 0$)

$$V^{\text{int}}(\mathbf{r}_1, \mathbf{r}_2, \ldots, \mathbf{r}_N) = \frac{1}{2}\sum_{i,j=1}^{N} V_{ij}(r_{ij}), \tag{1.202}$$

we can immediately verify that

$$\sum_{j=1}^{N}\mathbf{F}_{ij}(\mathbf{r}_i, \mathbf{r}_j) = -\boldsymbol{\nabla}_{\mathbf{r}_i} V^{\text{int}}(\mathbf{r}_1, \mathbf{r}_2, \ldots, \mathbf{r}_N), \tag{1.203}$$

as stipulated.

We thus have shown, by construction, that there exists a potential, *viz.*

$$V(\mathbf{r}_1, \mathbf{r}_2, \ldots, \mathbf{r}_N) = \sum_{j=1}^{N} V_j(\mathbf{r}_j) + \frac{1}{2}\sum_{i,j=1}^{N} V_{ij}(r_{ij}), \tag{1.204}$$

that indeed does satisfy (1.184) as assumed.

An example for such a potential is one for a system of charged, massive particles in a homogeneous gravitational field close to the surface of the Earth subject to an electrostatic field,

$$V(\mathbf{r}_1, \mathbf{r}_2, \ldots, \mathbf{r}_N) = \sum_{i=1}^{N} m_i g z_i + \sum_{i=1}^{N} q_i \Phi(\mathbf{r}_i) + \frac{1}{2}\sum_{i,j=1}^{N}{}' \frac{q_i q_j}{r_{ij}}, \tag{1.205}$$

where the gravitational field is $\mathbf{g} = -g\mathbf{e}_z$, the electrostatic potential is Φ, and the prime on the double sum indicates omission of the $i = j$ term. In this example, the external force on each particle has two parts, a gravitational and an electrostatic part. And the two-particle potential is the Coulomb potential.

1.5.4 *The ten integrals of motion*

The three conservations laws discussed in the preceding sections give rise to ten integration constants referred to as the *ten integrals of motion*:

- *Linear momentum conservation* determines the velocity $\mathbf{u}$ and the position $\mathbf{R}_0$ at $t = 0$ of the center of mass, corresponding to six constants.
- *Conservation of the angular momentum* $\mathbf{L}$ determines the spatial orientation of the *invariable plane* in which the motion takes place and the magnitude of the angular momentum (three constants).
- *Energy conservation* furnishes one constant, E.

If we know that the motion takes place in a plane, there are only $4+1+1 = 6$ integrals, and for a one-dimensional problem, there are $2 + 0 + 1 = 3$ integrals. The integrals of motion may be used to reduce the complexity of a particular problem.

For an N-body problem, we have $6N$ initial conditions (three position and three velocity components for each body). If all conservation laws apply, they yield 10 constants, leaving $6N - 10$ unused conditions. For the two-body problem, this leaves $12 - 10 = 2$ and for the three-body problem $18 - 10 = 8$, etc.

In the two-body case without external forces (see Sec. 1.8.1.1), the integrals of motion allow one to reduce the problem to a solvable equivalent one-dimensional problem that determines a complete solution of the original two-body case. A 'complete' solution is one where one obtains a single expression — either as an integral (that may have to be evaluated numerically) or in closed form[21] — that completely determines the solution for each coordinate individually, without coupling to the other coordinates.[22] For $N \geq 3$, such complete solutions cannot be obtained in general, and one must revert to approximation methods or to numerical integrations of the equations of motion.[23]

---

[21]We do not put any particular emphasis here on those solutions that can be expressed in 'closed' form in terms of known functions. 'Known' functions, after all, usually are defined in terms of more complicated expressions (power-series expansions, integrals, etc.); they have been given a name simply because they occur in many problems.

[22]More precisely, for a complete solution, it is not necessary that all equations of motion completely decouple — it is only necessary that all solutions can be obtained, either directly or recursively, in terms of the solution(s) of at least one completely decoupled equation.

[23]Let us emphasize that the lack of a 'complete' solution in the sense defined here does, of course, *not* mean that the corresponding problem is not solvable. It simply means

1.6   Inertial Frames of Reference

In Sec. 1.2.1, the notion of an *inertial* frame of reference was introduced in the context of Newton's Axioms. Under the simplifying assumptions of nonrelativistic kinematics, inertial frames of reference either are at rest or move with constant velocities with respect to the distant galaxies in the universe.[24] Inertial frames obey the *Galilean Relativity Principle*:[25]

> **Relativity Principle (Galilei):**
>
> (1) All inertial frames are equivalent.
> (2) Newton's equations are valid in all inertial frames.

$$(1.206)$$

In other words, a given physical phenomenon will be the same in all inertial frames. Formally, therefore, the corresponding Newtonian equations of motion must have the same form in all inertial frames. By contrast, in non-inertial frames the equations of motion acquire additional force terms, so-called *inertial forces*.

1.6.1   *Galilei transformations*

Defining an *event* as a point $\mathbf{r}$ in space together with an associated point in time, i.e.,

$$\text{event:} \quad (t, \mathbf{r}), \qquad (1.207)$$

the events of particular interest for a given physical phenomenon are the events $(t, \mathbf{r}(t))$ which are linked by the trajectories $\mathbf{r}(t)$ relevant to the phenomenon.[26]

Using two different inertial frames of reference, S and S' (see Fig. 1.10), the same event will have different representations in both frames. Since it

---

that one cannot escape solving it as a problem where each variable is coupled to at least one other variable. Usually, this requires a numerical approach.

[24]The 'distant galaxies in the universe' are invoked here because their motion is negligibly small over the time periods of typical experiments and, thus, they provide an idealized absolute frame of reference against which one can determine whether a given frame is an inertial one or not.

[25]Galileo Galilei (1564–1642) died in the year before Newton was born, so he did not know about Newton's axioms. Nevertheless, this principle given here in its modern formulation still bears his name to honor his contributions in establishing physics as an empirical science.

[26]The $(t, \mathbf{r})$ notation chosen here anticipates the notion of *spacetime* of special relativity (see Sec. 8.2).

is related to the same physical event, these representations must be related by a well-defined transformation, called a *Galilei transformation*,

$$(t', x_1', x_2', x_3')_{S'} \xleftarrow{\mathcal{G}} (t, x_1, x_2, x_3)_S , \tag{1.208}$$

where Cartesian coordinates are used for representing the frames.

According to the Relativity Principle, the Galilei transformation between two inertial frames of reference of the same physical event must leave the form of Newton's Axioms invariant. The requirement of *form invariance*, or *covariance*, therefore, may be used to derive the Galilei transformation.

First, since time is a mere parameter in classical mechanics, t and t' may at most differ by their zero-point setting, i.e.,

$$t' = t - t_0, \tag{1.209}$$

where $t_0 = const.$ This property is referred to as the *'homogeneity of time'* which means that time flows uniformly in all frames, independent of the location in space. Because of $dt' = dt$, a constant time dilation t_0 has no effect on Newton's equations.

Covariance of the 1st axiom

The space part of Eq. (1.208) may be written formally as

$$x_i' = \mathcal{G}_i(t, x_1, x_2, x_3), \qquad i = 1, 2, 3. \tag{1.210}$$

Applied to the required form invariance of Newton's first axiom, i.e.,

$$m\ddot{x}_i'(t') = 0 \xleftarrow{\mathcal{G}} m\ddot{x}_j(t) = 0, \tag{1.211}$$

this provides

$$0 = \ddot{x}_i' = \frac{\partial^2 \mathcal{G}_i}{\partial t^2} + 2\sum_{j=1}^{3} \frac{\partial^2 \mathcal{G}_i}{\partial x_j \partial t}\dot{x}_j + \sum_{k,j=1}^{3} \frac{\partial^2 \mathcal{G}_i}{\partial x_k \partial x_j}\dot{x}_k\dot{x}_j + \sum_{j=1}^{3} \frac{\partial \mathcal{G}_i}{\partial x_j}\ddot{x}_j. \tag{1.212}$$

The $\ddot{x}_j$ term vanishes anyway. Since the remaining three terms are linearly independent, their sum can only vanish if the corresponding second-order derivatives vanish individually. This means that $\mathcal{G}_i$ must be linear in x_j and t. The most general linear transformation may be written as

$$x_i' = \sum_{j=1}^{3} D_{ij}x_j - v_i't - d_i', \tag{1.213}$$

where D_{ij}, v_i', and d_i' are constants in space and time.

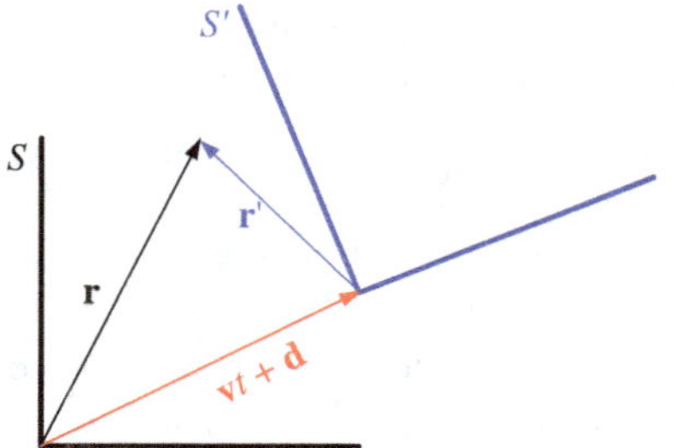

Fig. 1.10   Simplified two-dimensional sketch of the Galilei transformation relating the inertial frames S and S', whose origins are displaced by the uniformly time-dependent translation $\mathbf{v}t + \mathbf{d}$. The Cartesian axes of the frame S' are rotated with respect to those of S.

The meaning of the parameters v_i' and d_i' is found by putting $x_j = 0$. The ensuing relation $x_i' = -v_i't - d_i'$ describes the origin of S as seen from the origin of S' (using the clock of S). It shows that the origin of S' is moving with respect to S with a constant velocity $\mathbf{v} = \sum_i v_i' \mathbf{e}_i'$ (where $\mathbf{e}_i'$ are the fixed unit vectors of the S' inertial system), i.e., the total displacement of the origin of S' is $\mathbf{v}t + \mathbf{d}$, with the constant vector $\mathbf{d} = \sum_i d_i' \mathbf{e}_i'$ being the displacement at $t = 0$ (cf. Fig. 1.10).

The coefficients D_{ij} describe a possible constant rotation of the axes of the primed system with respect to the unprimed system, i.e.,

$$(\mathbf{e}_1', \mathbf{e}_2', \mathbf{e}_3')_{S'} \quad \overset{D}{\longleftarrow} \quad (\mathbf{e}_1, \mathbf{e}_2, \mathbf{e}_3)_S . \tag{1.214}$$

Rotations will be treated in detail in Sec. 5.1.1. Here we only mention that three-dimensional rotations are fully determined by three parameters,[27] and that the D_{ij} may be written in terms of the unit vectors as

$$D_{ij} = \mathbf{e}_i' \cdot \mathbf{e}_j = \cos(\mathbf{e}_i', \mathbf{e}_j). \tag{1.215}$$

The scalar product here is called a *direction cosine.*

In summary, the general Galilei transformations in Cartesian coordinates can be written as

$$t' = t - t_0, \tag{1.216a}$$

$$x_i' = \sum_{j=1}^{3} D_{ij}x_j - v_i't - d_i', \tag{1.216b}$$

which shows that they consist of the following individual transformation steps:

---

[27]This is intuitively obvious from the fact that a rotation can be determined by specifying a rotation axis (in terms of a unit vector along this axis: two parameters) and a rotation angle about this axis (one parameter). Another parametrization example are the three Euler angles (see Fig. 5.3).

- A constant temporal displacement by t_0;
- A constant spatial displacement by $\mathbf{d} = \sum_i d_i' \mathbf{e}_i'$;
- A uniform spatial displacement by $\mathbf{v}t$, where $\mathbf{v} = \sum_i v_i' \mathbf{e}_i'$;
- A rotation of the axes of S into the axes of S' described by D_{ij}.

Altogether, therefore, the Galilei transformations have ten parameters; they are directly related to the ten integrals of motion discussed in the previous Sec. 1.5.4.

Note that the spatial Galilei transformation, Eq. (1.216b), can just as well be written as

$$x_i' = \sum_{j=1}^{3} D_{ij}\left(x_j - v_j t - d_j\right), \tag{1.217}$$

where $v_j = \mathbf{v} \cdot \mathbf{e}_j$ and $d_j = \mathbf{d} \cdot \mathbf{e}_j$ [as compared to $v_i' = \mathbf{v} \cdot \mathbf{e}_i' = \sum_j D_{ij} v_j$ and $d_i' = \mathbf{d} \cdot \mathbf{e}_i' = \sum_j D_{ij} d_j$ for Eq. (1.216b)].

Under Galilei transformations the velocities and accelerations transform according to

$$\dot{x}_i'(t') = \frac{dx_i'}{dt'} = \frac{dx_i'}{dt} = \sum_{j=1}^{3} D_{ij}\dot{x}_j - v_i'$$

$$= \sum_{j=1}^{3} D_{ij}\left(\dot{x}_j - v_j\right), \tag{1.218}$$

$$\ddot{x}_i'(t') = \frac{d^2 x_i'}{dt'^2} = \frac{d^2 x_i'}{dt^2} = \sum_{j=1}^{3} D_{ij}\ddot{x}_j, \tag{1.219}$$

respectively. The latter relation, in particular, immediately gives back the covariance condition (1.211) upon multiplication by the mass m.

Note that the inverse of a given Cartesian Galilei transformation is found by rewriting Eq. (1.216b) as

$$\sum_{j=1}^{3} D_{ij} x_j = x_i' + v_i' t + d_i', \tag{1.220}$$

and multiplying by D_{ki}^{T} with a subsequent summation,

$$\sum_{i,j=1}^{3} D_{ki}^{\mathrm{T}} D_{ij} x_j = x_k = \sum_{i=1}^{3} D_{ki}^{\mathrm{T}}\left(x_i' + v_i' t + d_i'\right),$$

$$= \sum_{i=1}^{3} D_{ki}^{\mathrm{T}} x_i' + v_k t + d_k, \tag{1.221}$$

where the transposed rotation D^{T},

$$D_{ki}^{\mathrm{T}} = \mathbf{e}_k \cdot \mathbf{e}_i', \tag{1.222}$$

is the inverse of the rotation D, i.e.,

$$\sum_{i=1}^{3} D_{ki}^{\mathrm{T}} D_{ij} = \sum_{i=1}^{3} \mathbf{e}_k \cdot \mathbf{e}_i' \mathbf{e}_i' \cdot \mathbf{e}_j = \mathbf{e}_k \cdot \mathbf{1} \cdot \mathbf{e}_j = \mathbf{e}_k \cdot \mathbf{e}_j = \delta_{kj}, \qquad (1.223)$$

or

$$D^{\mathrm{T}} D = D D^{\mathrm{T}} = \mathbb{1}, \qquad (1.224)$$

where $\mathbb{1}$ is the unit matrix. The complete inverse Galilei transformations are therefore given by

$$t = t' + t_0, \qquad (1.225\text{a})$$

$$x_k = \sum_{i=1}^{3} D_{ki}^{\mathrm{T}} x_i' + v_k t' + \hat{d}_k, \qquad (1.225\text{b})$$

where $\hat{d}_k = d_k + v_k t_0$.

Covariance of the 2nd axiom

Next, we need to check if, and how, Newton's 2nd axiom remains form invariant under the transformations (1.216).

Writing the second axiom in components,

$$m\ddot{x}_j(t) = F_j(x, \dot{x}, t), \qquad (1.226)$$

where x and $\dot{x}$ generically stand for all components of the position and velocity, respectively, we first perform a rotation,

$$m\sum_{j=1}^{3} D_{ij} \ddot{x}_j(t) = \sum_{j=1}^{3} D_{ij} F_j(x, \dot{x}, t). \qquad (1.227)$$

Then, applying the inverse Galilei transformations (1.225) to the arguments of $F_j(x, \dot{x}, t)$ on the right-hand side, we may introduce a new function of x', $\dot{x}'$, and t' by

$$F_i'(x', \dot{x}', t') = \sum_{j=1}^{3} D_{ij} F_j(x, \dot{x}, t). \qquad (1.228)$$

Using (1.219), we thus have

$$m\ddot{x}_i'(t') = F_i'(x', \dot{x}', t') \; ; \qquad (1.229)$$

in other words, also the second axiom is covariant under Galilei transformations,

$$m\ddot{x}_i'(t') = F_i'(x', \dot{x}', t') \quad \overset{\mathcal{G}}{\longleftarrow} \quad m\ddot{x}_j(t) = F_j(x, \dot{x}, t). \qquad (1.230)$$

The force components $F_i'(x', \dot{x}', t')$ and $F_j(x, \dot{x}, t)$ in the two frames will be different functions of their respective variables, however, the physical forces $\mathbf{F}'$ and $\mathbf{F}$ themselves are identical. This is easily seen from

$$\mathbf{F}' = \sum_{i=1}^{3} \mathbf{e}_i' F_i' = \sum_{i,j=1}^{3} \mathbf{e}_i' D_{ij} F_j = \sum_{i,j=1}^{3} \mathbf{e}_i' \mathbf{e}_i' \cdot \mathbf{e}_j F_j = \sum_{j=1}^{3} \mathbf{e}_j F_j = \mathbf{F}, \quad (1.231)$$

where Eq. (1.228) was used. In view of this equality, the physical trajectories will be the same (albeit viewed from different frames), i.e., they will be *invariant* under the Galilei transformations (1.216).

Galilei transformations in terms of abstract vectors

The Galilei transformations (1.216), and their inverses (1.225), were derived here using Cartesian representations of the position vectors in S and S'. Written in terms of abstract vectors, however, they assume the simple forms (cf. Fig. 1.10)

$$t' = t - t_0, \tag{1.232a}$$
$$\mathbf{r}' = \mathbf{r} - \mathbf{v}t - \mathbf{d}, \tag{1.232b}$$

and

$$t = t' + t_0, \tag{1.233a}$$
$$\mathbf{r} = \mathbf{r}' + \mathbf{v}t' + \hat{\mathbf{d}}, \tag{1.233b}$$

resepctively, where $\hat{\mathbf{d}} = \mathbf{d} + \mathbf{v}t_0$. And the velocities and accelerations are related by

$$\dot{\mathbf{r}}' = \dot{\mathbf{r}} - \mathbf{v}, \tag{1.234}$$
$$\ddot{\mathbf{r}}' = \ddot{\mathbf{r}}. \tag{1.235}$$

The last relation, in particular, immediately provides the covariance of the 1st and 2nd axioms.

Instead of employing Cartesian representations from the start, we could of course just as well have started from the form invariance requirement of the first axiom and worked our way back from Eq. (1.235) to derive Eqs. (1.232) and (1.233). Note, however, that the possibility that the axes of the two frames may be rotated with respect to each other does not explicitly enter the abstract vector relations. Since this is relevant for many

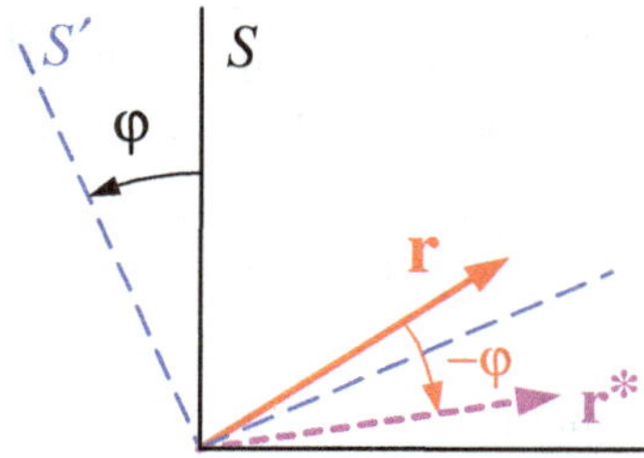

Fig. 1.11   Schematic two-dimensional visualization of passive and active rotations. The representation of the fixed vector **r** in the frame S' (rotated by an angle φ with respect to S) corresponds to a passive rotation D_φ. To obtain a vector **r*** with an identical representation in S, i.e., $\left(x_1^*, x_2^*, x_3^*\right)_S \equiv \left(x_1', x_2', x_3'\right)_{S'}$, the vector **r** needs to be actively rotated by an angle $-\varphi$. This active rotation $D_\varphi^{\mathrm{T}} = D_{-\varphi}$ thus is the inverse of the passive rotation D_φ.

practical considerations, a Cartesian representation was chosen here instead. Projecting onto the axes of the S' system, i.e.,

$$x_i' = \mathbf{e}_i' \cdot \mathbf{r}' = \mathbf{e}_i' \cdot \mathbf{r} - \underbrace{\mathbf{e}_i' \cdot \mathbf{v}}_{=v_i'}\, t - \underbrace{\mathbf{e}_i' \cdot \mathbf{d}}_{=d_i'}$$

$$= \mathbf{e}_i' \cdot \left(\sum_{j=1}^{3} x_j \mathbf{e}_j\right) - v_i' t - d_i'$$

$$= \sum_{j=1}^{3} \underbrace{\mathbf{e}_i' \cdot \mathbf{e}_j}_{=D_{ij}}\, x_j - v_i' t - d_i'$$

$$= \sum_{j=1}^{3} D_{ij}\, x_j - v_i' t - d_i', \tag{1.236}$$

we recover the explicit Galilei transformation (1.216b) from the abstract vector relation.

1.6.1.1   *Passive and active Galilei transformations*

In the previous subsections, the Galilei transformation was introduced by considering *one* physical system seen from *two* inertial frames. This is referred to as a *passive* Galilei transformation. In addition, we may also employ the Galilei transformations to relate *two* physical systems within *one* inertial frame of reference.

A simple example for such an *active* Galilei transformation would be a rotation of the physical system about the origin of the given frame. Figure 1.11 shows that the rotation matrix for such an active rotation would be just the inverse of the rotation matrix for a passive rotation if we would like to describe both on the same footing. In view of this, the active rotations are going to be denoted by D^{T} rather than D.

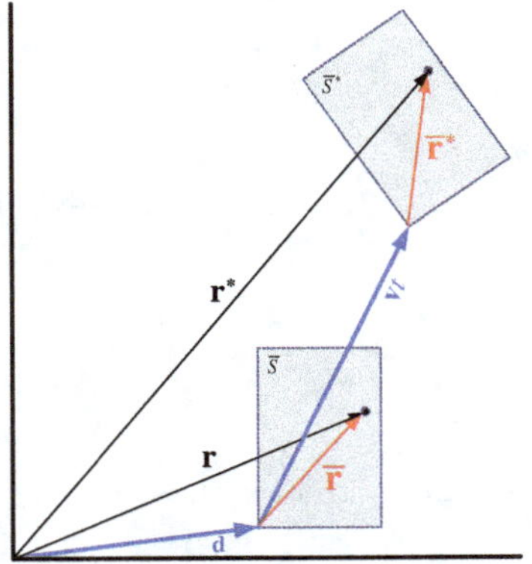

Fig. 1.12   Schematic illustration of an active Galilei transformation. The physical systems (shaded areas) are $\overline{S}$ and $\overline{S}^*$ before and after the transformation, respectively. Both are described within the same inertial frame S. Position vectors show the same point before (at $\mathbf{r}$) and after (at $\mathbf{r}^*$) the transformation. The vector $\mathbf{d}$ describes a point on the rotation axis at $t = 0$.

Labeling the active transformations by asterisks, they read

$$t \quad \longrightarrow \quad t^* = t + t_0, \tag{1.237a}$$

$$x_i \quad \longrightarrow \quad x_i^* = \sum_{j=1}^{3} D_{ij}^{\mathrm{T}} x_j + v_i t + \hat{d}_i^*, \tag{1.237b}$$

where *all* coordinates are now from *the same* inertial frame S, with $(t^*, x_1^*, x_2^*, x_3^*)$ displaced and rotated with respect to (t, x_1, x_2, x_3). Formally, the active transformation is seen here to look exactly like the inverse passive transformation (1.225).

To illustrate the physical meaning of the active transformations, it is convenient to introduce, at $t = 0$, a reference point within the physical system that lies on the rotation axis. Figure 1.12 shows the physical system $\overline{S}$ at $t = 0$ that is transformed into $\overline{S}^*$. At $t = 0$, this reference point lies at $\mathbf{d}$ and gets translated by $\mathbf{v}t$ when $\overline{S}$ goes over into $\overline{S}^*$. Relative to this reference point the position vectors $\overline{\mathbf{r}}$ in $\overline{S}$ and $\overline{\mathbf{r}}^*$ in $\overline{S}^*$ are related by a simple rotation,

$$\overline{x}_i^* = \sum_{j=1}^{3} D_{ij}^{\mathrm{T}} \overline{x}_j, \tag{1.238}$$

without a translation.

The meaning of the parameter $\hat{d}_i^*$ can be found from Fig. 1.12 via

$$x_i^* = d_i + v_i t + \overline{x}_i^*$$

$$= d_i + v_i t + \sum_{j=1}^{3} D_{ij}^{\mathrm{T}} \overline{x}_j$$

$$= d_i + v_i t + \sum_{j=1}^{3} D_{ij}^{\mathrm{T}} \left(x_j - d_j \right)$$

$$= \sum_{j=1}^{3} D_{ij}^{\mathrm{T}} x_j + v_i t + d_i - \sum_{j=1}^{3} D_{ij}^{\mathrm{T}} d_j, \qquad (1.239)$$

and therefore

$$\hat{d}_i^* = d_i - \sum_{j=1}^{3} D_{ij}^{\mathrm{T}} d_j, \qquad (1.240)$$

i.e., $\hat{d}_i^*$ does not have the simple geometric meaning that d_i' has for the passive Galilei transformation (1.216).

One easily verifies, following the previous procedure, that the form invariance of Newtons' Axioms remains true for active transformations. For the second axiom, in particular, one has

$$m\ddot{x}_i(t) = F_i(x, \dot{x}, t) \ \longrightarrow \ m\ddot{x}_i^*(t^*) = F_i^*(x^*, \dot{x}^*, t^*), \qquad (1.241)$$

where both equation relate to the same inertial system S. The force components, however, in general will *not* be the same, and therefore, in general, the trajectories of the transformed physical system will be different, despite the covariance of Newton's second law.

To illustrate this, imagine playing billiards on a table with a level top and on one with a skewed top. The two systems are related by an active Galilei transformation (a rotation) and for both systems Newton's Axioms are valid. Obviously, the trajectories of the balls on both table will be quite different since there is now an *external* force field (gravity) which cannot be rotated the same way as the system itself. In general, therefore, the actual external force at the transformed coordinates x^* will be different from the force resulting from the Galilei transformation of the external force at the old location,

$$F_i^{\mathrm{ext}}(x^*, \dot{x}^*, t^*) \ \neq \ F_i^{*\,\mathrm{ext}}(x^*, \dot{x}^*, t^*) \equiv \sum_{j=1}^{3} D_{ij}^{\mathrm{T}} F_j^{\mathrm{ext}}(x, \dot{x}, t), \qquad (1.242)$$

and therefore the total forces for both systems will be different. By contrast, if we played three-dimensional billiards inside a box somewhere in space, away from all external gravitational influences, there would be no external forces and then the trajectories would indeed be *invariant* under active Galilei transformations in the sense that relative to the (rotated) box, the forces would be identical,

$$F_i(x^*, \dot{x}^*, t^*) = F_i^*(x^*, \dot{x}^*, t^*), \qquad (1.243)$$

and therefore the trajectories would look identical. *Invariance* is a stronger requirement than *covariance* — it is satisfied when systems are *closed*. For invariance to obtain, it is sufficient to ascertain that the forces rotate the same way as the particle positions would rotate within the system.

1.6.2 *Closed systems*

A *closed* system is one which has no interaction with its environment. For a closed system of point masses, all external forces vanish.

Considering only conservative two-body forces, the internal forces on particle i are given by [cf. (1.201)]

$$\mathbf{F}_i(\mathbf{r}_1,\ldots,\mathbf{r}_N) = \sum_{j=1}^{N} \mathbf{F}_{ij}(\mathbf{r}_i,\mathbf{r}_j) = -\sum_{j=1}^{N} \frac{dV_{ij}(r_{ij})}{dr_{ij}} \frac{\mathbf{r}_i - \mathbf{r}_j}{|\mathbf{r}_i - \mathbf{r}_j|} \qquad (1.244)$$

(where all indices here pertain to particles and not to vector components, of course). The interparticle distance $\mathbf{r}_i - \mathbf{r}_j$ is invariant under uniform translations, i.e., it is subject only to pure rotations,

$$\mathbf{r}_i^* - \mathbf{r}_j^* = D^{\mathrm{T}}(\mathbf{r}_i - \mathbf{r}_j), \qquad (1.245)$$

where $\mathbf{r}' = D^{\mathrm{T}}\mathbf{r}$ is a symbolic notation for the rotation [cf. Eq. (1.18)]. Furthermore, its magnitude $r_{ij} = |\mathbf{r}_i - \mathbf{r}_j|$ is invariant under rotations [see Sec. 5.1.1 and, in particular, Eq. (5.11)] and therefore under active Galilei transformations,

$$r_{ij} \equiv |\mathbf{r}_i - \mathbf{r}_j| = |\mathbf{r}_i^* - \mathbf{r}_j^*| \equiv r_{ij}^*. \qquad (1.246)$$

Hence, the force on particle i transforms as

$$\begin{aligned}
\mathbf{F}_i^*(\mathbf{r}_1^*,\ldots,\mathbf{r}_N^*) &= D^{\mathrm{T}}\mathbf{F}_i(\mathbf{r}_1,\ldots,\mathbf{r}_N) \\
&= \sum_{j=1}^{N} D^{\mathrm{T}}\mathbf{F}_{ij}(\mathbf{r}_i,\mathbf{r}_j) \\
&= -\sum_{j=1}^{N} \frac{dV_{ij}(r_{ij})}{dr_{ij}} \frac{D^{\mathrm{T}}(\mathbf{r}_i - \mathbf{r}_j)}{|\mathbf{r}_i - \mathbf{r}_j|} \\
&= -\sum_{j=1}^{N} \frac{dV_{ij}(r_{ij}^*)}{dr_{ij}^*} \frac{\mathbf{r}_i^* - \mathbf{r}_j^*}{|\mathbf{r}_i^* - \mathbf{r}_j^*|} \\
&= \sum_{j=1}^{N} \mathbf{F}_{ij}(\mathbf{r}_i^*,\mathbf{r}_j^*) \\
&= \mathbf{F}_i(\mathbf{r}_1^*,\ldots,\mathbf{r}_N^*), \qquad (1.247)
\end{aligned}$$

i.e., showing indeed that the original force evaluated at the new space-time locations is equal to the transformed force at the new space-time locations, as required by (1.243). Conservative two-body forces thus leave the trajectories invariant under active Galilei transformations.

1.7 Accelerated Frames of Reference

Newton's Axioms are formulated in inertial systems. By contrast, non-inertial frames of reference are those which are accelerated relative to an inertial system — the acceleration produces additional frame-dependent force terms that lead to equations of motion of a form different from Newton's Axioms.[28] In the following, the necessary modifications for a linearly accelerated frame and for a rotating frame will be discussed. To simplify matters, the clocks in both frames are taken as synchronized so that $t' = t$.

1.7.1 *Linearly accelerated frame*

Consider the situation depicted in Fig. 1.10, but assume now that only S is an inertial system and S' is a system that is accelerated with respect to S. For a constant acceleration $\mathbf{a}$, the relation (1.232b) is modified to

$$\mathbf{r}' = \mathbf{r} - \tfrac{1}{2}\mathbf{a}t^2 - \mathbf{v}t - \mathbf{d}. \tag{1.248}$$

Given the validity of Newton's first law for a free motion in S,

$$m\ddot{\mathbf{r}} = 0 \quad \longrightarrow \quad m\ddot{\mathbf{r}}' = -m\mathbf{a}, \tag{1.249}$$

the description of the free motion in the accelerated frame S' is seen to acquire an additional force term $\mathbf{F}' = -m\mathbf{a}$, thus showing explicitly that the first law no longer holds true in S'. In the non-inertial frame, therefore, the equation of motion has the form of the second law, with a constant force field $\mathbf{F}' = -m\mathbf{a}$. Forces of this kind — which result from the *inertial* form $m\ddot{\mathbf{r}} = 0$ of the first law — are called *inertial forces*. They are also called *pseudoforces* since they are absent in inertial frames. Within the accelerated frame, however, pseudoforces appear as real (i.e., measurable) forces.

---

[28]Somewhat loosely speaking one often says that Newton's Laws are 'not valid' for non-inertial frames. What is meant is that the equations of Newton's Laws are 'not directly applicable' to non-inertial frames of reference. For such frames, as will be shown in this chapter, one needs to go through the additional step of transforming the equations of motion from an appropriate inertial frame to the non-inertial frame of interest. We emphasize, however, that Newton's Laws *always* form the basis for the description of nonrelativistic mechanics.

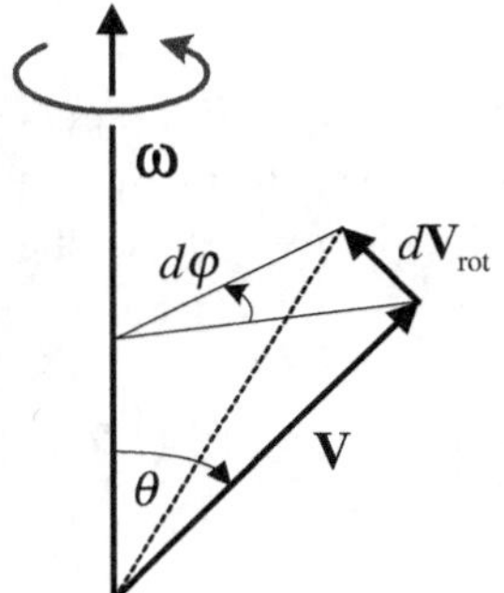

Fig. 1.13   Change $d\mathbf{V}_{\rm rot}$ of vector $\mathbf{V}$ as seen in frame S during infinitesimal rotation by $d\varphi$. The rotation increment satisfies $d\mathbf{V}_{\rm rot} \perp \mathbf{V}$ and $d\mathbf{V}_{\rm rot} \perp \boldsymbol{\omega}$ and $|d\mathbf{V}_{\rm rot}| = |\mathbf{V}|\sin\theta\, d\varphi$. (If the rotation axis coincides with the z axis, φ and θ correspond to the usual spherical coordinates.)

1.7.2   Rotating frame

Next, let us consider a frame S' rotating with a constant angular velocity

$$\boldsymbol{\omega} = \frac{d\boldsymbol{\varphi}}{dt} \tag{1.250}$$

whose origin coincides with that of an inertial frame S. The vectors $\boldsymbol{\omega}$ and $d\boldsymbol{\varphi}$ point in the direction of the rotation axis, and $d\varphi = |d\boldsymbol{\varphi}|$ is the angle of rotation during the time interval dt.[29]

Consider now a vector $\mathbf{V}$ and its changes during a time interval dt as observed from the two frames: Denoting its change in S' by $d'\mathbf{V}$, the *same* vector is seen in S as a rotating vector that changes according to

$$d\mathbf{V} = d'\mathbf{V} + d\mathbf{V}_{\rm rot}, \tag{1.251}$$

where (see Fig. 1.13)

$$d\mathbf{V}_{\rm rot} = d\boldsymbol{\varphi} \times \mathbf{V} = (\boldsymbol{\omega} \times \mathbf{V})\, dt \tag{1.252}$$

is the additional increment due to the rotation. Hence, the time derivatives in the two frames are related by

$$\boxed{\frac{d\mathbf{V}}{dt} = \frac{d'\mathbf{V}}{dt} + \boldsymbol{\omega} \times \mathbf{V},} \tag{1.253}$$

where the time derivative d'/dt on the right-hand side,

$$\frac{d'\mathbf{V}}{dt} = \left.\frac{d\mathbf{V}}{dt}\right|_{S'} = \dot{\mathbf{V}}', \tag{1.254}$$

affects only the quantities that change *within* S'. The notation $\dot{\mathbf{V}}'$ is defined by this equation.

---

[29]By convention, if the thumb of the right hand points in the direction of $d\boldsymbol{\varphi}$ (or $\boldsymbol{\omega}$), the fingers of the right hand curl in the direction of the rotation.

1.7.2.1   *Position and velocity*

Introducing Cartesian unit vectors $\mathbf{e}_i$ and $\mathbf{e}_i'(t)$ for both S and S', respectively, the latter are time-dependent, of course, when viewed from S, due to the rotation of S'. Since the origins of S and S' coincide, expressing the position vector in both frames produces

$$\mathbf{r} = \sum_{i=1}^{3} x_i(t)\mathbf{e}_i = \sum_{i=1}^{3} x_i'(t)\mathbf{e}_i'(t). \tag{1.255}$$

The velocity then is given by

$$\dot{\mathbf{r}} = \frac{d\mathbf{r}}{dt} = \sum_{i=1}^{3} \frac{dx_i'(t)}{dt}\mathbf{e}_i'(t) + \sum_{i=1}^{3} x_i'(t)\frac{d\mathbf{e}_i'(t)}{dt}, \tag{1.256}$$

where

$$\sum_{i=1}^{3} \frac{dx_i'(t)}{dt}\mathbf{e}_i'(t) = \frac{d'\mathbf{r}}{dt} = \left.\frac{d\mathbf{r}}{dt}\right|_{S'} = \dot{\mathbf{r}}' \tag{1.257}$$

is the change of the position vector relative to S'. Again, the last equality here is a definition of what is meant by $\dot{\mathbf{r}}'$ (as opposed to $\dot{\mathbf{r}}$). To evaluate $d\mathbf{e}_i'/dt$, we go back to Eq. (1.253) and find

$$\frac{d\mathbf{e}_i'(t)}{dt} = \boldsymbol{\omega} \times \mathbf{e}_i'(t) \tag{1.258}$$

since in S' the unit vectors $\mathbf{e}_i'$ are fixed, i.e., $d'\mathbf{e}_i' = 0$; hence,

$$\sum_{i=1}^{3} x_i'(t)\frac{d\mathbf{e}_i'(t)}{dt} = \boldsymbol{\omega} \times \sum_{i=1}^{3} \mathbf{e}_i'(t)x_i'(t) = \boldsymbol{\omega} \times \mathbf{r}. \tag{1.259}$$

The desired relation between the velocities in the inertial and rotating frames therefore reads

$$\boxed{\dot{\mathbf{r}} = \dot{\mathbf{r}}' + \boldsymbol{\omega} \times \mathbf{r}}, \tag{1.260}$$

which corresponds exactly to (1.253), of course.

1.7.2.2   *Equation of motion*

Writing Eq. (1.260) in components as

$$\dot{\mathbf{r}} = \sum_{i=1}^{3} \left(\dot{x}_i' + x_i' \,\boldsymbol{\omega}\times \right) \mathbf{e}_i', \tag{1.261}$$

the second derivative is found as

$$\ddot{\mathbf{r}} = \frac{d\dot{\mathbf{r}}}{dt} = \sum_{i=1}^{3} \left[\left(\ddot{x}'_i + \dot{x}'_i\, \boldsymbol{\omega}\times \right) \mathbf{e}'_i + \left(\dot{x}'_i + x'_i\, \boldsymbol{\omega}\times \right) \left(\boldsymbol{\omega} \times \mathbf{e}'_i \right) \right]$$

$$= \sum_{i=1}^{3} \left[\ddot{x}'_i \mathbf{e}'_i + 2\dot{x}'_i\, \boldsymbol{\omega} \times \mathbf{e}'_i + x'_i\, \boldsymbol{\omega} \times \left(\boldsymbol{\omega} \times \mathbf{e}'_i \right) \right], \qquad (1.262)$$

where the term

$$\sum_{i=1}^{3} \ddot{x}'_i \mathbf{e}'_i = \frac{d'\dot{\mathbf{r}}'}{dt} = \left. \frac{d\dot{\mathbf{r}}'}{dt} \right|_{S'} = \ddot{\mathbf{r}}' \qquad (1.263)$$

is the acceleration in S', with the notation $\ddot{\mathbf{r}}'$ chosen in analogy to (1.257). In vector notation, we thus have

$$\boxed{\ddot{\mathbf{r}} = \ddot{\mathbf{r}}' + 2\left(\boldsymbol{\omega} \times \dot{\mathbf{r}}' \right) + \boldsymbol{\omega} \times \left(\boldsymbol{\omega} \times \mathbf{r} \right)} \qquad (1.264)$$

Note that we arrive at the same result if we apply the generic equation (1.253) to Eq. (1.260), *viz.*

$$\ddot{\mathbf{r}} = \frac{d\dot{\mathbf{r}}}{dt} = \frac{d'\dot{\mathbf{r}}}{dt} + \boldsymbol{\omega} \times \dot{\mathbf{r}} = \frac{d'\left(\dot{\mathbf{r}}' + \boldsymbol{\omega} \times \mathbf{r} \right)}{dt} + \boldsymbol{\omega} \times \left(\dot{\mathbf{r}}' + \boldsymbol{\omega} \times \mathbf{r} \right)$$

$$= \ddot{\mathbf{r}}' + \boldsymbol{\omega} \times \dot{\mathbf{r}}' + \boldsymbol{\omega} \times \left(\dot{\mathbf{r}}' + \boldsymbol{\omega} \times \mathbf{r} \right). \qquad (1.265)$$

For a free particle in the inertial system S, the first axiom holds true, and therefore

$$m\ddot{\mathbf{r}} = 0 \quad \longrightarrow \quad m\ddot{\mathbf{r}}' = -2m\left(\boldsymbol{\omega} \times \dot{\mathbf{r}}' \right) - m\,\boldsymbol{\omega} \times \left(\boldsymbol{\omega} \times \mathbf{r} \right). \qquad (1.266)$$

Again, we see here that in the non-inertial rotating system S' the first law no longer holds; instead, inertial pseudoforces appear, the *Coriolis force*

$$\mathbf{F}_{\mathrm{C}} = -2m\left(\boldsymbol{\omega} \times \dot{\mathbf{r}}' \right), \qquad (1.267)$$

and the *centrifugal force*

$$\mathbf{F}_{\mathrm{c.f.}} = -m\,\boldsymbol{\omega} \times \left(\boldsymbol{\omega} \times \mathbf{r} \right). \qquad (1.268)$$

The Coriolis force is proportional to ω and the velocity of the point mass; it is perpendicular to the direction of the motion — it is difficult to walk a straight line on a merry-go-round. The centrifugal force is proportional to ω^2 and to the distance of the point mass from the axis of rotation. It points away from this axis — on a fast-rotating merry-go-round one must hold on to something to prevent being pushed toward the perimeter.

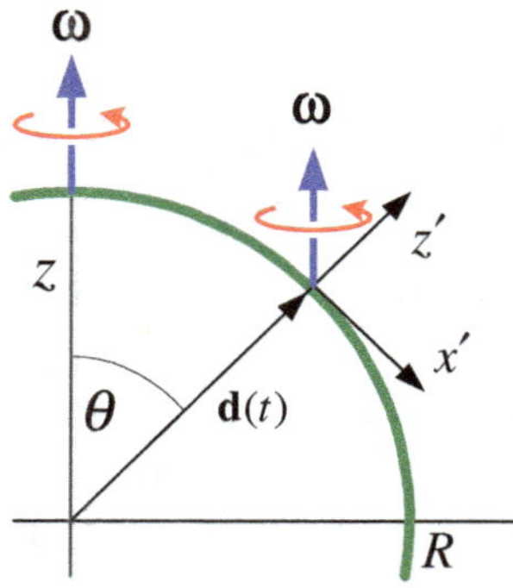

Fig. 1.14   A lab system on Earth's surface is an accelerated frame S', with coordinates x', y', z', where the x' axis is chosen to point south at all times. The origin of S' is rotating about the Earth's axis on a circle of radius $R\sin\theta$ with the angular velocity $\boldsymbol{\omega}$. S' itself rotates about an axis parallel to Earth's rotation axis with the same angular velocity $\boldsymbol{\omega}$.

If the body experiences a force $\mathbf{F}$ in the inertial frame, the inertial forces must be added to produce an effective force in the rotating frame,

$$\mathbf{F}_{\text{eff}} = \mathbf{F} - 2m\left(\boldsymbol{\omega} \times \dot{\mathbf{r}}'\right) - m\,\boldsymbol{\omega} \times (\boldsymbol{\omega} \times \mathbf{r}). \tag{1.269}$$

Even though the corresponding equation of motion,

$$m\ddot{\mathbf{r}}' = \mathbf{F}_{\text{eff}}, \tag{1.270}$$

looks like the second law, strictly speaking this is *not* the second law because of the appearance of the frame-dependent inertial forces — the second law is only valid in inertial frames.

1.7.2.3   *Laboratory system on rotating Earth*

A laboratory on Earth's surface (see Fig. 1.14) corresponds to an accelerated frame of reference S'. Let us take the center of mass of the Earth as the origin of an (approximate[30]) inertial frame S. A position $\mathbf{r}$ in S is related to the same position $\mathbf{r}'$ in S' by

$$\mathbf{r} = \mathbf{r}' + \mathbf{d}, \tag{1.271}$$

where the origin of S', described by $\mathbf{d}$, rotates with an angular velocity

$$\boldsymbol{\omega} = \omega\,\mathbf{e}_z = \omega\left(-\mathbf{e}'_x \sin\theta + \mathbf{e}'_z \cos\theta\right), \tag{1.272}$$

which is the same rate with which S' rotates about an axis through its origin parallel to Earth's axis.

With

$$\frac{d^2\mathbf{r}}{dt^2} = \frac{d'^2\mathbf{r}}{dt^2} + 2\boldsymbol{\omega} \times \frac{d'}{dt}\mathbf{r} + \boldsymbol{\omega} \times (\boldsymbol{\omega} \times \mathbf{r}) \qquad \text{and} \qquad \frac{d'}{dt}\mathbf{d} = 0, \tag{1.273}$$

---

[30]This will be a good approximation if the time period for an experiment is short enough so that Earth's trajectory may be approximated by a straight line.

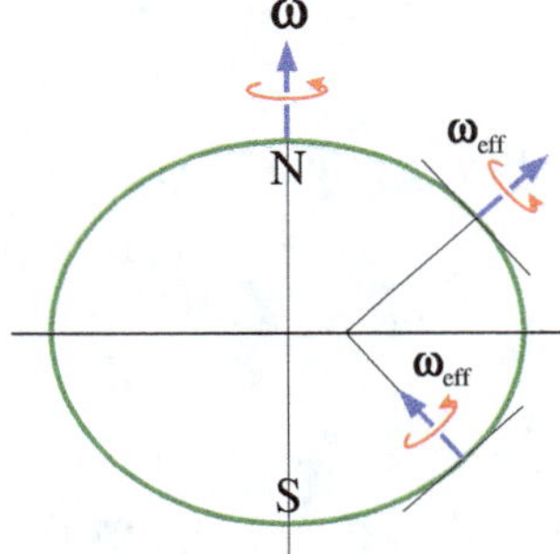

Fig. 1.15   Orientations of effective angular velocities $\boldsymbol{\omega}_{\text{eff}}$ of laboratory systems on the northern and southern hemispheres in relation to Earth's angular velocity $\boldsymbol{\omega}$, according to Eq. (1.281). The flattening of Earth is greatly exaggerated here.

we then find the equation of motion in S',

$$m\frac{d'^2\mathbf{r}'}{dt^2} \equiv m\ddot{\mathbf{r}}' = \hat{\mathbf{F}} + m\hat{\mathbf{g}} - m\ddot{\mathbf{d}} - 2m\left(\boldsymbol{\omega}\times\dot{\mathbf{r}}'\right) - m\,\boldsymbol{\omega}\times\left(\boldsymbol{\omega}\times\mathbf{r}'\right), \quad (1.274)$$

where the equation of motion in S was written as $m\ddot{\mathbf{r}} = \hat{\mathbf{F}} + m\hat{\mathbf{g}}$, i.e., the applied force was split into the gravitational force $m\hat{\mathbf{g}}$ and a remainder $\hat{\mathbf{F}}$. The lab frame (at the tip of $\mathbf{d}$) moves with Earth's rotation in a circle of radius $R\sin\theta$. In S, it is subject to a centripetal acceleration toward the rotation axis,

$$\ddot{\mathbf{d}} = \boldsymbol{\omega}\times\left(\boldsymbol{\omega}\times\mathbf{d}\right) = -\omega^2 R\sin\theta\left(\mathbf{e}'_x\cos\theta + \mathbf{e}'_z\sin\theta\right), \quad (1.275)$$

which corresponds to the centrifugal pseudoforce $-m\ddot{\mathbf{d}}$ in the lab frame equation (1.274).

A rotating ball of liquid is flattened into an ellipsoid under the influence of the gravitational force $m\hat{\mathbf{g}}$ and the centrifugal force $-m\ddot{\mathbf{d}}$ such that in equilibrium the sum of the two forces is perpendicular to the surface everywhere. (If this were not the case, there would be residual forces displacing mass elements in the surface.) The flattening of Earth from a sphere into what is called a *geoid* corresponds to this situation to a very good approximation (see Fig. 1.15). The resulting force is therefore approximately perpendicular to the surface of the Earth,

$$m(\hat{\mathbf{g}} - \ddot{\mathbf{d}}) \approx -mg\,\mathbf{e}'_z, \quad (1.276)$$

where $\mathbf{e}'_z$ is the direction of a plumb line on the flattened Earth and g is the effective gravitational acceleration. Its value depends on the latitude; at the equator, it is about 0.53% less than at the poles.

Earth's angular velocity $\omega = 2\pi/24\text{h}$ is small to begin with; in many applications, therefore, we have $\omega \ll |\dot{\mathbf{r}}'|/|\mathbf{r}'|$ and the centrifugal force term $-m\,\boldsymbol{\omega}\times\left(\boldsymbol{\omega}\times\mathbf{r}'\right)$ can then be neglected against the Coriolis force. In good approximation, we may then write

$$m\ddot{\mathbf{r}}' = \hat{\mathbf{F}} - mg\,\mathbf{e}'_z - 2m\left(\boldsymbol{\omega}\times\dot{\mathbf{r}}'\right), \quad (1.277)$$

or

$$m\ddot{x}' = \hat{F}_{x'} + 2m\omega\,\dot{y}'\cos\theta, \qquad (1.278a)$$

$$m\ddot{y}' = \hat{F}_{y'} - 2m\omega\left(\dot{x}'\cos\theta + \dot{z}'\sin\theta\right), \qquad (1.278b)$$

$$m\ddot{z}' = \hat{F}_{z'} - mg + 2m\omega\,\dot{y}'\sin\theta. \qquad (1.278c)$$

For horizontal motion ($\dot{z}' \approx 0$), the first two equations can be considered separately,

$$m\ddot{x}' = \hat{F}_{x'} + 2m\omega\,\dot{y}'\cos\theta, \qquad (1.279a)$$

$$m\ddot{y}' = \hat{F}_{y'} - 2m\omega\,\dot{x}'\cos\theta, \qquad (1.279b)$$

where the inertial force components are just those of a Coriolis force

$$\mathbf{F}_{\mathrm{C,eff}} = -2m\left(\boldsymbol{\omega}_{\mathrm{eff}} \times \dot{\mathbf{r}}'\right) \qquad (1.280)$$

for a rotation about the z'-axis with an effective angular frequency

$$\boldsymbol{\omega}_{\mathrm{eff}} = \omega\cos\theta\,\mathbf{e}'_z. \qquad (1.281)$$

The motion thus corresponds to that on a merry-go-round that rotates with ω_{eff}. This frequency depends on the geographical latitude $\pi/2 - \theta$ and trajectories are deflected according to the direction of the effective Coriolis force: On the northern hemisphere ($\cos\theta > 0$), it points to the right of the direction of motion and on the southern hemisphere ($\cos\theta < 0$) to the left (see also Fig. 1.15).

Foucault pendulum

A (mathematical) pendulum of mass m and length ℓ swinging horizontally with a small amplitude in Earth's gravitational field can be described by the equations of motion (1.279), with force components $\hat{F}_{x'} = -mgx'/\ell$ and $\hat{F}_{y'} = -mgy'/\ell$. On the northern hemisphere, the pendulum changes its plane of oscillation with an angular velocity $-\boldsymbol{\omega}_{\mathrm{eff}}$ (i.e., in a clockwise motion) relative to the ground due to the Coriolis force (1.280).

Hurricane

On the northern hemisphere, hurricanes swirl in an anti-clockwise direction because of the Coriolis force. Hurricanes have a low-pressure center; air coming into this center is deflected by the Coriolis force to produce what is called the *cyclonic* anti-clockwise motion around the center, hence the technical name *tropical cyclone* for a hurricane. However, while this effect indeed initiates the spinning motion, it is not responsible for the high winds associated with a hurricane. Those are caused by thermodynamical effects that feed energy into the cyclonic motion.

1.8 The Central-Force Problem

Central forces, i.e., forces of the form

$$\mathbf{F} = f(r)\frac{\mathbf{r}}{r}, \tag{1.282}$$

were introduced in Sec. 1.4.2.1. It was shown there that they conserve the angular momentum, which gives rise to the area law (1.137). Although it was not shown there explicitly, they are, of course, also conservative: There exists a potential $V(r)$, with total differential

$$dV(r) = -\mathbf{F} \cdot d\mathbf{r} = -f(r)\frac{\mathbf{r} \cdot d\mathbf{r}}{r} = -f(r)\,dr, \tag{1.283}$$

or

$$f(r) = -\frac{dV(r)}{dr}. \tag{1.284}$$

The brief treatment of central forces given so far has one body moving in a generic fixed-center force field. In this section, we will consider the general aspects of the problem of two bodies interacting via mutual central forces. As an application, we will then treat the motion of two point masses subject to gravity; this will allow us to describe the motion of a planet around the Sun, commonly called the *Kepler problem*.

1.8.1 *The two-body problem*

Consider a system of two point particles, with masses m_1 and m_2, subject to internal conservative forces only. The forces are taken to act along the line connecting the two masses, thus satisfying the collinearity constraint (1.57).

1.8.1.1 *Reduction to the equivalent one-body problem*

In an inertial frame, with position vectors $\mathbf{r}_i$, as depicted in Fig. 1.16, the individual equations of motion of the two bodies are

$$m_1\ddot{\mathbf{r}}_1 = \mathbf{F}_{12} \qquad \text{and} \qquad m_2\ddot{\mathbf{r}}_2 = \mathbf{F}_{21}. \tag{1.285}$$

According to Sec. 1.5, the internal forces can be written as

$$\mathbf{F}_{12} = f(|\mathbf{r}_1 - \mathbf{r}_2|)\frac{\mathbf{r}_1 - \mathbf{r}_2}{|\mathbf{r}_1 - \mathbf{r}_2|} = -\mathbf{F}_{21}, \tag{1.286}$$

with

$$f(|\mathbf{r}_1 - \mathbf{r}_2|) = -\frac{dV(|\mathbf{r}_1 - \mathbf{r}_2|)}{d|\mathbf{r}_1 - \mathbf{r}_2|}, \tag{1.287}$$

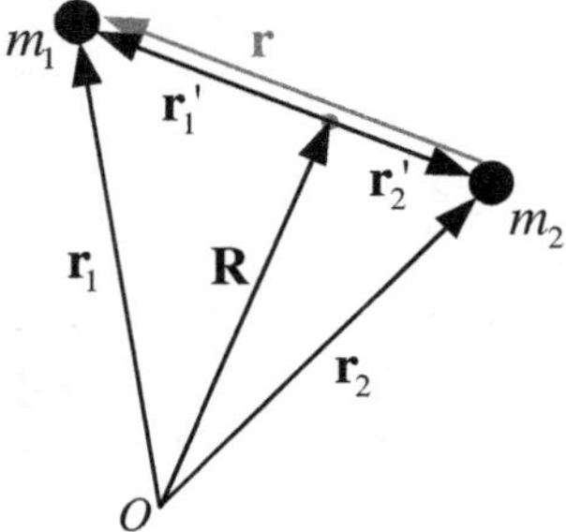

Fig. 1.16   Coordinates for the two-body problem.

which depend only on the relative position vector and its magnitude,

$$\mathbf{r} = \mathbf{r}_1 - \mathbf{r}_2, \quad \text{and} \quad r = |\mathbf{r}_1 - \mathbf{r}_2|. \tag{1.288}$$

From (1.285), it then follows that

$$\ddot{\mathbf{r}} = \ddot{\mathbf{r}}_1 - \ddot{\mathbf{r}}_2 = \frac{\mathbf{F}_{12}}{m_1} - \frac{\mathbf{F}_{21}}{m_2} = \left(\frac{1}{m_1} + \frac{1}{m_2} \right) \mathbf{F}_{12}, \tag{1.289}$$

or

$$\boxed{m\ddot{\mathbf{r}} = \mathbf{F} \equiv f(r)\frac{\mathbf{r}}{r}}, \tag{1.290}$$

where $\mathbf{F} \equiv \mathbf{F}_{12}$ and

$$m = \frac{m_1 m_2}{m_1 + m_2}, \quad \text{or} \quad \frac{1}{m} = \frac{1}{m_1} + \frac{1}{m_2}, \tag{1.291}$$

is the *reduced mass*.

Equation (1.290) looks exactly like the equation of motion for a single (fictitious) body of mass m sitting at the position of the mass m_1 in a central fixed-center force field emanating from the position of mass m_2.[31] Indeed, the solution of this *equivalent one-body problem* provides a complete solution of the two-body problem (1.285).

To see this, we consider the center of mass (CM),

$$\mathbf{R} = \frac{m_1 \mathbf{r}_1 + m_2 \mathbf{r}_2}{m_1 + m_2}, \tag{1.292}$$

and note that, since there are no external forces,

$$M\ddot{\mathbf{R}} = m_1 \ddot{\mathbf{r}}_1 + m_2 \ddot{\mathbf{r}}_2$$

$$= \mathbf{F}_{12} + \mathbf{F}_{21}$$

$$= 0, \tag{1.293}$$

____

[31]Of course, one may just as well formulate the problem with the roles of m_1 and m_2 exchanged.

with $M = m_1 + m_2$ being the total mass. Hence, the CM moves uniformly with constant velocity $\mathbf{u}$, i.e.,

$$\mathbf{R}(t) = \mathbf{R}_0 + \mathbf{u}t, \tag{1.294}$$

where $\mathbf{R}_0$ is a constant vector determined by the initial conditions. The center-of-mass system (CMS) is then another inertial frame, with position vectors

$$\mathbf{r}_1' = \frac{m_2}{M}\mathbf{r} = \mathbf{r}_1 - \mathbf{R}, \qquad \mathbf{r}_2' = -\frac{m_1}{M}\mathbf{r} = \mathbf{r}_2 - \mathbf{R}, \tag{1.295}$$

that are, up to mass factors, given by the relative vector

$$\mathbf{r} = \mathbf{r}_1 - \mathbf{r}_2 = \mathbf{r}_1' - \mathbf{r}_2'. \tag{1.296}$$

The solution of Eq. (1.290) for $\mathbf{r}$, therefore, provides complete solutions

$$\mathbf{r}_1(t) = \mathbf{R}_0 + \mathbf{u}t + \frac{m}{m_1}\mathbf{r}(t), \tag{1.297a}$$

$$\mathbf{r}_2(t) = \mathbf{R}_0 + \mathbf{u}t - \frac{m}{m_2}\mathbf{r}(t), \tag{1.297b}$$

for the original two-body problem (1.285).

Note also that the total kinetic energy T can be written as

$$\begin{aligned} T &= \frac{1}{2}m\dot{\mathbf{r}}_1^2 + \frac{1}{2}m\dot{\mathbf{r}}_2^2 \\ &= \frac{1}{2}M\mathbf{u}^2 + T'. \end{aligned} \tag{1.298}$$

where

$$\begin{aligned} T' &= \frac{1}{2}m_1\dot{\mathbf{r}}_1'^2 + \frac{1}{2}m_2\dot{\mathbf{r}}_2'^2 \\ &= \frac{1}{2}m\dot{\mathbf{r}}^2, \end{aligned} \tag{1.299}$$

i.e., it is the sum of the kinetic energy of the center of mass and the kinetic energy T' of the motion about the center of mass,

For the total angular momentum $\mathbf{L}$, we have

$$\begin{aligned} \mathbf{L} &= (\mathbf{r}_1 \times m_1\dot{\mathbf{r}}_1) + (\mathbf{r}_2 \times m_2\dot{\mathbf{r}}_2) \\ &= (\mathbf{R} \times M\mathbf{u}) + \boldsymbol{\ell}, \end{aligned} \tag{1.300}$$

where the first term is the angular momentum of the center of mass with respect to the original origin and

$$\begin{aligned} \boldsymbol{\ell} &= (\mathbf{r}_1' \times m_1\dot{\mathbf{r}}_1') + (\mathbf{r}_2' \times m_2\dot{\mathbf{r}}_2') \\ &= \mathbf{r} \times m\dot{\mathbf{r}} \end{aligned} \tag{1.301}$$

is the total angular momentum with respect to the CM which in turn is exactly the same as the angular momentum of a single particle with mass m and position vector $\mathbf{r}$ in an inertial frame.

Therefore, separating off the trivial center-of-mass motion, the original two-body problem can indeed be treated completely equivalently to that of a single body — with mass m, kinetic energy $m\dot{\mathbf{r}}^2/2$, and angular momentum $\boldsymbol{\ell}$ — subject to a fixed-center force $\mathbf{F} = f(r)\mathbf{e}_r$.

1.8.1.2   *Reduction to the one-dimensional problem*

In view of angular momentum conservation, $\boldsymbol{\ell} = const$, the one-body central-force problem can be further reduced to an equivalent *one-dimensional* problem, according to the results of Sec. 1.4.2.1, with equations of motion

$$m\ddot{r} = \hat{f} \equiv f + \frac{\ell^2}{mr^3}, \tag{1.302}$$

$$\dot{\varphi} = \frac{\ell}{mr^2} \tag{1.303}$$

[cf. Eqs. (1.134) and (1.135)]. As before, the motion takes place in the xy-plane, i.e., $\boldsymbol{\ell} = \ell\mathbf{e}_z$, and φ is the azimuthal angle. Since the centrifugal part of the force function can be written as a derivative,

$$\frac{\ell^2}{mr^3} = -\frac{d}{dr}\left(\frac{\ell^2}{2mr^2}\right), \tag{1.304}$$

we may introduce

$$V_{\mathrm{cb}}(r) = \frac{\ell^2}{2mr^2}, \tag{1.305}$$

called the *centrifugal barrier*, and write

$$\hat{f} = -\frac{dV_{\mathrm{eff}}(r)}{dr}, \tag{1.306}$$

with an *effective potential,*

$$V_{\mathrm{eff}}(r) = V(r) + V_{\mathrm{cb}}(r), \tag{1.307}$$

that is the sum of the original two-body potential $V(r)$ and the centrifugal barrier term.

In view of

$$\boldsymbol{\ell}^2 = \ell^2 = (\mathbf{r} \times m\dot{\mathbf{r}}) \cdot (\mathbf{r} \times m\dot{\mathbf{r}}) = m^2 r^2 \left(\dot{\mathbf{r}}^2 - \dot{r}^2\right), \tag{1.308}$$

the kinetic energy becomes

$$\frac{1}{2}m\dot{\mathbf{r}}^2 = \frac{1}{2}m\dot{r}^2 + \frac{\ell^2}{2mr^2} = \frac{1}{2}m\dot{r}^2 + V_{\mathrm{cb}}(r), \tag{1.309}$$

which in turn allows for the (conserved) total energy to be written as in the one-dimensional case, i.e.,

$$E = \frac{1}{2}m\dot{\mathbf{r}}^2 + V(r) = \frac{1}{2}m\dot{r}^2 + V_{\text{eff}}(r). \qquad (1.310)$$

As in the one-dimensional treatments of Sec. 1.3.1, solving for $\dot{r} = dr/dt$ leads to

$$dt = \frac{dr}{\sqrt{\frac{2}{m}\left[E - V_{\text{eff}}(r)\right]}} \qquad (1.311)$$

in which the t dependence on the left is separated from the r dependence on the right. Integrating this relation with appropriate initial conditions produces $t(r)$, which may then be inverted to obtain the radial solution $r = r(t)$. From Eq. (1.303), we also have

$$d\varphi = \frac{\ell}{mr^2}dt, \qquad (1.312)$$

which provides $\varphi = \varphi(t)$. Both solutions together then completely determine the motion in the xy-plane in terms of polar coordinates $r = r(t)$ and $\varphi = \varphi(t)$.

Note that we may also use Eq. (1.312) to eliminate the time dependence in Eq. (1.311), i.e.,

$$\begin{aligned}
d\varphi &= \frac{\ell}{mr^2}\frac{dr}{\sqrt{\frac{2}{m}\left[E - V(r) - \frac{\ell^2}{2mr^2}\right]}} \\
&= \frac{1}{r^2}\frac{dr}{\sqrt{\frac{2mE}{\ell^2} - \frac{2m}{\ell^2}V(r) - \frac{1}{r^2}}},
\end{aligned} \qquad (1.313)$$

and then obtain an orbital equation $r = r(\varphi)$ by inverting $\varphi = \varphi(r)$.

Without considering specific choices for the potential $V(r)$, the behavior of the solutions can only be discussed qualitatively by noting that physical solutions must obey

$$E - V_{\text{eff}}(r) = \frac{1}{2}m\dot{r}^2 \geq 0, \qquad (1.314)$$

similar to the considerations in Sec. 1.3.1. An application of this criterion will be given in the subsequent Sec. 1.8.2 for the Kepler problem.

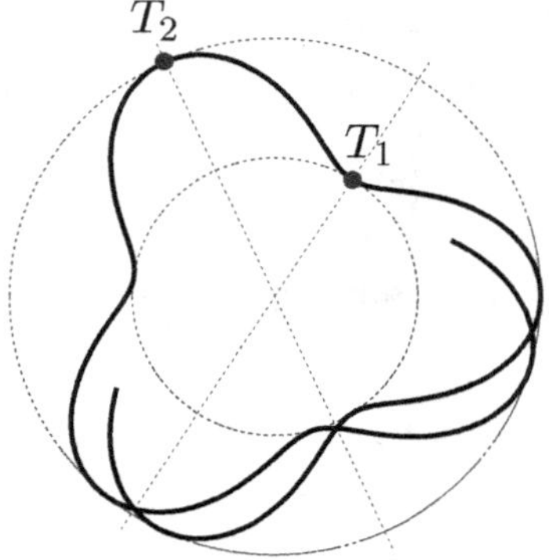

Fig. 1.17   Generic example of the symmetry exhibited by a bound orbit due to central forces. The plot shows the symmetry axes for two adjacent orbital turning points T_1 and T_2.

Symmetries

In Eq. (1.311), we took the positive root of $\dot{r}^2$, i.e., we assumed that r increases as t increases. Indeed, to be able to obtain $r(t)$ by inversion, a monotonic functional behavior $t(r)$ is even required. In reality, of course, this is not necessarily the case. However, the solution for decreasing r with increasing t can be immediately written in terms of the solution for Eq. (1.311) and, therefore, no separate treatment is necessary. Specifically, writing the differential equation (1.302) in terms of the angle φ, via (1.312), and substituting $u = 1/r$, one finds immediately

$$\frac{d^2 u}{d\varphi^2} + u + \frac{m}{\ell^2}\frac{dV(\frac{1}{u})}{du} = 0. \tag{1.315}$$

This equation of motion is explicitly seen to be invariant under the symmetry operation $\varphi - \varphi_0 \to \varphi_0 - \varphi$, where $\varphi_0 = const$. The solution, therefore, is symmetric about any point φ_0 for which the initial conditions are the same in $\varphi - \varphi_0$ and $\varphi_0 - \varphi$. This is the case for the *turning points* of the motion (defined by $\dot{r} = 0$), where

$$u(\varphi_0) = u_0 \qquad \text{and} \qquad \left.\frac{du}{d\varphi}\right|_{\varphi=\varphi_0} = 0. \tag{1.316}$$

At these points an increasing r goes over into a decreasing r, and vice versa. We conclude then that *the orbit is invariant under reflection about the line connecting the origin with a turning point*. Thus, calculating the orbit between two adjacent turning points, we can then determine the neighboring orbit portion by reflections about the symmetry lines; this procedure may be repeated — indefinitely if necessary — to obtain the complete orbit (see Fig. 1.17). It is sufficient, therefore, to consider Eq. (1.311) to obtain a complete solution for this problem by taking, for example, the closest-approach point to the origin as a starting point.

Power-law potentials

Let us consider potentials which are integer powers of r, i.e.,

$$V(r) = \frac{a}{r^n},\tag{1.317}$$

where a is a constant (that may be positive or negative) and n is an integer. Using the substitution $u = 1/r$, Eq. (1.313) can be turned into the standard form

$$d\varphi = -\frac{du}{\sqrt{A - Bu^n - u^2}},\tag{1.318}$$

with $A = 2mE/\ell^2$ and $B = 2ma/\ell^2$. Solutions in terms of well-known functions can be obtained for some specific values of n:

$$\text{trigonometric functions:} \qquad n = -2,\ 1,\ 2,\tag{1.319}$$
$$\text{elliptic functions:} \qquad n = -6,\ -4,\ -1,\ 3,\ 4,\ 6.\tag{1.320}$$

With the exception of solutions for some fractional values of n, in most other cases one must resort to numerical integrations of (1.318) to obtain the orbital solution.

1.8.2 The Kepler problem

A case of special interest is the potential

$$V(r) = \frac{-\alpha}{r}, \qquad \text{with} \quad \alpha > 0,\tag{1.321}$$

since this subsumes the attraction of two massive bodies by the gravitational force

$$\mathbf{F} \equiv \mathbf{F}_{12} = -G\frac{m_1 m_2}{r^2}\frac{\mathbf{r}_1 - \mathbf{r}_2}{r}.\tag{1.322}$$

In other words, the scalar force function for this case is given by

$$f(r) = \frac{-\alpha}{r^2}, \qquad \text{with} \quad \alpha = Gm_1 m_2,\tag{1.323}$$

where G is the gravitational constant.

Specifically, this will allow us to describe the motion of the planets around the Sun.[32] This is usually referred to as the *Kepler problem*, in honor of Johannes Kepler (1571–1630), who discovered the laws of planetary motion described below. The mathematical derivation of these laws was one

---

[32]This assumes that the perturbing influence between planets can be neglected and that the Sun is the only 'driving' force in the solar system. As long as the time periods of interest are not too long, this is a very good approximation since the Sun is 1000 times more massive than the next largest body in the solar system, Jupiter.

of the great successes of Newtonian mechanics at its inception in the 17th century.

To solve this case explicitly, let us go back to Eq. (1.313) and employ a slightly different substitution, for reasons which will become obvious in Sec. 1.8.2.2,

$$u = \frac{1}{r_0} - \frac{1}{r},$$

(1.324)

with

$$r_0 = \frac{\ell^2}{m\alpha},$$

(1.325)

where r_0 is the distance at which the effective potential $V_{\text{eff}}(r)$ has its minimum. We then find

$$d\varphi = \frac{\ell}{m} \frac{du}{\sqrt{\frac{2}{m}\left[\left(E + \frac{m\alpha^2}{2\ell^2}\right) - \frac{\ell^2}{2m}u^2\right]}} = \frac{du}{\sqrt{\frac{\varepsilon^2}{r_0^2} - u^2}},$$

(1.326)

where

$$\boxed{\varepsilon = \sqrt{1 + \frac{2\ell^2}{m\alpha^2}E}}.$$

(1.327)

This integral can be found in standard tables as

$$\varphi - \phi + \frac{\pi}{2} = \arcsin \frac{ur_0}{\varepsilon} = \arcsin\left(\frac{1}{\varepsilon} - \frac{r_0}{\varepsilon r}\right),$$

(1.328)

where ϕ is the integration constant; the angle $\pi/2$ was added here explicitly to facilitate comparison with results of Sec. 1.8.2.3 below.

Inversion then provides

$$\boxed{r(\varphi) = \frac{r_0}{1 - \varepsilon \cos(\varphi - \phi)}}.$$

(1.329)

This is the parametrization of a *conic section*, where the origin, $r = 0$, coincides with one of the foci (see Fig. 1.18). The parameter ε is the *numerical eccentricity* of the conic section that determines the type of section and its value is closely related to the available energy E via (1.327):

$$\text{hyperbola:} \qquad \varepsilon > 1, \qquad E > 0, \tag{1.330a}$$

$$\text{parabola:} \qquad \varepsilon = 1, \qquad E = 0, \tag{1.330b}$$

$$\text{ellipse:} \qquad 0 < \varepsilon < 1, \qquad -\frac{m\alpha^2}{2\ell^2} < E < 0, \tag{1.330c}$$

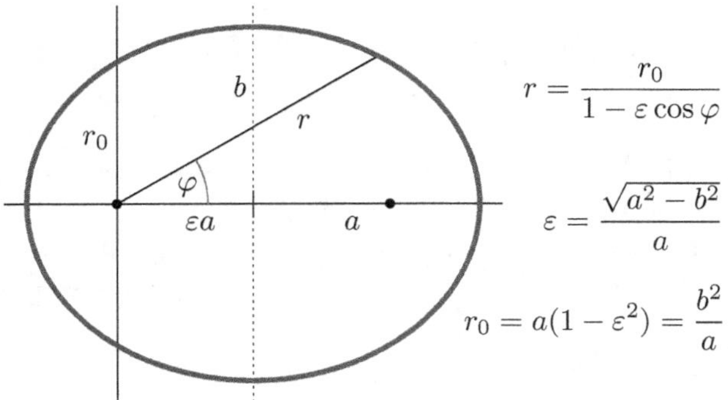

Fig. 1.18 Geometric relationship between the semimajor axis a and semiminor axis b of the ellipse and the parameters of the conic section (1.329). Solid dots indicate the two focal points. Replacing φ by $\varphi - \phi$, one obtains (1.329), where ϕ is the angle by which the ellipse is rotated about the focal point on the left, which is the force center.

$$\text{circle:} \qquad \varepsilon = 0, \qquad E = -\frac{m\alpha^2}{2\ell^2}. \qquad (1.330\text{d})$$

Geometrically, for bound orbits ($\varepsilon < 1$), ε is the ratio of the distance d between a focus and the center of the ellipse and its semimajor axis a, i.e., $\varepsilon = d/a$ (see Fig. 1.18).

In fact, one may qualitatively discuss the behavior of the orbits by noting, as previously mentioned, that physical solutions must satisfy

$$E - V_{\text{eff}}(r) = E + \frac{\alpha}{r} - \frac{\ell^2}{2mr^2} = \frac{1}{2}m\dot{r}^2 \geq 0. \qquad (1.331)$$

These considerations are summarized in Fig. 1.19 for the present Kepler case.[33]

1.8.2.1 *Kepler's laws*

For negative total energies, $\varepsilon < 1$, and the trajectories described by (1.329) are bound orbits. More specifically, the orbits are *ellipses*, with the force center being one of the foci. This is the famous *first law* by Kepler: *The planets move in ellipses, with the Sun at one focal point.* His *second law* we have encountered already; it was the area law of Eq. (1.137): *The radius vector sweeps out equal areas in equal times.* For Kepler's third law, we use

---

[33]The findings summarized in this figure are generically true for any attractive potential V that satisfies $\lim_{r \to 0} r^2 V = 0$ and $\lim_{r \to \infty} 1/r^2 V = 0$, i.e., for which the centrifugal barrier rises faster for $r \to 0$ and falls off faster for $r \to \infty$ than the potential.

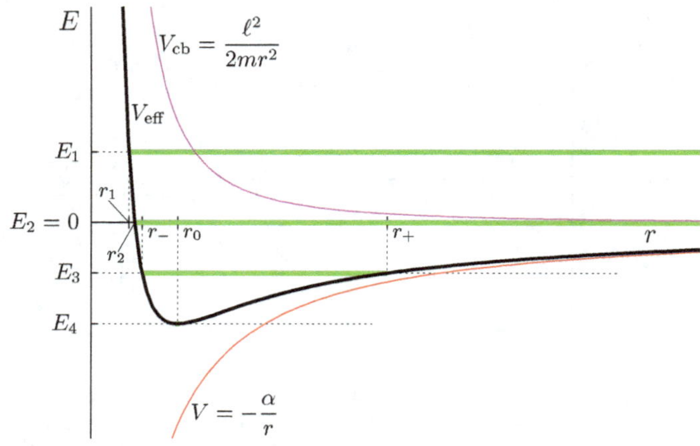

Fig. 1.19   Plots of the gravitational potential $V = -\alpha/r$, the centrifugal barrier $V_{\mathrm{cb}} = \ell^2/2mr^2$, and the resulting effective potential $V_{\mathrm{eff}} = V + V_{\mathrm{cb}}$. The allowed regions of motion satisfying the condition $E - V_{\mathrm{eff}} = m\dot{r}^2/2 \geq 0$ are shaded for four different energies E_i: For $E_1 > 0$, the motion is an unbounded hyperbola, with $r \geq r_1$; for $E_2 = 0$, the motion is a parabola, also unbounded, with $r \geq r_2 = r_0/2$; for the energy E_3 (with $0 > E_3 > E_4$), the motion is an ellipse bounded between minimum and maximum values of r of $r_- = r_0/(1+\varepsilon)$ and $r_+ = r_0/(1-\varepsilon)$, respectively; and for the energy E_4 (which coincides with the minimum of the effective potential), the motion is a circle of radius $r_0 = \ell^2/m\alpha$. For energies $E < E_4$, no physical solutions are possible.

(1.303) in the form

$$\frac{mr^2}{\ell} \, d\varphi = dt \tag{1.332}$$

and integrate around one complete orbit, i.e.,

$$2\sqrt{\frac{m}{\alpha}} \, r_0^{3/2} \int_0^{\pi} \frac{d\varphi}{(1 - \varepsilon \cos \varphi)^2} = T, \tag{1.333}$$

where T is the orbital period, and the initial condition $\phi = 0$ was chosen to simplify the calculation. This yields

$$2\pi \sqrt{\frac{m}{\alpha}} \left(\frac{r_0}{1 - \varepsilon^2} \right)^{3/2} = 2\pi \sqrt{\frac{m}{\alpha}} \, a^{3/2} = T, \tag{1.334}$$

with the ellipse's *semimajor axis*

$$a = \frac{r_0}{1 - \varepsilon^2}. \tag{1.335}$$

Written in the form

$$\frac{T^2}{a^3} = 4\pi^2 \frac{m}{\alpha} = \frac{4\pi^2}{G(m_1 + m_2)} = const, \tag{1.336}$$

the *third law* then reads: *The ratio of the square of the period T and the cube of the semimajor axis a is constant.* This means, in particular, that the actual shape of the orbit does not matter: a circle with radius a will have the same period T as an extremely elongated ellipse with a semimajor axis a of the same value.

Kepler's laws were purely empirical, based on astronomical observations — Newtonian mechanics had not been invented yet! We see here that his three laws follow from the fact that gravity is an *inverse-square* central force. The second law — the area law — follows from angular-momentum conservation alone and therefore is generically true for all central forces. The first and third laws, however, depend on the specific form of the force function, $f(r) = -\alpha/r^2$. The first law postulates the orbits to be ellipses and the third law obtains only for orbits of this geometrical form. Elliptical orbits are distinguished by the fact that *they are closed*, i.e., after one revolution the body returns to the same point in space with conditions exactly matching the initial conditions, and the motion repeats itself indefinitely. It is remarkable that there exist *only two* force laws with stable closed orbits, the inverse-square law and the linear-force law — Hooke's law — for the harmonic oscillator.[34] Indeed, in the subsequent section 1.8.2.2 it will be shown that these two force laws are very closely related to each other.

The left-hand side of Fig. 1.20 depicts the closed-orbit Kepler ellipse; the figure shows that the angle ϕ in the conic section (1.329) defines the orientation of the ellipse in space. The fact that this orientation is stable is due to an underlying conserved quantity, the Runge–Lenz vector, that will be discussed in Sec. 1.8.2.3.

The heliocentric frame of reference

It should be mentioned that Kepler's laws were formulated (about 400 years ago!) with the implicit assumption that the Sun provided a fixed force center for planetary motion which — in modern language — we would call an inertial frame. We know now, of course, that such a *heliocentric* frame is not an inertial frame, since Newton's third law means that the same force that is exerted by the Sun on a planet, this planet will exert on the Sun, thus making the heliocentric frame of reference an accelerated one, necessitating inertial pseudoforces to account for that acceleration.

---

[34]It should be noted that several force laws provide closed orbits that are stable only if the initial conditions are within a specific, narrow range. However, only the inverse-square law and Hooke's law provide stable orbits for any set of initial conditions corresponding to bound orbits.

To briefly discuss this in more detail let us consider the specific example of the Earth-Sun system, where $m_1 \equiv m_E$ and $m_2 \equiv m_S$ are the Earth's and Sun's masses, respectively. Denoting quantities from the heliocentric frame with asterisks and the necessary pseudoforce by $\mathbf{F}'$, the equations of motion of the heliocentric Kepler problem would be

$$m_E \ddot{\mathbf{r}}^* = \mathbf{F}^* \qquad \longrightarrow \qquad m_E \ddot{\mathbf{r}} = \mathbf{F} + \mathbf{F}' = \mathbf{F}^*, \tag{1.337}$$

where $\mathbf{r} \equiv \mathbf{r}^*$ since the relative position is already identical to the position vector in the heliocentric frame. Note that Eq. (1.290) can be rewritten as

$$m_E \ddot{\mathbf{r}} = \frac{m_E + m_S}{m_S}\mathbf{F} = \mathbf{F} + \frac{m_E}{m_S}\mathbf{F} = \mathbf{F} + \mathbf{F}', \tag{1.338}$$

where the identification

$$\mathbf{F}' = \frac{m_E}{m_S}\mathbf{F} \tag{1.339}$$

for the pseudoforce is made by comparison with the preceding equation (1.337). Since this is still completely equivalent to Eq. (1.290), the solution $\mathbf{r}$ is the same, of course. It shows that the heliocentric pseudoforce $\mathbf{F}'$ is actually quite small because of the factor $m_E/m_S \ll 1$. Neglecting $\mathbf{F}'$, therefore, would be an acceptable approximation in most cases. Nevertheless, even if neglected, *all of Kepler's laws would remain valid* since this remaining approximate force is still an inverse-square central force. The only quantitative change would occur for the third law, where the constant on the right-hand side of Eq. (1.336) would become slightly larger by a term of the order $m_E/m_S \approx 3 \cdot 10^{-6}$.

1.8.2.2   *Relation to the simple harmonic oscillator*

It was mentioned in the discussion above that the linear force of Hooke's law is the only force other than the inverse-square force that provides stable orbits. The very close relation between these two seemingly quite different forces can be seen from Eq. (1.326). Indeed, replacing the free parameters E and ℓ of the Kepler case by defining a new 'total energy'

$$\hat{E} \equiv E + \frac{m\alpha^2}{2\ell^2} \tag{1.340}$$

and a 'spring constant'

$$k \equiv \frac{\ell^2}{m}, \tag{1.341}$$

Eq. (1.326) can be recast in the form

$$d\varphi = \omega_0 d\tau = \frac{\omega_0\, du}{\sqrt{\frac{2}{m}\left[\hat{E} - \frac{1}{2}ku^2\right]}}, \qquad (1.342)$$

which looks exactly like the equation for a simple harmonic oscillator[35] with angular frequency $\omega_0 \equiv \sqrt{k/m}$! The 'time' variable τ is defined here by this equation as a simple rescaling of the angle variable φ.

This remarkable result is of course no accident. The isogonal mapping (1.324),

$$u = \frac{1}{r_0} - \frac{1}{r},$$

that ties together both of these force laws exhibits an underlying symmetry: In addition to the usual rotational invariance in three dimension, which leads to a conserved angular momentum, the Kepler problem (and therefore also the simple harmonic oscillator) is rotationally invariant in *four* dimensions. In group-theory language, rotational invariance in three dimensions is governed by the group $SO(3)$ (see Sec. 5.1.1), and in four dimensions by $SO(4)$. The latter is a larger group that contains $SO(3)$ as a subgroup, i.e., $SO(4) \supset SO(3)$.

1.8.2.3 The Runge–Lenz vector

The $SO(4)$ symmetry just discussed manifests itself most prominently by the fact that, in addition to the angular-momentum vector $\boldsymbol{\ell}$, there is another conserved vector in the Kepler case, the *Runge–Lenz vector* $\boldsymbol{\varepsilon}$ defined by

$$\boldsymbol{\varepsilon} = \frac{\mathbf{r}}{r} - \frac{1}{m\alpha}\left(\mathbf{p} \times \boldsymbol{\ell}\right), \qquad (1.343)$$

with $\mathbf{p} = m\dot{\mathbf{r}}$. This vector is perpendicular to the angular momentum,

$$\boldsymbol{\varepsilon} \cdot \boldsymbol{\ell} = 0, \qquad (1.344)$$

and thus lies in the orbital plane. With $\boldsymbol{\ell} = \ell\mathbf{e}_z$, its polar representation is

$$\boldsymbol{\varepsilon} = \left(1 - \frac{\ell^2}{m\alpha r}\right)\mathbf{e}_r + \frac{\ell\dot{r}}{\alpha}\mathbf{e}_\varphi. \qquad (1.345)$$

and, assuming a central force, one easily finds that its time derivative,

$$\frac{d\boldsymbol{\varepsilon}}{dt} = \left[\frac{\alpha}{r^2} + f(r)\right]\frac{\ell}{m\alpha}\mathbf{e}_\varphi, \qquad (1.346)$$

vanishes for the Kepler case where $f(r) = -\alpha/r^2$.

---

[35]See Eq. (1.84) in Sec. 1.3.2.

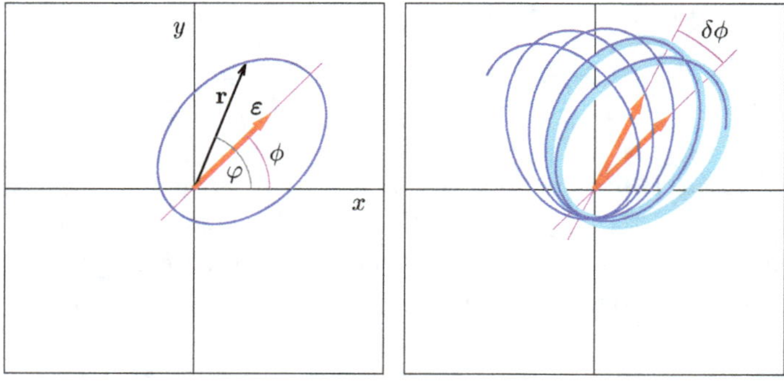

Fig. 1.20   *(Left)* Runge–Lenz vector $\boldsymbol{\varepsilon}$ in relation to the spatial orientation of the Kepler ellipse. For an inverse-square central force, the orbit is closed and the ellipse is stationary in space, with $\boldsymbol{\varepsilon}$ pointing toward the aphelion point of the orbit. *(Right)* Precessing of the Kepler ellipse under the influence of a perturbing force. In addition to the continuous line of the true orbit, the plot shows that the perturbing force changes the orientation of the Kepler ellipses (shaded), and their associated Runge–Lenz vectors, by an angle increment $\delta\phi$ between two orbits.

Both $\boldsymbol{\ell}$ and $\boldsymbol{\varepsilon}$ completely determine the Kepler orbit: $\boldsymbol{\ell}$ is perpendicular to the plane of orbit, i.e., it defines the orbit plane, and $\boldsymbol{\varepsilon}$ determines the orientation of the stationary Kepler ellipse in this plane. The latter fact is easily seen by taking the scalar product of $\boldsymbol{\varepsilon}$ and the position vector $\mathbf{r} = r\mathbf{e}_r$,

$$\mathbf{r} \cdot \boldsymbol{\varepsilon} = r\varepsilon \cos(\varphi - \phi) = r\left(1 - \frac{\ell^2}{m\alpha r}\right), \qquad (1.347)$$

where φ and ϕ are the azimuthal angles of $\mathbf{r}$ and $\boldsymbol{\varepsilon}$, respectively. Solving this for r immediately yields the previously obtained polar equation for the orbit,

$$r(\varphi) = \frac{\ell^2}{m\alpha} \frac{1}{1 - \varepsilon \cos(\varphi - \phi)}. \qquad (1.348)$$

It is seen here that the spatial orientation of this ellipse is determined by $\boldsymbol{\varepsilon}$'s azimuthal angle ϕ (see Fig. 1.20). Moreover, the eccentricity ε is found to be identical to the magnitude of the Runge–Lenz vector,

$$\varepsilon = |\boldsymbol{\varepsilon}|. \qquad (1.349)$$

It is for this reason $\boldsymbol{\varepsilon}$ is sometimes called the *eccentricity vector.*[36]

---

[36]The story of how the names of Runge and Lenz became associated with $\boldsymbol{\varepsilon}$ — neither of whom actually discovered this conserved quantity — is a very interesting piece of science history (for details, see [2] and references therein). It was discovered in 1710 by J. Hermann [[(Giornale de Letterati D'Italia **2**, 447–467 (1710)] and — giving credit to all pertinent contributions in chronological order — should perhaps more appropriately be called Hermann-Bernoulli-Laplace-Hamilton-Runge-Lenz-Pauli vector.

The present derivation shows that *relative to* $\boldsymbol{\varepsilon}$ *the orbit equation always has the form of a conic section*, irrespective of whether $\boldsymbol{\varepsilon}$ is conserved or not. In fact, for force functions $f(r)$ different from $-\alpha/r^2$, one may use Eq. (1.346) as a differential equation for $\boldsymbol{\varepsilon}$ to determine both $\phi = \phi(t)$ and $\varepsilon = \varepsilon(t)$ as functions of time [or of φ via Eq. (1.332)]. For perturbations of the pure Kepler case, this time dependence causes the orbital ellipse to precess so that relative to the changing $\boldsymbol{\varepsilon}$, the orbit remains a conic section, as illustrated on the right-hand side of Fig. 1.20. For celestial mechanics, in particular, differential equations of motion for the perturbed Runge-Lenz vectors of celestial bodies based on generalizations of Eq. (1.346) for arbitrary forces provide a tool for the numerical determination of orbits that is several orders more efficient than using standard Newtonian techniques.[37]

1.8.3 *Newton's cannonball*

As an argument why the force that keeps the Moon in its (nearly) circular orbit about the Earth and the force that makes things fall to the ground on Earth is the same, Newton offered the following gedankenexperiment:[38] Consider a cannonball launched horizontally from a high mountaintop, neglecting all influences other than gravity. For small launch velocities v_0, the ball will drop to the ground after a short flight. The larger v_0 the farther the ball will travel until eventually, at sufficiently high velocities, it will circumnavigate the globe in a closed orbit, just like the Moon orbits the Earth.

As a generalization of Newton's original problem and as an illustration of the kinematics resulting for Kepler-type problems, let us consider a cannonball of mass m launched at an arbitrary angle β with respect to the vertical, as shown in Fig. 1.21, assuming that the launch velocity v_0 is sufficiently large and the launch point is atop a mountain sufficiently high so that the ball can go into an elliptical orbit.

We choose a coordinate system in the orbital xy-plane centered at Earth's center of mass (see Fig. 1.21). In view of a cannonball's small mass

---

[37]See E.A. Bartnik, H. Haberzettl, and W. Sandhas, *"Equations of motion using the dynamical evolution of the Runge-Lenz vector"*, Astrophys. J. **334**, 517 (1988).

[38]In science, the German word *Gedankenexperiment* (literally *thought experiment*, from *Gedanke = thought*) is often used to describe experiments carried out in our thoughts, instead of in reality, to test or illustrate certain assumptions about physical theories. To lead to valid conclusions, the experiments must be such that they could be undertaken in reality *in principle* if one could overcome whatever technical obstacles exist to actually carry them out successfully. Sometimes thought experiments envisioned now may lead to actual experiments at a later time when they become technically feasible.

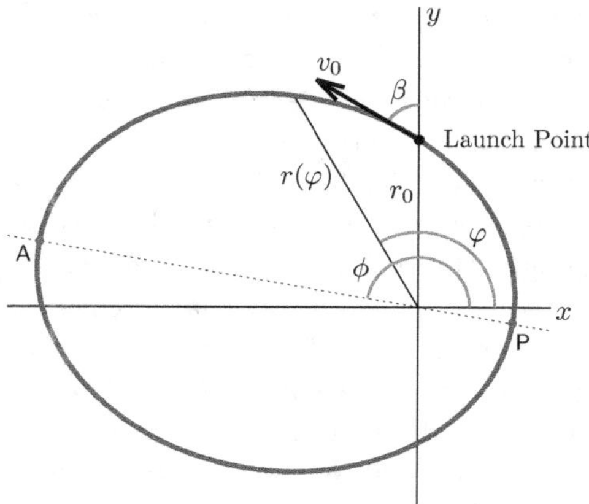

Fig. 1.21   Coordinates used here for the description of Newton's cannonball problem. The y-axis is defined by the location of the launch point and the orbital and apogee angles φ and ϕ, respectively, are measured from the x-axis. The dotted line connects the orbit's perigee P with the apogee A.

m compared to Earth's vastly larger mass of $m_E \approx 6 \times 10^{24}$ kg, the difference between the two-body center-of-mass frame and the geocentric frame is negligibly small. For all practical purposes, therefore, this is an inertial frame (neglecting in addition, of course, the influence of all other celestial bodies in the solar system). The reduced mass of this two-body system, therefore, is practically equal to the ball's mass m, i.e, $m\,m_E/(m+m_E) \approx m$, and we may take over all relations derived above for two-body systems as being equally valid for the cannonball's geocentric orbit. Furthermore, let us put the launch point on the y-axis, at a distance r_0 from the center of the Earth.[39] We also assume — without lack of generality — that the cannonball is launched in a counter-clockwise direction, as shown in the diagram 1.21.[40]

It is convenient for the following to introduce the escape velocity v_e as a parameter, which is the lowest velocity that will permit a body launched at height r_0 to escape the influence of Earth's gravitational field, become unbound, and travel to infinity.[41] According to (1.330), this will happen

---

[39]See the note below at the end of this subsection how this choice affects the result.

[40]The diagram depicts conditions where the ball is launched in the direction of the apogee; it is possible, of course, to find conditions where the ball is launched towards the perigee. In fact, for every launch angle β towards the apogee (for which $\beta < \pi/2$), there exists a similar ellipse, obtained by a reflection at the y axis, that corresponds to a launch angle $\pi - \beta$ towards the perigee, with the same initial velocity v_0.

[41]On the surface of the Earth, the escape velocity is about 11.2 km/s. Of course, the escape velocity *relative* to the surface of the Earth is smaller or larger depending respectively on whether the launch happens in an easterly or westerly direction since one needs to add to or subtract from the launch velocity the radial speed of the launch position due to Earth's rotation. It is more cost-effective, therefore, to launch spacecraft in an easterly direction as close to the equator as possible.

when the orbital energy E is zero, in other words, when the kinetic energy is equal in magnitude to the potential energy,

$$\frac{m}{2}v_{\mathrm{e}}^2 = \frac{\alpha}{r_0} \quad \Rightarrow \quad v_{\mathrm{e}} = \sqrt{\frac{2Gm_E}{r_0}} \quad \text{for} \quad \alpha = Gmm_E, \tag{1.350}$$

For the present problem, the parameters of the Kepler ellipse (1.329) then are readily found as

$$\alpha = \frac{m}{2}v_{\mathrm{e}}^2 r_0 \quad \text{and} \quad \ell = r_0(mv_\perp) = mr_0v_0 \sin\beta, \tag{1.351}$$

where $v_\perp = v_0 \sin\beta$ is the velocity perpendicular to the radius vector at launch time, and

$$\varepsilon = \sqrt{1 + 4\sin^2\beta \left(\frac{v_0}{v_{\mathrm{e}}}\right)^2 \left[\left(\frac{v_0}{v_{\mathrm{e}}}\right)^2 - 1\right]}. \tag{1.352}$$

Squaring the last relation and solving the resulting quadratic equation for the squared velocity ratio produces

$$\left(\frac{v_0}{v_{\mathrm{e}}}\right)^2 = \frac{1}{2} - s\frac{\sqrt{\varepsilon^2 - \cos^2\beta}}{2\sin\beta}, \tag{1.353}$$

where the sign $s = \pm 1$ depends on whether the ball is launched towards the apogee ($s = -1$, $\beta < \pi/2$) or the perigee ($s = +1$, $\beta > \pi/2$) as follows from inspecting the diagram. This corresponds to the sign of $\cos\phi = s|\cos\phi|$ since $s = -1$ for $\pi/2 < \phi < 3\pi/2$, and $s = +1$ otherwise (ϕ is taken here modulo 2π). Launching the cannonball horizontally, as in Newton's original gedankenexperiment, yields

$$\left(\frac{v_0}{v_{\mathrm{e}}}\right)^2 = \frac{1 \pm \varepsilon}{2}, \quad \text{for} \quad \beta = \frac{\pi}{2}, \tag{1.354}$$

which shows that $v_0 = v_{\mathrm{e}}/\sqrt{2}$ for a circular orbit (where $\varepsilon = 0$). Velocities above or below that value correspond to having the launch point right at the perigee or the apogee, respectively, of an elliptical orbit.

The main task is now to express the direction of the apogee, ϕ, in terms of the launch angle β and the other initial parameters, v_0 and r_0.

The preceding results allow us to rewrite the conic section (1.329) as

$$\frac{r_0}{r} = \left(\frac{v_{\mathrm{e}}}{v_0}\right)^2 \frac{1 - \varepsilon\cos(\varphi - \phi)}{2\sin^2\beta}. \tag{1.355}$$

To calculate ϕ, put $\varphi = 90°$, where $r = r_0$, i.e.,

$$1 = \left(\frac{v_e}{v_0}\right)^2 \frac{1 - \varepsilon \sin \phi}{2 \sin^2 \beta}, \tag{1.356}$$

or

$$\sin \phi = \frac{1}{\varepsilon}\left[1 - 2\sin^2 \beta \left(\frac{v_0}{v_e}\right)^2\right]. \tag{1.357}$$

Moreover,

$$\cos \phi = s\sqrt{1 - \sin^2 \phi}$$

$$= -\frac{2\cos \beta \sin \beta}{\varepsilon}\left(\frac{v_0}{v_e}\right)^2, \tag{1.358}$$

where $s|\cos \beta| = -\cos \beta$ was used. We can now write

$$\varepsilon \cos(\varphi - \phi) = \varepsilon\,(\cos \varphi \cos \phi + \sin \varphi \sin \phi)$$

$$= -\sin \varphi + 2\sin \beta \left(\frac{v_0}{v_e}\right)^2 \cos(\varphi - \beta). \tag{1.359}$$

Inserting this into Eq. (1.355), we immediately obtain

$$\frac{r_0}{r} = \left(\frac{v_e}{v_0}\right)^2 \frac{1 - \sin \varphi}{2\sin^2 \beta} + \frac{\cos(\varphi - \beta)}{\sin \beta}, \tag{1.360}$$

or

$$r(\varphi) = \frac{2 r_0 v_0^2 \sin^2 \beta}{v_e^2 \left(1 - \sin \varphi\right) + 2 v_0^2 \sin \beta \cos(\varphi - \beta)} \tag{1.361}$$

which is the desired result expressing the orbit entirely in terms of β, r_0, v_0, and v_e (instead of ϕ, ℓ, α, and m of the original parametrization).

Note that the specific form (1.360) given here depends on how we chose to orient the coordinate axes with respect to the launch point. For example, if the launch point is taken to be on the x-axis, instead of on the y-axis as in (1.360), we would need to rotate the coordinate system by $90°$ and the orbital angle would then be

$$\varphi \quad \longrightarrow \quad \theta = \varphi - \frac{\pi}{2}, \tag{1.362}$$

leading to

$$\frac{r_0}{r} = \left(\frac{v_e}{v_0}\right)^2 \frac{1 - \cos\theta}{2\sin^2\beta} - \frac{\sin(\theta - \beta)}{\sin\beta}$$

$$= \left(\frac{v_e}{v_0}\right)^2 \left(\frac{\sin\frac{\theta}{2}}{\sin\beta}\right)^2 - \frac{\sin(\theta - \beta)}{\sin\beta} \tag{1.363}$$

or

$$r(\theta) = \frac{r_0 v_0^2 \sin^2\beta}{v_e^2 \sin^2\frac{\theta}{2} - v_0^2 \sin\beta \sin(\theta - \beta)}. \tag{1.364}$$

In the usual Kepler-ellipse parametrization (1.329), this frame dependence does not arise because both φ and ϕ are measured with respect to the same reference axis and thus the argument difference of $\cos(\varphi - \phi)$ is unaffected by any rotation of the xy coordinates.

1.8.4 *The restricted three-body problem*

For the two-body case, we have seen that the central-force problem can be reduced to a one-dimensional effective one-body problem that may even be fully integrable for many cases. The situation becomes much more complex if we now add a third body. This classical *three-body problem* in general does not have a complete solution.[42]

Here, we will consider the *restricted three-body problem of celestial mechanics*. This is a rather simplified version of a full three-body problem, but it will suffice to show some of the complexities. This is one of the classic problems of celestial mechanics and it concerns the motion of a small body in the force fields of two much larger bodies. Specifically, let us consider an asteroid, with negligibly small mass μ, moving subject to the gravitational attraction of the Sun (mass m_S) and of Jupiter (mass m_J). In view of the smallness of μ, the asteroid has no influence on the motion of the Sun and Jupiter, and therefore both move in unperturbed Kepler orbits that are found as a solution of the Sun–Jupiter two-body problem along the lines discussed above.[43]

---

[42]This does *not* mean that the problem does not have a solution — after all, three-body systems do exist in nature! It only means that a *complete* solution in the sense defined in Sec. 1.5.4 does not exist in general; complete solutions have only been found for special configurations. Of course, the problem can be solved numerically. For a discussion of some aspects of the three-body case, see Ref. [1].

[43]This presumes, of course, that the influence of all other celestial bodies can be neglected. Since the Sun and Jupiter are by far the most massive objects in the solar system, this is a reasonable assumption for many applications.

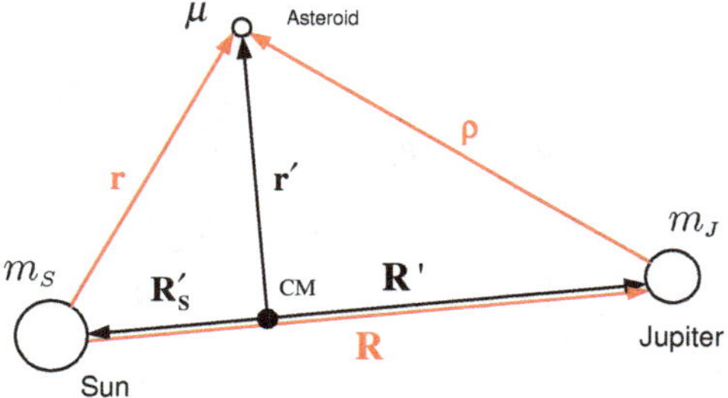

Fig. 1.22   Asteroid in the gravitational field of the Sun and Jupiter. The asteroid's mass μ is taken as negligibly small compared to the other two masses.

With the variables defined as in Fig. 1.22, the force on the asteroid is given in the (inertial) center-of-mass system of the Sun and Jupiter by

$$\mu\ddot{\mathbf{r}}' = -G\mu m_S \frac{\mathbf{r}}{r^3} - G\mu m_J \frac{\boldsymbol{\rho}}{\rho^3}. \tag{1.365}$$

Using

$$\mathbf{r}' = \mathbf{r} + \mathbf{R}'_S \tag{1.366}$$

and the fact that the force on the Sun due to Jupiter is given by

$$m_S \ddot{\mathbf{R}}'_S = G m_S m_J \frac{\mathbf{R}}{R^3}, \tag{1.367}$$

we can eliminate $\ddot{\mathbf{R}}'_S$ and derive

$$\mu\ddot{\mathbf{r}} = -\alpha \frac{\mathbf{r}}{r^3} + \mathbf{F}_J(\mathbf{r}) \equiv \mathbf{F}(\mathbf{r}) \tag{1.368}$$

where

$$\mathbf{F}_J(\mathbf{r}) = -\alpha k \left[\frac{\boldsymbol{\rho}}{\rho^3} + \frac{\mathbf{R}}{R^3} \right], \tag{1.369}$$

with $\alpha = G\mu m_S$ and $\boldsymbol{\rho} = \mathbf{r} - \mathbf{R}$. $\mathbf{F}$ is the force on the asteroid in the heliocentric frame. In the absence of Jupiter, and in view of the smallness of μ, this could be considered an inertial frame, with the first term on the right-hand side of Eq. (1.368) providing the usual inverse-square force for the unperturbed Kepler orbits. The perturbing force $\mathbf{F}_J$ due to Jupiter contains two terms; the first is Jupiter's direct gravitational force on the asteroid, and the second, $-\alpha k \mathbf{R}/R^3$, is an inertial pseudoforce that appears because the heliocentric frame is not an inertial frame. The entire strength of Jupiter's influence is suppressed by the mass ratio $k = m_J/m_S \approx 1/1000$.

To further discuss this, let us — as is customary for the restricted problem — assume that the asteroid's initial conditions are such that it moves in the plane of the Kepler orbits of the Sun and Jupiter. The force $\mathbf{F}$ then will only have radial and azimuthal components,

$$\mathbf{F}(\mathbf{r}) = F_r \mathbf{e}_r + F_\varphi \mathbf{e}_\varphi. \tag{1.370}$$

In detail, one finds

$$F_r = -\frac{\alpha}{r^2} - \frac{\alpha k}{r^2} \left[\frac{r^3}{\rho^3} + \frac{r^2 \mathbf{R} \cdot \mathbf{e}_r}{R^3} \left(1 - \frac{R^3}{\rho^3} \right) \right], \tag{1.371a}$$

$$F_\varphi = -\frac{\alpha k}{r^2} \left[\frac{r^2 \mathbf{R} \cdot \mathbf{e}_\varphi}{R^3} \left(1 - \frac{R^3}{\rho^3} \right) \right]. \tag{1.371b}$$

This result is written so that the terms in square brackets are dimensionless quantities. The first term of F_r would correspond to the unperturbed Kepler problem; all other force contributions are suppressed by the factor k. The equations of motion then can be written as

$$\mu \ddot{r} = \frac{\ell^2}{\mu r^3} + F_r, \tag{1.372a}$$

$$\dot{\ell} = r F_\varphi, \tag{1.372b}$$

$$\dot{\varphi} = \frac{\ell}{\mu r^2}, \tag{1.372c}$$

where the angular momentum ℓ is defined by the last equation [see also (1.374)]. This clearly is no longer a central-force problem and we cannot decouple these equations, as we could do it in the equivalent one-body formulation of the two-body case (where F_φ was absent, giving rise to angular momentum conservation). It is not possible, therefore, to obtain a complete solution. The equations can only be integrated numerically as a coupled set of differential equations,

Even though this is a simplified version of a full celestial three-body problem, and even though the perturbation may not seem to be large, this force may lead to rather complex orbits for the asteroid, with sometimes very violent changes of the asteroid's orbit during close encounters with Jupiter. These effects stem from the ρ-dependent dimensionless factors in the square brackets of (1.371). Averaged over one Jupiter revolution, they are of order one for most of the time. However, during close encounters, when $r \approx R$, one has $\rho \ll r$ and these factors then get so large that they compensate the small mass ratio k.

One may further simplify the problem by assuming that Jupiter's orbit around the Sun is a perfect circle,[44] with a constant angular velocity ω_J.

---

[44] Jupiter's eccentricity is $\varepsilon \approx 0.05$, so this is not a bad approximation to get semi-quantitatively reliable results.

This *circular restricted three-body problem* can be shown to have another constant of motion, the *Jacobi integral*,

$$C = E_0 + kU - \omega_J \ell = const, \qquad (1.373)$$

where $\ell = |\boldsymbol{\ell}|$ is the magnitude of the asteroid's heliocentric angular momentum

$$\boldsymbol{\ell} = \mathbf{r} \times \mu \dot{\mathbf{r}}, \qquad (1.374)$$

and

$$E_0 = \frac{1}{2}\mu\dot{\mathbf{r}}^2 - \frac{\alpha}{r} \qquad (1.375)$$

is the energy of the asteroid's unperturbed Kepler orbit around the Sun, and

$$kU = -\int \mathbf{F}_J(\mathbf{r}) \cdot d\mathbf{r} = -\frac{\alpha k}{\rho} + \alpha k \frac{\mathbf{r} \cdot \mathbf{R}}{R^3} \qquad (1.376)$$

is the potential energy due to Jupiter's perturbing force. Without going into further details, the existence of this conserved quantity can be related to the fact that the asteroid's motion exhibits certain symmetries relative to the line connecting the Sun and Jupiter which rotates with constant angular velocity ω_J.

Even the simplified problem as outlined here has no complete solutions. It is quite instructive, however, to perform numerical simulations of this restricted three-body system. This may, for example, be used to model the flight path of a spacecraft close to a planet (provided the planet's eccentricity is small enough to warrant this approach).

1.8.5   *Scattering in a central-force field*

The examples discussed in the previous sections — in particular, the Kepler problem — concern the bound motion of a body with negative total energy. For positive energies, however, the trajectories are unbound. For the inverse-square gravitational force, for example, these unbound trajectories are hyperbolas, with a limiting trajectory of a parabola for zero energy; this follows from Eq. (1.330) with (1.327).

The interaction of two bodies in an unbound relative motion allows us to describe a *scattering* process (also called a *collision* process): Coming from infinity, a body with mass m_{p} (the *projectile*) approaches another body with mass m_{t} (the *target*); at close proximity, the motion is affected by the interaction between the two bodies. In the rest frame of the target, the projectile has no potential energy at infinity, only (positive) kinetic energy. Since therefore the total energy of such a process is always positive, the projectile particle cannot be captured into a bound orbit (with negative total energy) without violating energy conservation. Hence, the projectile *scatters* off the target and moves again away to infinity, usually in a direction different from the incident direction. 'Infinity' here refers to distances so far away from the target that the projectile-target interaction can be ignored.

This generic scenario of a scattering process applies, for example, to the collisions of billiard balls, to the trajectories of celestial objects that enter the Sun's gravitational field from outside the solar system, to the scattering of nuclei off each other, and to many, many more processes of physical interest.

The collision outlined in this manner is a two-body process. In the equivalent one-body formulation of the two-body problem given in Sec. (1.8.1.1), the two-body collision process corresponds to the unbound motion of a (fictitious) body, with the reduced mass $m = m_\mathrm{p} m_\mathrm{t}/(m_\mathrm{p} + m_\mathrm{t})$, in the relative interaction potential. The true two-body collision process is thus reduced to the scattering of a single (fictitious) projectile 'particle' on a 'target' described in terms of a fixed-center force field.

In practice, the scattering process often involves a stream of many identical projectile particles that collide with a known configuration of many identical target bodies. For example, in the classic Rutherford experiment discussed below, a beam of alpha particles is scattered off a target of gold nuclei in the form of a thin gold foil. If the projectile particles do not interact with each other and if the process can be understood in terms of each projectile particle scattering just once off one of the target bodies, then the two-body picture may be employed to describe the entire process by suitably summing over the individual two-body collisions. Scattering experiments of this type, therefore, are treated by considering the scattering of a stream of identical (fictitious) particles in a central potential field. The terms 'particle', 'target', etc., in the subsequent discussion of scattering processes are to be understood in this sense.

1.8.5.1 *Cross section*

Consider an homogeneous *incident beam*, i.e., a stream of incoming identical particles of mass m such that the density of particles over the entire cross-sectional area A of the beam is uniform; we define a *current*, or *flux*,

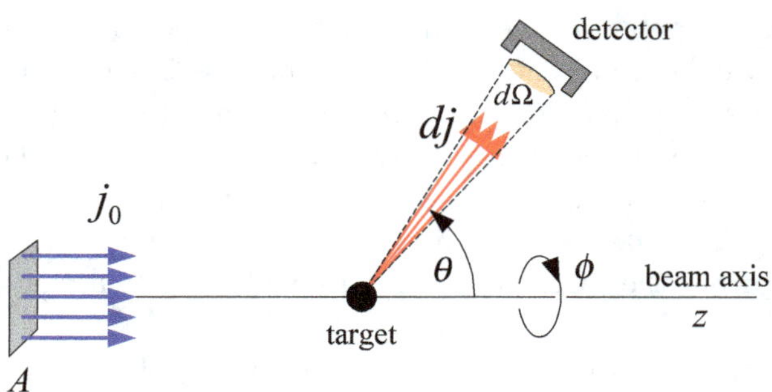

Fig. 1.23   Generic setup of a scattering experiment. The incident beam described by
the particle current j_0 is aimed at the target. A fraction dj/j_0 is scattered under an
angle θ off the beam axis. This fraction is measured by a detector covering a solid angle
$d\Omega$. If the interaction between beam and target is cylindrically symmetric about the
beam axis (by convention taken to be the z-axis), the outcome of such an experiment is
independent of the azimuthal angle ϕ.

density by

$$j_0 = \frac{N_0}{\Delta t\, A} = \frac{\text{number of incoming particles}}{\text{time} \cdot \text{area}}. \tag{1.377}$$

The direction of the beam, by convention, is taken as the positive z-
direction. This beam is aimed at the target (sitting at $z = 0$) and a
part dj of the incident current is deflected away from the beam axis by
the *scattering angle θ* (see Fig. 1.23).

We consider here only central forces; the original spherical symmetry of
such forces is broken by the beam-axis direction, but a cylindrical symmetry
about this axis remains. The scattered part dj, therefore, only depends on
the polar angle θ, and it remains constant for changes of the azimuthal
angle ϕ, i.e., $dj = dj(\theta)$.

The scattered particles are measured with a detector whose detection
aperture covers a solid angle $d\Omega$. If there are $\Delta N(\theta)$ particles detected per
time Δt, the corresponding *scattering-current density* is

$$dj(\theta) = \frac{\Delta N(\theta)}{\Delta t\, d\Omega} = \frac{\text{number of particles deflected by } \theta}{\text{time} \cdot \text{solid angle}}. \tag{1.378}$$

For fixed $d\Omega$, this density is independent of how far away from the scattering center the detector is placed (indeed, it is independent of whether
a measurement takes place at all); it only depends on how many particles
are scattered into the solid angle $d\Omega$ subtended around the direction θ.

The ratio

$$\frac{dj(\theta)}{j_0} = \frac{\text{number of scattered particles per time and solid angle}}{\text{incident current density}} \tag{1.379}$$

describes how much of the incident flux ends up in this particular solid angle $d\Omega$. This measure for the effectiveness of the scattering process at this scattering angle is called the *differential scattering cross section*, and it is denoted by

$$\boxed{\frac{d\sigma}{d\Omega} = \frac{dj(\theta)}{j_0}}. \tag{1.380}$$

The units of the differential cross section are *area/steradian*. Integrating $d\sigma/d\Omega$ over all directions produces the *total cross section*,

$$\sigma = \int d\Omega \frac{d\sigma}{d\Omega} = \int_0^{2\pi} d\phi \int_0^{\pi} d\theta \, \sin\theta \, \frac{d\sigma}{d\Omega} = 2\pi \int_0^{\pi} d\theta \, \sin\theta \, \frac{d\sigma}{d\Omega}, \tag{1.381}$$

with units of an *area*. The product σj_0 is equal to the total number of particles scattered per unit time. The total cross section σ, therefore, corresponds to the total effective area that produces scattered particles. For example, for the hard-sphere scattering considered below (see Fig. 1.25), σ is equal to the cross-sectional circular area seen by the incident beam, i.e., $\sigma = \pi R^2$, where R is the radius of the sphere. For the scattering of a nucleon off a heavy nucleus (with a radius $R \sim 6 \cdot 10^{-13}$ cm), we expect a cross section $\sigma \sim \pi R^2 \approx 10^{-24}$ cm^2, and for the scattering of two atoms $(R \sim 2 \cdot 10^{-8}$ cm) off each other, $\sigma \sim 10^{-15}$ cm^2. These are indeed found to be valid orders of magnitudes for the cross sections of such processes. In fact, 10^{-24} cm^2 is a convenient unit for nuclear-physics cross sections called one *barn*.

The concept of an homogeneous incident particle beam takes care of the incoming flux of many identical particles. The cross sections defined here presume that the scattering process takes place on one single particle. As mentioned already, in practice, the target usually consists of a large number of identical particles. For a *thin* target (so that each incident particle scatters at most once before it leaves the target area), the number of detected particles is directly proportional to the number of target particles 'illuminated' by the incident beam. Therefore, to arrive at the cross sections as defined here, the measured rates must be divided by that number. With these assumptions, we may now employ the two-body formalism of the previous sections and derive more explicit expressions for the cross section.

Each particle within the incident beam has the same initial velocity (at infinity) and, hence, carries the same initial energy

$$E = \frac{m}{2} v_0^2. \tag{1.382}$$

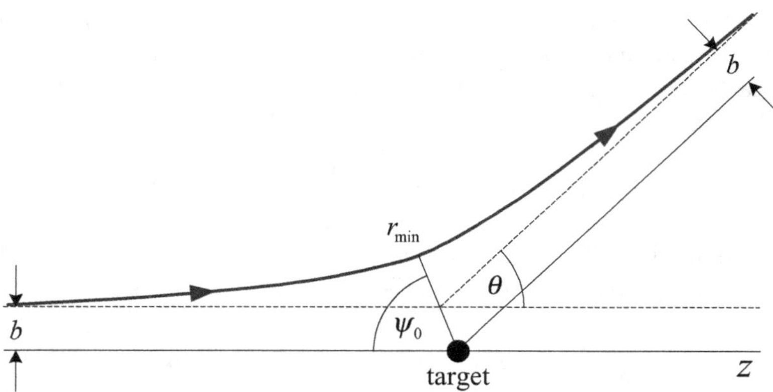

Fig. 1.24   Trajectory of scattered particle for a repulsive central force. The trajectory is symmetric about the line connecting the target and the closest-approach position r_{min}. Asymptotically the particle moves along lines that are a distance b parallel to lines that go through the target at the origin.

All particles are taken here to travel parallel to each other. In view of the finite width of the beam, only the central beam axis goes right through the target at the origin. In general, at infinity, a particle will be offset from this axis by a distance b, called the *impact parameter* (see Fig. 1.24). The particle's (conserved) angular momentum with respect to the origin is then given by

$$\ell = m v_0 b = b\sqrt{2mE}. \tag{1.383}$$

We know from the general result (1.313) that, for a given central potential, E and ℓ are the only parameters that enter the calculation of the trajectories. The scattering angle θ, that follows directly from the outgoing asymptotic trajectory, is then completely specified by the values of E and b. For the present case, we have[45]

$$\theta = \theta(b) \quad \text{or} \quad b = b(\theta) \quad (E \text{ fixed}), \tag{1.384}$$

where E is the given energy of the scattering process.

The number of particles that enter the scattering region per unit time with an impact parameter in the range b to $b + db$ is given by $j_0 2\pi b\,|db|$. Since particles do not get lost, this current must be equal to the particles

---

[45]It is assumed here that $\theta(b)$ can be inverted; for some potentials (e.g., $V \sim 1/r^2$), this is not possible. If $\theta(b)$ is a nonmonotonic function, it must be divided into piecewise monotonic branches that can be inverted separately, $b \rightarrow b_i = b_i(\theta)$. The differential cross section (1.385) is then replaced by a corresponding sum over these branches. See [1] for more details.

scattered per unit time into the solid angle $d\Omega = 2\pi \sin\theta\,|d\theta|$ corresponding to the angular range θ to $\theta + d\theta$. Hence,

$$j_0 2\pi b\,|db| = dj\,d\Omega \qquad \Longrightarrow \qquad \boxed{\frac{d\sigma}{d\Omega} = \frac{b(\theta)}{\sin\theta}\left|\frac{db(\theta)}{d\theta}\right|}, \qquad (1.385)$$

which thus reduces the determination of the differential cross section to finding the function $b(\theta)$. Only the magnitudes of the increments db and $d\theta$ enter here since $d\sigma/d\Omega$ is positive by definition, but db and $d\theta$ may have opposite signs (see the subsequent example for scattering off a hard sphere); $\sin\theta$ is positive since the scattering angle can only vary between 0 and π.

Scattering off a hard sphere

As a first example, we consider the elastic scattering of point masses off a fixed hard sphere of radius R (see Fig. 1.25). For the incident angle γ,

$$\sin\gamma = \frac{b}{R}. \qquad (1.386)$$

Assuming ideal reflection, the scattering angle is

$$\theta = \pi - 2\gamma = \pi - 2\arcsin\frac{b}{R}, \qquad (1.387)$$

which translates into

$$b = R\sin\frac{\pi - \theta}{2} = R\cos\frac{\theta}{2}. \qquad (1.388)$$

Hence

$$\frac{d\sigma}{d\Omega} = \frac{b}{\sin\theta}\left|\frac{db(\theta)}{d\theta}\right| = \frac{b}{\sin\theta}\left(\frac{R}{2}\sin\frac{\theta}{2}\right) = \frac{R^2}{4} \qquad (1.389)$$

and

$$\sigma = \int \frac{d\sigma}{d\Omega}\,d\Omega = 2\pi\int_0^\pi \frac{d\sigma}{d\Omega}\sin\theta\,d\theta = \pi R^2, \qquad (1.390)$$

i.e., the total cross section is equal to the circular area 'seen' by the incident beam.

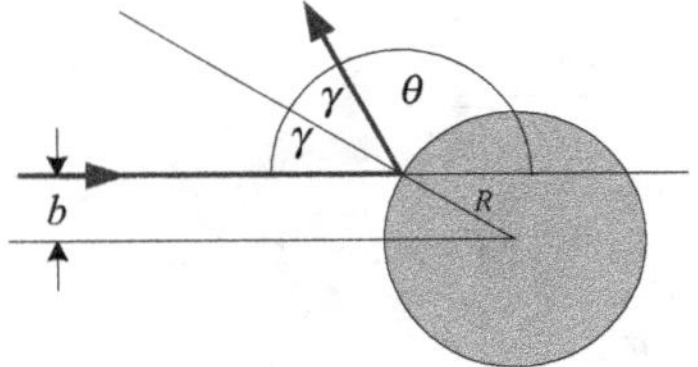

Fig. 1.25   Elastic scattering off an ideally reflecting hard sphere.

1.8.5.2   *Scattering trajectories*

The trajectory of the scattered particle follows from the general result (1.313), with the angle φ, defined in the (fixed) plane of the trajectory, with respect to the positive z axis. For the following considerations, we assume a repulsive force, leading to the generic picture shown in Fig. 1.24.[46] It is convenient to rewrite (1.313) in the form

$$d\psi = -\frac{b}{r^2}\frac{dr}{\sqrt{1 - \frac{b^2}{r^2} - \frac{V(r)}{E}}}, \tag{1.391}$$

where (1.383) was used and $\psi = \pi - \varphi$. The angle ψ is defined with respect to the *negative* z-axis; it varies according to $0 \leq \psi \leq \pi - \theta$, describing the progress of the scattered particle, with the limits $\psi = 0$ and $\psi = \pi - \theta$ corresponding to the incident particle coming from infinity or moving away to infinity, respectively.

The overall minus sign in (1.391) means that this equation is valid only during the time when the particle approaches the target, when $dr < 0$. The angle ψ then varies between $\psi = 0$ and $\psi = \psi_0$, when the particle is closest to the target (see Fig. 1.24). The trajectory is symmetric about this *periapsis* (closest approach) angle ψ_0. This follows from the spherical symmetry of the interaction and the fact that the description would be the same if time ran backward.[47] It is sufficient, therefore, to calculate the approach part of the trajectory according to Eq. (1.391) to determine the entire scattering solution.

The minimal periapsis distance $r_{\min}$ (see Fig. 1.24) follows from the condition

$$\frac{dr}{d\psi} = -\frac{r^2}{b}\sqrt{1 - \frac{b^2}{r^2} - \frac{V(r)}{E}} = 0, \tag{1.392}$$

or

$$\frac{b^2}{r_{\min}^2} + \frac{V(r_{\min})}{E} = 1. \tag{1.393}$$

The associated periapsis angle ψ_0 is related to the scattering angle by

$$\theta = \pi - 2\psi_0. \tag{1.394}$$

Calculating ψ_0, therefore, will provide an explicit expression for θ; one has

$$\theta(b) = \pi - 2\int_{r_{\min}}^{\infty} \frac{b}{r^2}\frac{dr}{\sqrt{1 - \frac{b^2}{r^2} - \frac{V(r)}{E}}}, \tag{1.395}$$

---

[46]Scattering for attractive forces can just as easily be described along the lines presented here.

[47]See also the symmetry considerations regarding Eq. (1.313), on page 64.

where the integral is equal to ψ_0. [Note that (1.393) means that the integrand is singular at the lower limit of the integration; the integral, however, exists.] In general, this expression for $\theta(b)$ has to be evaluated numerically. This determines $\theta(b)$ and hence $b(\theta)$, which allows us to calculate the differential cross section according to (1.385).

1.8.5.3 *Rutherford scattering*

For the $1/r$ potential, the preceding expressions can be evaluated explicitly. This concerns a famous historical experiment, the scattering of alpha particles (helium nuclei) off gold nuclei performed by Ernest Rutherford (1871–1937) and his undergraduate student, Ernest Marsden, in 1910. The outcome of the experiment provided confirmation of Rutherford's nuclear model of the atom, i.e., that the atom behaves like a tiny planetary system, with negatively charged light electrons orbiting a tightly packed, positively charged heavy central core, the nucleus. In this experiment, the potential acting between the two nuclei is a repulsive Coulomb potential,

$$V(r) = \frac{Z_1 Z_2 e^2}{r}, \tag{1.396}$$

where Z_1 and Z_2 are the charge numbers of the nuclei (for alpha particles and gold, the respective numbers are $Z_1 = 2$ and $Z_2 = 79$); e is the elementary charge.

For the repulsive Coulomb potential, we can take over the expressions obtained for the Kepler problem without modifications, except for the fact that the potential strength α of (1.321) is now negative, i.e., in the expressions of Sec. 1.8.2 the replacement

$$\alpha = -\beta, \qquad \text{with} \qquad \beta = Z_1 Z_2 e^2, \tag{1.397}$$

should be used to exhibit this negative sign explicitly.

Integration of (1.391) provides the trajectory

$$r(\psi) = \frac{2b^2 E}{\beta} \frac{1}{\varepsilon \sin(\psi + \frac{\theta}{2}) - 1}, \qquad \text{with} \quad 0 \le \psi \le \pi - \theta, \tag{1.398}$$

which explicitly contains the scattering angle θ. This result can also be read off Eq. (1.329), using Eq. (1.383), $\varphi = \pi - \psi$, and $\phi = (\pi + \theta)/2$, where the latter is the direction of the associated Runge–Lenz vector. This expression describes one branch of a *hyperbola*, with an eccentricity parameter [cf. Eq. (1.327)]

$$\varepsilon = \sqrt{1 + \left(\frac{2bE}{\beta}\right)^2} = \frac{1}{\sin \frac{\theta}{2}}, \tag{1.399}$$

where the last equality follows from the fact that at $\psi = 0$, the denominator of (1.398) must vanish since this corresponds to the particle coming in from infinity.

To determine the impact parameter as a function of the scattering angle, consider the y-projection of (1.398) at $\psi = 0$, i.e.,

$$y = \frac{2b^2 E}{\beta} \frac{\sin\psi}{\varepsilon \sin(\psi + \frac{\theta}{2}) - 1} \xrightarrow{\psi=0} b = \frac{2b^2 E}{\beta} \tan\frac{\theta}{2}, \qquad (1.400)$$

or

$$b(\theta) = \frac{\beta}{2E} \cot\frac{\theta}{2}. \qquad (1.401)$$

For the cross section, one then finds, via (1.385),

$$\boxed{\frac{d\sigma}{d\Omega} = \left(\frac{\beta}{4E}\right)^2 \frac{1}{\sin^4\frac{\theta}{2}} \qquad \text{Rutherford scattering}} \qquad (1.402)$$

This is the famous equation verified experimentally by Lord Rutherford that corroborated his nuclear model of the atom.

With hindsight, one may wonder why the strong nuclear force (not known to Rutherford) did not considerably modify the outcome of this experiment. Using (1.393), one finds that for $\ell = 0$ (i.e., $b = 0$), the minimum distance between the helium and gold nuclei is

$$r_{\min} = \frac{\beta}{E} = \frac{Z_1 Z_2 e^2}{E} \approx 30\text{–}60 \text{ fm} \qquad (1.403)$$

for energies of $E = 4\text{–}8$ MeV available to Rutherford. This distance is well beyond the range where the strong nuclear interaction can be felt. For $\ell > 0$, $r_{\min}$ is even larger. Coulomb effects due to the electron shell of the gold atoms in the target foil are several orders of magnitude smaller and can be ignored as well. It is justified, therefore, to use only the Coulomb interaction between the nuclei to describe this experiment.

The differential cross section for Rutherford scattering is singular for $\theta = 0$. As a consequence, the total cross section σ actually diverges. Physically, this finding results from the extremely long range of the Coulomb potential, i.e., it is not really possible to consider an incoming or outgoing particle to be free of the interaction even very far away from the scattering center. Therefore, even particles with large impact parameter still feel the force of the scattering center, contributing more and more to the cross section for $\theta = 0$ as b becomes larger and larger. In an actual experiment, however, the pure Coulomb potential, for large distances, becomes screened by the electrons surrounding the nucleus. This effective potential, falling off faster than $1/r$ at large distances, then provides finite cross sections.

The present classical result for Rutherford scattering is found to remain true even when the problem is treated — as it ought to — completely quantum-mechanically (with relativistic corrections ignored). While this agreement is certainly fortuitous from Rutherford's point of view (he did not know anything about quantum mechanics, since it had not been invented yet when he derived the corresponding cross section), it is *not* an accident, as it is often called, but follows from the fact that the Coulomb potential is an *inherently macroscopic*, and therefore classical, potential, where the usual quantum-mechanical treatment of a collision process must be modified, leading to one of the most delicate mathematical problems in nonrelativistic quantum scattering theory. In quantum mechanics, where one uses the concept of *wave functions* for the description of particle properties, the scattering information is carried out of the interaction region (at $r = 0$) by an outgoing spherical wave that falls off like $1/r$. For the Coulomb potential, this is the same fall-off behavior as the interaction itself. This means that even at macroscopic distances (where the detector is located), the scattering information cannot be considered to be independent of the interaction. All quantum-mechanical modification that are necessary to treat the special case of the Coulomb interaction can be traced back to this interference.

1.8.5.4 *Transformation to laboratory coordinates*

In view of the equivalence between the effective one-body formulation and the true two-body problem, the scattering formalism developed in the preceding sections provides a complete solution of the problem. What remains to be done now is to relate the scattering angle θ of the equivalent one-body formalism to the scattering angle θ_L actually measured in the laboratory. In the language of the two-body problem, the scattering angle θ is directly related to the angle $2\psi_0 = \pi - \theta$ between the initial and final directions of the relative vector between the two particles in their center-of-mass (CM) coordinates (see Fig. 1.26). However, the observed scattering angle θ_L in an experiment is the angle between the initial and final directions of the projectile as measured in the lab frame. The two angles are only identical if the target remains stationary after the collision. In general, this is not the case since the projectile will transfer momentum to the target during the collision. Therefore, to relate the present results to an actual experimental

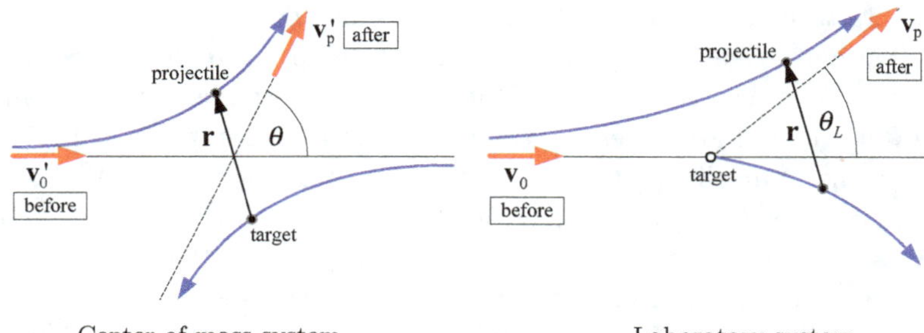

Fig. 1.26   Schematic picture of the scattering of two particles in the center-of-mass system and the laboratory system. In the lab system, the target is at rest initially. The relative position vector **r** is the same in both systems; during the scattering process, it rotates by an angle $\pi - \theta$. Also shown are the directions of the projectile's velocity vector before and after the collision.

measurement taking place in the laboratory, we need to find the relation between θ and θ_L.

Both lab and CM systems are inertial systems. They are related by Galilei transformations written in terms of the CM vector $\mathbf{R}(t)$ in the lab system (see Sec. 1.8.1.1). The lab position $\mathbf{r}_i$ of particle i is related to its CM position $\mathbf{r}'_i$ by

$$\mathbf{r}_i(t) = \mathbf{R}(t) + \mathbf{r}'_i(t), \tag{1.404}$$

with the prime denoting CM quantities. The corresponding velocities are related by

$$\mathbf{v}_i(t) = \mathbf{u} + \mathbf{v}'_i(t), \tag{1.405}$$

where $\mathbf{u} = \dot{\mathbf{R}} = const$ is the constant velocity of the CM in the lab system. Since the target is initially at rest in the lab frame, the initial velocity $\mathbf{v}_0$ of the projectile in that frame is identical to the initial *relative* velocity of the particles, i.e., v_0 of Eq. (1.382) is given by $v_0 = |\mathbf{v}_0|$. The velocity $\mathbf{u}$ of the CM is directly related to this relative velocity via total-momentum conservation; one finds in the lab frame

$$(m_\mathrm{p} + m_\mathrm{t})\mathbf{u} = m_\mathrm{p}\mathbf{v}_0 \quad \Longrightarrow \quad \mathbf{u} = \frac{m_\mathrm{p}}{m_\mathrm{p} + m_\mathrm{t}}\mathbf{v}_0, \tag{1.406}$$

where m_p and m_t are the projectile and target masses, respectively.

The direction of the CM velocity $\mathbf{u}$ is the direction of the beam axis; this direction is the *same* in both frames. The scattering angles in both frames, therefore, are defined with respect to this direction (see Fig. 1.26). Using Eq. (1.405) for the projectile's velocities long after the collision has taken

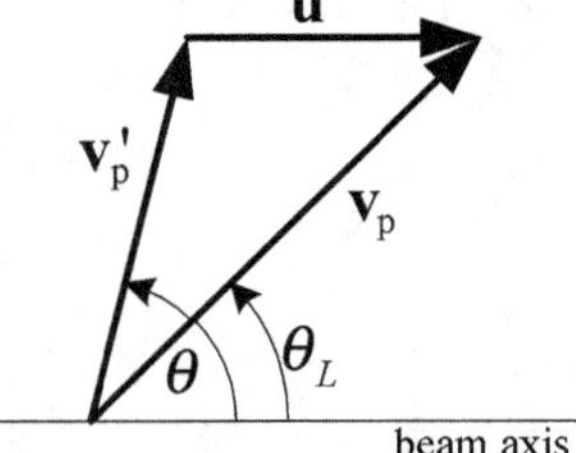

Fig. 1.27 Relation between the projectile velocities in the CM and lab systems and the associated scattering angles long *after* the scattering has taken place (see also Fig. 1.26). The beam direction is the same in both frames; it is along the CM velocity **u**.

place, the relationship of the velocity vectors is shown in Fig. 1.27. One directly reads off

$$\boxed{\tan \theta_L = \frac{\sin \theta}{\cos \theta + \nu}}, \tag{1.407}$$

or alternatively

$$\boxed{\cos \theta_L = \frac{\cos \theta + \nu}{\sqrt{1 + 2\nu \cos \theta + \nu^2}}}. \tag{1.408}$$

The parameter ν is given by the velocity ratio

$$\nu = \frac{|\mathbf{u}|}{|\mathbf{v}_p'|} = \frac{u}{v_p'} = \frac{m_{\mathrm{p}}}{m_{\mathrm{p}} + m_{\mathrm{t}}} \frac{v_0}{v_p'}. \tag{1.409}$$

It follows by differentiating Eq. (1.295) that

$$v_{\mathrm{p}}' = \frac{m_{\mathrm{t}}}{m_{\mathrm{t}} + m_{\mathrm{p}}} v_{\mathrm{rel}}, \tag{1.410}$$

where v_{rel} is the relative velocity *after* the collision. Hence, we have

$$\nu = \frac{m_{\mathrm{p}}}{m_{\mathrm{t}}} \frac{v_0}{v_{\mathrm{rel}}}. \tag{1.411}$$

This general result holds true for all scattering processes. Within the present context of potential scattering, there is no energy loss during the collision, i.e., the scattering is purely *elastic*, and this can be simplified further.[48] Energy conservation $E = mv_0^2/2 = mv_{\mathrm{rel}}^2/2$ holds true for potential scattering, and thus the relative velocities before and after the collision are equal, $v_0 = v_{\mathrm{rel}}$; hence

$$\boxed{\nu = \frac{m_{\mathrm{p}}}{m_{\mathrm{t}}} \quad \text{(elastic scattering)}}. \tag{1.412}$$

To derive a relation between the cross sections, we note that the number of particles scattered into the solid angle $d\Omega_L$ around θ_L in the lab frame

---

[48]For further discussion of inelastic scattering, see, e.g., [1].

must be equal to those scattered during the same time into the correspond-ing solid angle $d\Omega$ around θ in the CM frame, i.e.,

$$dj(\theta)\,d\Omega(\theta) = dj_L(\theta_L)\,d\Omega_L(\theta_L),\tag{1.413}$$

or

$$j_0\,\frac{d\sigma}{d\Omega}\,2\pi\,\sin\theta\,|d\theta| = j_{L,0}\,\frac{d\sigma_L}{d\Omega_L}\,2\pi\,\sin\theta_L\,|d\theta_L|.\tag{1.414}$$

Since the time is the same in all frames in the context of classical mechanics, the incoming flux densities j_0 and $j_{L,0}$ can only differ by the number of incoming particles per unit time [see Definition (1.377)]. However, since for every 'particle' of the equivalent one-body problem, there is one, and only one, projectile-target pair of the two-body problem, the two are actually identical; hence, $j_0 = j_{L,0}$. We, therefore, have

$$\frac{d\sigma_L}{d\Omega_L} = \frac{d\sigma}{d\Omega}\,\frac{\sin\theta}{\sin\theta_L}\,\left|\frac{d\theta}{d\theta_L}\right| = \frac{d\sigma}{d\Omega}\,\left|\frac{d\cos\theta}{d\cos\theta_L}\right|,\tag{1.415}$$

which, using (1.408), leads to the result

$$\boxed{\frac{d\sigma_L}{d\Omega_L} = \frac{d\sigma}{d\Omega}\,\frac{\left(1 + 2\nu\cos\theta + \nu^2\right)^{3/2}}{1 + \nu\cos\theta}}\,.\tag{1.416}$$

The cross sections in general depend on the energy of the projectile (the Rutherford cross section, for example, contains a factor $1/E^2$); this energy must be converted as well. The projectile energy of the equivalent one-body system is given by, cf. (1.382),

$$E = \frac{1}{2}\,\frac{m_t m_p}{m_t + m_p}\,v_0^2 = \frac{m_t}{m_t + m_p}\,\left[\frac{m_p}{2}\,v_0^2\right].\tag{1.417}$$

Since v_0, as explained above, is also the initial velocity of the lab projectile, the quantity in the square bracket is the kinetic energy E_L of the projectile in the lab frame, i.e.,

$$\boxed{E = \frac{m_t}{m_t + m_p}\,E_L}\,.\tag{1.418}$$

Let us consider two special cases. If the target mass is very much larger than the projectile mass, one has $\nu \approx 0$ and one finds then

$$\theta_L \approx \theta, \qquad \frac{d\sigma_L}{d\Omega_L} \approx \frac{d\sigma}{d\Omega}, \qquad E \approx E_L \qquad (m_t \gg m_p).\tag{1.419}$$

This case is approximatively true for the original Rutherford experiment of scattering alpha particles off gold nuclei. Corrections are of the order of a few percent. For the second example of equal masses, one finds

$$\tan\theta_L = \frac{\sin\theta}{\cos\theta + 1} = \tan\frac{\theta}{2} \qquad (m_p = m_t),\tag{1.420}$$

hence

$$\theta_L = \frac{\theta}{2}, \quad \frac{d\sigma_L}{d\Omega_L} = 4\cos\theta_L \frac{d\sigma(2\theta_L)}{d\Omega}, \quad E = \frac{E_L}{2} \qquad (m_p = m_t). \quad (1.421)$$

The lab scattering angle θ_L cannot exceed $\pi/2$ for this case since $\theta \leq \pi$. For the collision of two billiard balls, for example, this suggests that the maximum deflection of a cue ball would be $\pi/2$. Of course, this completely ignores spin effects, friction, etc., which in practice may modify this idealized result quite substantially.

Chapter 2

Lagrangian Mechanics

The elementary formulation of mechanics given in the previous chapter presumes that the forces are known. In practice, however, one very often encounters situations where some of the forces are not known *a priori*. This typically involves constraints that limit the allowed motion such that, in many instances, one may not be able to formulate the corresponding constraining forces independent of the dynamical situation. In such cases, one needs to adapt Newton's second law so that the determination of the constraining forces becomes an integral part of the solution of the equations of motion. The formal framework to achieve this is furnished by the Lagrangian approach to mechanics, named after Joseph Louis Lagrange.[49]

2.1   Lagrange Equations of the First Kind

First, we will reformulate Newton's equations of motion by explicitly accounting for the constraints such that determining their *a priori* unknown underlying forces will become an integral part of the solution procedure.

---

[49] Joseph Louis Lagrange (1736–1813) is usually considered a French national, but he was actually born in Turin, Italy, as Giuseppe Lodovico Lagrangia, according to his birth certificate. Turin originally belonged to the French Duchy of Savoy, but in 1720, well before Lagrange's birth, it became the capital of the Italian Kingdom of Sardinia. Lagrange had studied mathematics and spent his early professional life in Turin. Later, in 1766, he succeeded Euler as Director of Mathematics at the Berlin Academy of Sciences, one of the most prestigious appointments available at the time. His main treatment of analytical mechanics, *Traité de mécanique analytique*, was completed while still in Berlin, but published in Paris, in 1788, after he had accepted a position there, in 1787. Jean-Baptiste Joseph Fourier (1768–1830), one of his contemporaries in Paris, complained that his lectures were delivered with a strong Italian accent and were too abstract for most students. Lagrange stayed in Paris for the rest of his life.

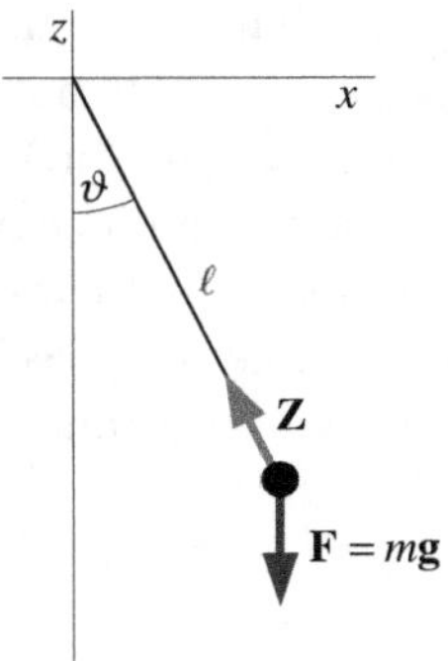

Fig. 2.1   The forces acting on the mass m of the planar pendulum are the gravity $\mathbf{F}$ and an *a priori* unknown constraining force $\mathbf{Z}$ exerted by the string.

2.1.1   *Constraints*

If the motion of a particle is confined to, for example, a surface, the particle's trajectory $\mathbf{r} = \mathbf{r}(t)$ must satisfy a constraint that can be expressed in the form[50]

$$g(\mathbf{r}, t) = g(x, y, z, t) = 0, \tag{2.1}$$

where the function g parameterizes the surface and the explicit time dependence allows for the fact that the surface itself may be moving or otherwise change form. If the motion is confined to a particular curve, the constraint would take the form

$$g_1(x, y, z, t) = 0 \qquad \text{and} \qquad g_2(x, y, z, t) = 0, \tag{2.2}$$

where the intersection of the two surfaces g_1, g_2 defines the curve. An example for the latter case is the trajectory of a planar pendulum of length ℓ confined to move in the xz-plane (see Fig. 2.1), whose constraint equations are

$$x^2 + y^2 + z^2 - \ell^2 = 0 \qquad \text{and} \qquad y = 0, \tag{2.3}$$

where the first equation constrains the motion to a spherical shell of radius ℓ and the second constraint takes a vertical slice through that shell leaving a circle of radius ℓ in the xz-plane. For Newton's second axiom, we have

$$m\ddot{\mathbf{r}} = \mathbf{F} + \mathbf{Z}, \tag{2.4}$$

where in addition to the gravitational force, $\mathbf{F} = m\mathbf{g}$, we must introduce a *constraining force* $\mathbf{Z}$ that accounts for the fact that the string exerts a

---

[50]To avoid possible confusion: The meaning here is that there is a function $g = g(x, y, z, t)$, and for those positions $\mathbf{r}(t) = x(t)\mathbf{e}_x + y(t)\mathbf{e}_y + z(t)\mathbf{e}_z$ that describe points on the trajectory, the condition $g(x, y, z, t) = 0$ must be met. In other words g is *not* the function that vanishes identically [however, see the discussion related to Eq. (2.72)].

force on the mass that brings about the constrained motion. The problem is that while we do know the constraint conditions (2.3), *a priori* we do not know the constraining force $\mathbf{Z}$. As a matter of fact, the constraining force will, in general, depend dynamically on the circumstances of the motion.

Clearly, for a single particle, the maximum number of non-trivial constraint equations is two, since three equations would determine all three coordinates (x, y, z) completely. For n particles, we may have l constraints of the form

$$g_k(\mathbf{r}_1, \mathbf{r}_2, \ldots, \mathbf{r}_n, t) = 0, \qquad k = 1, 2, \ldots, l, \tag{2.5}$$

depending on all particle positions $\mathbf{r}_i$, where l is limited to $l \leq 3n - 1$.

Constraints that may be expressed in the manner discussed above, i.e., as equations that connect the particle coordinates (and that possibly depend explicitly on the time), are called *holonomic*. Constraints that cannot be written in this way are called *nonholonomic*. Examples for the latter are constraint equations that contain the velocities of the particles, which can be written as

$$f_k(\mathbf{r}_1, \ldots, \mathbf{r}_n, \dot{\mathbf{r}}_1, \ldots, \dot{\mathbf{r}}_n, t) = 0, \quad k = 1, 2, \ldots, l \tag{2.6}$$

and *cannot* be reduced to the form (2.5). Other nonholonomic constraints involve inequalities, rather than equalities. An example would be a particle confined to the inside of a sphere, where the constraint is $R^2 - x^2 - y^2 - z^2 \geq 0$. If the constraint equations contain the time explicitly, the conditions are called *rheonomous*; if they are time-independent, they are called *scleronomous*. Most of the standard examples for constraints are holonomic, however, most of the constrained phenomena in nature are nonholonomic.

In the following discussion, we will at first assume that all constraints can be expressed either by Eq. (2.5) or by (2.6). The discussion of constraints based on inequalities will take place in Sec. 2.2.2.9.

2.1.2 *Forces of constraint*

For n particles subject to l constraint conditions, the equations of motions are

$$\boxed{m_i \ddot{\mathbf{r}}_i = \mathbf{F}_i + \mathbf{Z}_i, \qquad i = 1, \ldots, n} \tag{2.7}$$

The applied forces $\mathbf{F}_i$ are given, but the *forces of constraint* $\mathbf{Z}_i$ generally are unknown *a priori* and need to be determined in accordance with the constraint conditions.

Let us first restrict the discussion to holonomic constraints, as in (2.5), and consider *instantaneous* infinitesimal displacements $\delta\mathbf{r}_i$ of the n particle positions $\mathbf{r}_i$ chosen such that they are compatible with the constraints, i.e.,

$$g_k(\mathbf{r}_1,\ldots,\mathbf{r}_n,t) \equiv g_k(\mathbf{r},t) = 0, \qquad (2.8a)$$

$$g_k(\mathbf{r}_1 + \delta\mathbf{r}_1,\ldots,\mathbf{r}_n + \delta\mathbf{r}_n,t) \equiv g_k(\mathbf{r} + \delta\mathbf{r},t) = 0, \qquad (2.8b)$$

for all $k = 1, 2, \ldots, l$, where

$$\mathbf{r} = (\mathbf{r}_1,\ldots,\mathbf{r}_n), \qquad (2.9)$$

etc., generically stands for the entire list of variables on the right. The $\delta\mathbf{r}_i$ are called *virtual displacements*; this is in contradistinction to *real displacements* $d\mathbf{r}_i$ that *always* take place during a non-zero time interval dt. The virtuality of the displacements thus ensures that there will be no changes of the forces or the constraints. Loosely speaking, virtual displacements allow one to probe the topological properties of the spatial region surrounding a point on the trajectory at a given moment in time in a static manner without affecting the dynamic evolution of the trajectory.

The virtual change δg_k of each constraint condition itself then vanishes,

$$\delta g_k = g_k(\mathbf{r} + \delta\mathbf{r},t) - g_k(\mathbf{r},t) = 0 \qquad (2.10)$$

or, by performing a Taylor expansion on the right-hand side,

$$\sum_{j=1}^{n} \boldsymbol{\nabla}_j g_k(\mathbf{r},t) \cdot \delta\mathbf{r}_j = 0, \qquad k = 1, 2, \ldots, l, \qquad (2.11)$$

where $\boldsymbol{\nabla}_j \equiv \boldsymbol{\nabla}_{\mathbf{r}_j}$ is the gradient with respect to the position of particle j. As in the one-particle case, for the j-th particle the constraint $g_k(\mathbf{r},t) = 0$ defines a surface if we keep the other coordinates $\mathbf{r}_i$, $i \neq j$, fixed. By construction, the displacement $\delta\mathbf{r}_j$ is compatible with the constraints and must lie within the respective surfaces. Topologically, this means that the gradients $\boldsymbol{\nabla}_j g_k$ are perpendicular to the constraining surfaces for this particle.[51]

---

[51]To see this more clearly note that we may uniquely decompose the gradients as

$$\boldsymbol{\nabla}_j g_k = \mathbf{j}\,(\mathbf{j} \cdot \boldsymbol{\nabla}_j g_k) + \mathbf{j}_\perp (\mathbf{j}_\perp \cdot \boldsymbol{\nabla}_j g_k),$$

where $\mathbf{j} = \delta\mathbf{r}_j/|\delta\mathbf{r}_j|$ is the unit vector in the direction of $\delta\mathbf{r}_j$ and

$$\mathbf{j}_\perp = \frac{\boldsymbol{\nabla}_j g_k - \mathbf{j}\,(\mathbf{j} \cdot \boldsymbol{\nabla}_j g_k)}{\sqrt{(\boldsymbol{\nabla}_j g_k)^2 - (\mathbf{j} \cdot \boldsymbol{\nabla}_j g_k)^2}}$$

is its orthogonal complement. The first term of the gradient decomposition contains the derivative $\mathbf{j} \cdot \boldsymbol{\nabla}_j g_k$ in the direction of $\delta\mathbf{r}_j$. Since for virtual displacements $\delta\mathbf{r}_j$ the constraint equation $g_k = 0$ remains true, the function g_k does not change in this direction and hence $\mathbf{j} \cdot \boldsymbol{\nabla}_j g_k = 0$, which in turn implies $\boldsymbol{\nabla}_j g_k \perp \delta\mathbf{r}_j$.

The set of all l constraints defines a $(3n - l)$-dimensional hypersurface in the $3n$-dimensional configuration space of all n particles via the intersection of all individual surfaces, and the set of all $\boldsymbol{\nabla}_j g_k$, with $k = 1, \ldots, l$, forms an l-dimensional basis for any vector that is perpendicular to the instantaneous hypersurface constraining particle j.

If, instead, we consider a *real displacement* $d\mathbf{r}_i$, such that

$$g_k(\mathbf{r}, t) = 0, \qquad g_k(\mathbf{r} + d\mathbf{r}, t + dt) = 0, \tag{2.12}$$

we also have $dg_k = 0$, analogously to (2.10), and thus

$$\sum_{j=1}^{n} \boldsymbol{\nabla}_j g_k(\mathbf{r}, t) \cdot d\mathbf{r}_j + \frac{\partial g_k(\mathbf{r}, t)}{\partial t} dt = 0, \qquad k = 1, \ldots, l. \tag{2.13}$$

This shows that real displacements lie only within a *given* hypersurface if the constraint conditions are scleronomous, $\partial g_k / \partial t = 0$, i.e., if the hypersurface is at rest. By contrast, real displacements *connect* one constraint hypersurface at time t with another constraint hypersurface at time $t + dt$. Note that Eq. (2.13) may also be written as

$$\sum_{j=1}^{n} \boldsymbol{\nabla}_j g_k(\mathbf{r}, t) \cdot \dot{\mathbf{r}}_j + \frac{\partial g_k(\mathbf{r}, t)}{\partial t} \equiv \frac{dg_k(\mathbf{r}, t)}{dt} = 0, \qquad k = 1, \ldots, l. \tag{2.14}$$

Which one of these two forms is used is a matter of convenience. However, it should be emphasized that these sets of equations are *completely equivalent* to introducing the holonomic constraints as it was done originally in Eq. (2.5). The present form, however, allows us now to generalize the approach to include nonholonomic conditions.

Equation (2.13) is generalized by allowing for arbitrary functions $\mathbf{h}_{kj} = \mathbf{h}_{kj}(\mathbf{r}, t)$ and $h_{k0} = h_{k0}(\mathbf{r}, t)$ that specify the constraints in the form

$$\sum_{j=1}^{n} \mathbf{h}_{kj} \cdot d\mathbf{r}_j + h_{k0}dt = 0, \quad k = 1, \ldots, l, \tag{2.15}$$

or, equivalently, by

$$\sum_{j=1}^{n} \mathbf{h}_{kj} \cdot \dot{\mathbf{r}}_j + h_{k0} = 0, \quad k = 1, \ldots, l, \tag{2.16}$$

which is of the more general form (2.6). In other words, the $\mathbf{h}_{kj}$ and h_{k0} need not be related to common constraint functions g_k. However, they must satisfy

$$\sum_{j=1}^{n} \mathbf{h}_{kj} \cdot \delta\mathbf{r}_j = 0, \tag{2.17}$$

Table 2.1   Classification of constraints expressed generically by Eq. (2.15).

	Rheonomous	*Scleronomous*
Nonholonomic	$\mathbf{h}_{kj}$ and h_{k0} arbitrary	$\dfrac{\partial \mathbf{h}_{kj}}{\partial t} = 0$ and $h_{k0} = 0$
Holonomic	$\mathbf{h}_{kj} = \boldsymbol{\nabla}_j g_k$ and $h_{k0} = \dfrac{\partial g_k}{\partial t}$	$\mathbf{h}_{kj} = \boldsymbol{\nabla}_j g_k,$ with $\dfrac{\partial \mathbf{h}_{kj}}{\partial t} = 0$ and $h_{k0} = \dfrac{\partial g_k}{\partial t} = 0$

similar to (2.11), i.e., the vectors $\mathbf{h}_{kj}$, for $k = 1, \ldots, l$ must form an l-dimensional basis for any vector that is perpendicular to the instantaneous constraint hypersurface of particle j.

Nonholonomic constraints are such that Eq. (2.15) *cannot* be integrated, i.e., for constraints of this type, the left-hand side of (2.15) is not a perfect differential[52] and thus cannot be recast in the form (2.13). The properties of nonholonomic constraints will be discussed in more detail in Sec. 2.2. The various types of constraints are summarized in Table 2.1.

A constraining force $\mathbf{Z}_j$ for particle j should not affect motion *within* the particle's instantaneous constraint hypersurface and therefore should be perpendicular to it. Clearly, this is *not* true if one wishes to express friction phenomena as constraints as well because friction forces, in particular kinetic friction forces, act along trajectories within constraining hypersurfaces. However, mechanical friction is a macroscopic phenomenon and the associated forces are explicitly known; they can therefore always simply be incorporated into the applied forces $\mathbf{F}_j$. Hence, for the following, we will only consider topological constraining forces perpendicular to the hypersurfaces.[53]

---

[52]Expressions of the type $\sum_i a_i \, d\xi_i$ are also called *Pfaffian forms*. If there exists a function f such that $a_i = \partial f / \partial \xi_i$ for all a_i, the Pfaffian form is equal to the perfect differential df. See also footnote 19.

[53]It is indeed possible to treat friction forces as constraints whose virtual work (see Sec. 2.1.4) does not vanish; see R. Kalaba and F. Udwadia, "Analytical dynamics with constraint forces that do work in virtual displacements", Applied Mathematics and Computation **121**, 211 (2001). However, the resulting equations of motion can always be equivalently reformulated as explicitly adding the known friction forces to the applied forces without affecting the treatment of the remaining topological constraint forces.

We may then write $\mathbf{Z}_j$ as a linear combination of the corresponding perpendicular basis vectors,

$$\mathbf{Z}_j = \sum_{k=1}^{l} \lambda_k \, \mathbf{h}_{kj}, \tag{2.18}$$

with unknown coefficient functions λ_k, called *Lagrange multipliers*.[54] For holonomic conditions, this reads

$$\mathbf{Z}_j = \sum_{k=1}^{l} \lambda_k \boldsymbol{\nabla}_j g_k. \tag{2.19}$$

2.1.3   *Equations of motion with constraints*

We therefore find that all constraints are satisfied by the following *Lagrange equations of the first kind*:

$$m_j \ddot{\mathbf{r}}_j - \mathbf{F}_j - \sum_{k=1}^{l} \lambda_k \mathbf{h}_{kj} = 0, \qquad j = 1, \dots, n, \tag{2.20a}$$

$$\sum_{j=1}^{n} \mathbf{h}_{kj} \cdot d\mathbf{r}_j + h_{k0} dt = 0, \qquad k = 1, \dots, l. \tag{2.20b}$$

This set of $(3n + l)$ equations — $3n$ second-order differential equations and l algebraic constraint conditions — for the $3n$ Cartesian components (x_j, y_j, z_j) of the n particles and the l Lagrange multipliers λ_k completely determine the trajectories subject to all constraints.

It is customary to renumber the components, etc., according to

$$\mathbf{r}_j = (x_j, y_j, z_j) \quad \longrightarrow \quad (x_{3j-2}, x_{3j-1}, x_{3j}), \tag{2.21a}$$

$$\mathbf{F}_j = (F_{j,x}, F_{j,y}, F_{j,z}) \quad \longrightarrow \quad (F_{3j-2}, F_{3j-1}, F_{3j}), \tag{2.21b}$$

$$\mathbf{h}_{kj} = (h_{kj,x}, h_{kj,y}, h_{kj,z}) \quad \longrightarrow \quad (h_{k,3j-2}, h_{k,3j-1}, h_{k,3j}), \tag{2.21c}$$

$$m_j \quad \longrightarrow \quad m_{3j-2} = m_{3j-1} = m_{3j}, \tag{2.21d}$$

---

[54]Note that the Lagrange multipliers λ_k cannot depend on the particle index j since this would violate Newton's third axiom. To illustrate this, consider a two-body system constrained such that the relative distance is kept fixed, $g(\mathbf{r}_1, \mathbf{r}_2) = |\mathbf{r}| - d = 0$, with $\mathbf{r} = \mathbf{r}_1 - \mathbf{r}_2$. Allowing for λ to depend on the particle index i, the constraint forces are $\mathbf{Z}_i = \lambda_i \boldsymbol{\nabla}_i g$. One then has $\mathbf{Z}_1 + \mathbf{Z}_2 = (\lambda_1 - \lambda_2)\, \mathbf{e}_r$; hence $\lambda_1 = \lambda_2$ since Newton's third law requires $\mathbf{Z}_1 + \mathbf{Z}_2 = 0$.

and write the Lagrange equations as

$$m_i\ddot{x}_i - F_i - \sum_{k=1}^{l}\lambda_k h_{ki} = 0, \qquad i = 1,\ldots,3n, \tag{2.22a}$$

$$\sum_{j=1}^{3n} h_{kj}dx_j + h_{k0}dt = 0, \qquad k = 1,\ldots,l. \tag{2.22b}$$

Specifically *for holonomic conditions*, this reads

$$m_i\ddot{x}_i - F_i - \sum_{k=1}^{l}\lambda_k \frac{\partial g_k}{\partial x_i} = 0, \qquad i = 1,\ldots,3n, \tag{2.23a}$$

$$g_k(x_1,\ldots,x_{3n},t) = 0, \tag{2.23b}$$

or $$\sum_{j=1}^{3n} \frac{\partial g_k}{\partial x_j}\dot{x}_j + \frac{\partial g_k}{\partial t} \equiv \frac{dg_k}{dt} = 0, \qquad k = 1,\ldots,l, \tag{2.23c}$$

where the last two equations are equivalent alternative ways of specifying the holonomic conditions.

2.1.3.1 *Constructing the forces of constraint*

In the original formulation of Newton's axioms all forces are presumed to be known. The Lagrange equations derived here provide a *nontrivial* extension to Newton's second law to cases were *a priori* unknown constraining forces are required. Newton's second law is still valid, of course, but the Lagrange formalism allows one to construct the necessary constraining forces *before* they enter Newton's second equation. The construction proceeds along the following lines. First, take the total time derivative of the constraint equation,

$$\frac{d}{dt}\left(\sum_{j=1}^{3n} h_{kj}\dot{x}_j + h_{k0}\right) = \sum_{j=1}^{3n} h_{kj}\ddot{x}_j + \sum_{j=1}^{3n} \dot{h}_{kj}\dot{x}_j + \dot{h}_{k0} = 0, \tag{2.24}$$

and eliminate $\ddot{x}_j$ with the help of (2.22a), producing[55]

$$\sum_{k=1}^{l}\left(\sum_{j=1}^{3n} \frac{h_{rj}h_{kj}}{m_j}\right)\lambda_k = -\sum_{j=1}^{3n} \frac{h_{rj}F_j}{m_j} - \sum_{j=1}^{3n} \dot{h}_{rj}\dot{x}_j - \dot{h}_{r0}, \tag{2.25}$$

---

[55]It is assumed here that there appear no higher time derivatives than $\dot{x}_i$ in the components F_j of the applied forces.

where r runs through all l constraints, i.e., this is a set of l linear inhomogeneous equations for the l unknowns λ_k. These equations therefore completely determine

$$\lambda_k = \lambda_k(x, \dot{x}, t), \tag{2.26}$$

where x and $\dot{x}$ generically stand for the lists of positions $(x_1, \ldots, x_{3n})$ and velocities $(\dot{x}_1, \ldots, \dot{x}_{3n})$, respectively, and thus produce well-defined components

$$Z_i = \sum_{k=1}^{l} \lambda_k(x, \dot{x}, t)\, h_{ki}(x, t) \tag{2.27}$$

of the forces of constraint. The equations of motion,

$$m_i \ddot{x}_i = F_i(x, \dot{x}, t) + Z_i(x, \dot{x}, t), \tag{2.28}$$

now are written in terms of *known* functions of x, $\dot{x}$, and t on the right-hand side. We can therefore now apply the usual methods of the Newtonian approach to this set of equations to obtain the solutions $x = x(t)$. The integration constants are to be determined by the initial conditions *and* the constraint conditions.

2.1.4   *Virtual work: The d'Alembert Principle*

Note that

$$\sum_{j=1}^{n} \mathbf{Z}_j \cdot \delta\mathbf{r}_j = \sum_{k=1}^{l} \lambda_k \sum_{j=1}^{n} \mathbf{h}_{kj} \cdot \delta\mathbf{r}_j = 0 \tag{2.29}$$

because of (2.17). The quantity on the left is called the *net virtual work of the forces of constraint*. The result here says that this virtual work vanishes.

Historically, this was one of the basic assumptions entering the treatment of constraints. Using the *net total virtual work* of all particles,

$$\delta W = \sum_{i=1}^{n} \left(\mathbf{F}_i + \mathbf{Z}_i \right) \cdot \delta\mathbf{r}_i = \sum_{i=1}^{n} m_i \ddot{\mathbf{r}}_i \cdot \delta\mathbf{r}_i, \tag{2.30}$$

and rewriting this as

$$\sum_{i=1}^{n} \left(\mathbf{F}_i - m_i \ddot{\mathbf{r}}_i \right) \cdot \delta\mathbf{r}_i = -\sum_{i=1}^{n} \mathbf{Z}_i \cdot \delta\mathbf{r}_i, \tag{2.31}$$

it was *postulated* that the net virtual work of the forces of constraint on the right vanish. The resulting condition,

$$\boxed{\sum_{i=1}^{n} \left(\mathbf{F}_i - m_i \ddot{\mathbf{r}}_i \right) \cdot \delta\mathbf{r}_i = 0}, \tag{2.32}$$

is called the *d'Alembert Principle*.[56] It states that *the motion of a system of particles will be such that the virtual work of the part of the force that does not cause acceleration vanishes at any instant in time.* (This force, $\mathbf{F}_i - m_i\ddot{\mathbf{r}}_i$, is sometimes called the 'lost' force.) This is a nontrivial statement since the virtual displacements $\delta\mathbf{r}_i$ must satisfy the constraint conditions and are therefore *not* independent.

The essence of the d'Alembert Principle is the assumption that the forces of constraints must be perpendicular to hypersurfaces within which the motion takes place and hence do not contribute to the particle's acceleration. The Lagrangian approach, therefore, is completely equivalent to using d'Alembert's Principle as a starting point.

2.1.5 *Energy conservation*

Let us consider the question whether energy is conserved in the presence of forces of constraint. Assuming that the applied forces are conservative, i.e.,

$$F_i = -\frac{\partial V}{\partial x_i}, \quad i = 1, \ldots, 3n, \tag{2.33}$$

where the potential V is time-independent, we find from Eq. (2.22) upon multiplication by $\dot{x}_i$ and summation over all particles

$$\begin{aligned}
0 &= \sum_{i=1}^{3n} m_i \ddot{x}_i \dot{x}_i - \sum_{i=1}^{3n} F_i \dot{x}_i - \sum_{i=1}^{3n}\sum_{k=1}^{l} \lambda_k h_{ki} \dot{x}_i \\
&= \underbrace{\frac{d}{dt}\left(\sum_{i=1}^{3n} \frac{m_i}{2}\dot{x}_i^2\right)}_{=dT/dt} + \underbrace{\sum_{i=1}^{3n} \frac{\partial V}{\partial x_i}\dot{x}_i}_{=dV/dt} - \sum_{k=1}^{l} \lambda_k \underbrace{\left(\sum_{i=1}^{3n} h_{ki}\dot{x}_i\right)}_{=-h_{k0}}.
\end{aligned} \tag{2.34}$$

In other words

$$\frac{d}{dt}(T+V) = -\sum_{k=1}^{l} \lambda_k h_{k0}, \tag{2.35}$$

which specifically for holonomic conditions reads

$$\frac{d}{dt}(T+V) = -\sum_{k=1}^{l} \lambda_k \frac{\partial g_k}{\partial t}. \tag{2.36}$$

---

[56]After Jean-Baptiste le Rond d'Alembert (1717–1783). The principle appeared in his *Traité de dynamique*, published in 1743, that provided a notion of force much improved over Newton's *Principia*.

For energy conservation to hold, the right-hand sides here must vanish. We find therefore: *The total energy is conserved if the applied forces are conservative and the constraints are scleronomous.*

Note that irrespective of whether we have holonomic or nonholonomic constraints, using (2.22b) and (2.23c), the right-hand sides of Eqs. (2.35) and (2.36) can be cast in the generic form

$$\left. \begin{aligned} -\sum_{k=1}^{l} \lambda_k h_{k0} &= \sum_{i=1}^{n} \left(\sum_{k=1}^{l} \lambda_k \mathbf{h}_{ki} \right) \cdot \dot{\mathbf{r}}_i \\ -\sum_{k=1}^{l} \lambda_k \frac{\partial g_k}{\partial t} &= \sum_{i=1}^{n} \left(\sum_{k=1}^{l} \lambda_k \boldsymbol{\nabla}_i g_k \right) \cdot \dot{\mathbf{r}}_i \end{aligned} \right\} = \sum_{i=1}^{n} \mathbf{Z}_i \cdot \dot{\mathbf{r}}_i. \tag{2.37}$$

In other words, we may generically write[57]

$$\frac{d}{dt}(T + V) = \sum_{i=1}^{n} \mathbf{Z}_i \cdot \dot{\mathbf{r}}_i, \tag{2.38}$$

which shows that time-*dependent* constraining forces effectively act like dissipative forces that change the energy of the system, similar to what was found in Eq. (1.148). In the scleronomous case, they all act perpendicular to the fixed constraining hypersurface and thus the scalar products with velocities $\dot{\mathbf{r}}_i$ vanish because the velocities all lie within their respective surfaces.

2.1.6 *Applications*

To illustrate the formalism, let us consider a few examples.

2.1.6.1 *Planar pendulum*

As a first application, let us solve the case of the pendulum depicted in Fig. (2.1), whose holonomic constraints are given in (2.3). It should be noted, however, that this is done mainly for the purpose of illustration — there are simpler ways to solve this particular problem.

With the constraint functions

$$g_1 = y \qquad \text{and} \qquad g_2 = x^2 + y^2 + z^2 - \ell^2, \tag{2.39}$$

the forces of constraint are

$$\mathbf{Z}_1 = \lambda_1 \boldsymbol{\nabla} g_1 = \lambda_1 \boldsymbol{\nabla} y$$
$$= \lambda_1 \mathbf{e}_y, \tag{2.40a}$$

---

[57]Of course, this particular result can also be obtained directly starting from the basic equations of motion (2.7), without the detour of the details of the Lagrange formalism.

$$\mathbf{Z}_2 = \lambda_2 \nabla g_2 = \lambda_2 \nabla (x^2 + y^2 + z^2 - l^2)$$
$$= 2\lambda_2(x\mathbf{e}_x + y\mathbf{e}_y + z\mathbf{e}_z), \tag{2.40b}$$

resulting in Lagrange equations

$$m\ddot{x} = 2\lambda_2 x, \tag{2.41a}$$
$$m\ddot{y} = \lambda_1 + 2\lambda_2 y, \tag{2.41b}$$
$$m\ddot{z} = -mg + 2\lambda_2 z, \tag{2.41c}$$
$$y = 0, \tag{2.41d}$$
$$x^2 + y^2 + z^2 = l^2. \tag{2.41e}$$

For the y component, we have $y = 0$, of course, and therefore $\lambda_1 = 0$ and $\mathbf{Z}_1 = 0$. To solve the remaining two equations, let

$$x = l \sin \vartheta, \quad \text{and} \quad z = -l \cos \vartheta, \tag{2.42}$$

where the angle ϑ is given in Fig. (2.1). With $y = 0$, this satisfies the last constraint equation (2.41e) and produces

$$x\text{-comp.:} \qquad ml(\ddot{\vartheta} \cos \vartheta - \dot{\vartheta}^2 \sin \vartheta) = 2\lambda_2 l \sin \vartheta, \tag{2.43a}$$
$$z\text{-comp.:} \qquad ml(\ddot{\vartheta} \sin \vartheta + \dot{\vartheta}^2 \cos \vartheta) = -mg - 2\lambda_2 l \cos \vartheta. \tag{2.43b}$$

Multiplying the first equation by $\cos \vartheta$ and the second by $\sin \vartheta$ and adding the equations and then multiplying the first equation by $\sin \vartheta$ and the second by $\cos \vartheta$ and subtracting the equations yields

$$l\ddot{\vartheta} = -g \sin \vartheta, \tag{2.44a}$$
$$2\lambda_2 l = -mg \cos \vartheta - ml\dot{\vartheta}^2. \tag{2.44b}$$

The first equation here is the well-known second-order differential equation for the pendulum in terms of the angle ϑ measuring displacement from the vertical equilibrium position. With the Lagrange multiplier λ_2 from the second equation, the constraining force $\mathbf{Z}_2$ becomes

$$\mathbf{Z}_2 = 2\lambda_2 l \underbrace{\left(\sin \vartheta \, \mathbf{e}_x - \cos \vartheta \, \mathbf{e}_z \right)}_{\text{radial unit vector } \mathbf{e}_r}$$
$$= -\left(mg \cos \vartheta + ml\dot{\vartheta}^2 \right)\mathbf{e}_r. \tag{2.45}$$

This constraining force has the direction indicated in Fig. 2.1 (where $\mathbf{Z} \equiv \mathbf{Z}_1 + \mathbf{Z}_2 = \mathbf{Z}_2$); the first term here compensates the radial component of gravity and the second term, $-ml\dot{\vartheta}^2 \mathbf{e}_r$, provides the centripetal acceleration to keep the mass on a circular path.

To better understand the dynamical roles played by the constraining forces in this example, it should be emphasized that $\mathbf{Z}_2$ is non-zero because the applied force, gravity, in general has a component perpendicular to the constraint trajectory and in addition there is a dynamically generated force, the centrifugal force, which also is perpendicular; both need to be compensated by the corresponding force of constraint $\mathbf{Z}_2$. The constraint to the xz-plane, by contrast, is rather trivial since the applied force, gravity, has no component along y, i.e., $y = 0$ means that we cannot choose any initial conditions that would lead the particle out of the initial xz-plane and therefore there is no need for any explicit constraining force and that is why $\mathbf{Z}_1 = 0$. Usually, of course, this trivial constraint is taken care of by setting up the problem from the very beginning as a two-dimensional problem in x and z only.

2.1.6.2 *Bead on rotating rod*

In the next example, we follow the construction procedure explained in Sec. 2.1.3.1 that provides the constraining force in a form that is immediately applicable to Newton's second law as if it were a known applied force. We consider a rod of length $2L$ that is rotating horizontally about its midpoint with constant angular velocity ω. A bead of mass m is sliding frictionless along the rod (see Fig. 2.2). Initially, the bead is a distance ℓ from the midpoint, moving with a velocity v_0.

Since the motion takes place in a horizontal plane, taken as the xy-plane, we can ignore motion in the z-direction. The constraining condition for the bead to stay on the rod is given by

$$g = \varphi - \omega t = \arctan \frac{y}{x} - \omega t = 0, \tag{2.46}$$

where $x = x(t)$ and $y = y(t)$ describe the motion in the x and y directions, respectively.

The Lagrange equations of the 1st kind are then given by

$$m\ddot{\mathbf{r}} = \lambda \, \boldsymbol{\nabla} g. \tag{2.47}$$

Using polar coordinates according to $x = r \cos \varphi$ and $y = r \sin \varphi$, the position vector can be written as

$$\mathbf{r} = x(t)\, \mathbf{e}_x + y(t)\, \mathbf{e}_y = r(t)\, \mathbf{e}_r \tag{2.48}$$

while the bead is on the rod. In polar coordinates, the gradient is given by

$$\boldsymbol{\nabla} = \mathbf{e}_r \frac{\partial}{\partial r} + \mathbf{e}_\varphi \frac{1}{r} \frac{\partial}{\partial \varphi}, \tag{2.49}$$

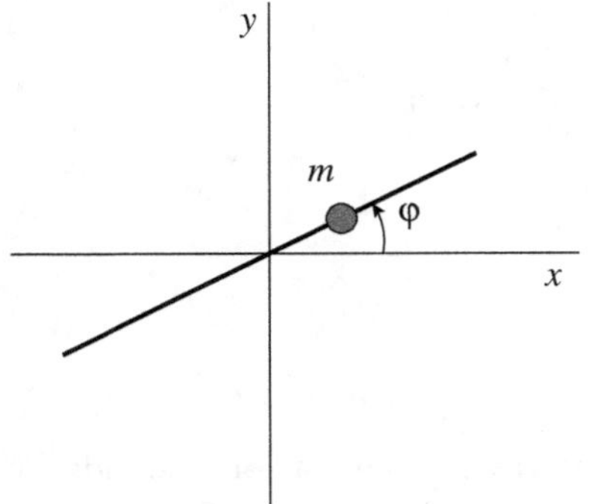

Fig. 2.2   A point mass m sliding without friction along a rod of length $2L$ rotating about its midpoint. The bead's distance from the midpoint is taken as positive if it is on the same side on which it was initially; otherwise it is counted as negative.

and hence we get for the two components of the equation of motion (2.47)

$$m\ddot{r} - mr\dot{\varphi}^2 = 0, \qquad mr\ddot{\varphi} + 2m\dot{r}\dot{\varphi} = \frac{\lambda}{r}. \tag{2.50}$$

Differentiating the constraint equation (2.46) yields

$$\frac{d^2 g}{dt^2} = \ddot{\varphi} = 0, \tag{2.51}$$

producing

$$\lambda = \lambda(r, \dot{r}, \dot{\varphi}) = 2mr\dot{r}\dot{\varphi}, \tag{2.52}$$

which is now of the same form as an applied force. Putting this into (2.50) then leads to standard equations of motion

$$m\ddot{r} - mr\dot{\varphi}^2 = 0, \qquad mr\ddot{\varphi} = 0. \tag{2.53}$$

The second equation has the trivial (unstable) solution $r(t) = 0$; for $r \neq 0$, it is solved by the constraint equation itself,

$$\varphi(t) = \omega t. \tag{2.54}$$

Putting this into the first equation,

$$\ddot{r} - \omega^2 r = 0, \tag{2.55}$$

this describes a one-dimensional motion along the radial direction, with the general solution

$$r(t) = A\,e^{\omega t} + B\,e^{-\omega t}, \tag{2.56}$$

where A and B are determined by the initial conditions

$$r(0) = \ell = A + B, \qquad \left.\frac{dr(t)}{dt}\right|_{t=0} = v_0 = (A - B)\omega. \tag{2.57}$$

Hence

$$r(t) = \left(\frac{\ell}{2} + \frac{v_0}{2\omega} \right) e^{\omega t} + \left(\frac{\ell}{2} - \frac{v_0}{2\omega} \right) e^{-\omega t}$$

$$= \ell \cosh \omega t + \frac{v_0}{\omega} \sinh \omega t, \tag{2.58}$$

and for the constraining force one finds

$$\mathbf{Z} = \lambda \, \boldsymbol{\nabla} g = 2m\dot{r}\omega \mathbf{e}_\varphi$$

$$= 2m\omega^2 \mathbf{e}_\varphi \left(\ell \sinh \omega t + \frac{v_0}{\omega} \cosh \omega t \right). \tag{2.59}$$

Note that for $v_0 > -\omega\ell$, the bead will eventually fall off on the side of the rod on which it sits initially (which by way of $\ell > 0$ is defined as the positive side of the rod). This is obvious if $v_0 \geq 0$ since the bead will simply continue sliding towards the perimeter in accelerated motion. If $v_0 < 0$, it will initially move towards the midpoint, but it will never reach the midpoint and ultimately the centrifugal force will reverse the motion and the bead will then also slide off at the rod's positive end. For $v_0 = -\omega\ell$, Eq. (2.58) shows that the exponentially rising term is cancelled and the motion is governed completely by the exponentially falling term. In other words, the bead will slowly approach the midpoint and will ultimately (after an infinitely long time) come to rest at the midpoint. And for $v_0 < -\omega\ell$, the bead will decelerate towards the midpoint, overshoot it and then begin accelerating towards the other end of the rod where it will eventually fall off. Hence, given the right initial conditions, the bead can remain on the rod for an infinitely long time (albeit in an extremely unstable equilibrium).

2.2   Lagrange Equations of the Second Kind

If one is only interested in the trajectories, but not in the explicit details of the constraining forces, it would be useful to reformulate the preceding approach such that the forces of constraint are eliminated altogether from the equations. As will be shown in the following, this can indeed be done for holonomic constraints, leading to Lagrange equations of the second kind.

2.2.1   *Generalized coordinates*

In view of l constraints, the $3n$ Cartesian coordinates are not independent of each other. In other words, the positions of the particles are completely specified if

$$N = 3n - l \tag{2.60}$$

Cartesian components are given, i.e., the l constraints leave only N *degrees of freedom* for the trajectories of the n bodies. For example, for the planar pendulum treated in Sec. 2.1.6, the two constraint equations reduce the problem to one degree of freedom described by the angle ϑ. This angle determines the coordinates according to Eq. (2.42) and it thus satisfies the constraint equation (2.41e).

In general, the degrees of freedom can be described by N *generalized coordinates*,

$$q_1, \ q_2, \ \ldots, \ q_N, \tag{2.61}$$

which completely describe the position of the particles according to

$$x_i = x_i(q_1, q_2, \ldots, q_N, t), \qquad i = 1, \ldots, 3n. \tag{2.62}$$

The velocity components are then given by

$$\dot{x}_i = \sum_{r=1}^{N} \frac{\partial x_i}{\partial q_r} \dot{q}_r + \frac{\partial x_i}{\partial t} = \dot{x}_i(q_1, \ldots, q_N, \dot{q}_1, \ldots, \dot{q}_N, t), \tag{2.63}$$

$$i = 1, \ldots, 3n,$$

where the $\dot{q}_r$ are referred to as the *generalized velocities*. As we shall see in the examples below, the generalized coordinates and velocities need not have the respective dimensions of their Cartesian counterparts.

For the following, it will be convenient to introduce the short-hand notation

$$x = (x_1, \ldots, x_{3n}), \qquad\qquad \dot{x} = (\dot{x}_1, \ldots, \dot{x}_{3n}), \tag{2.64a}$$

$$q = (q_1, \ldots, q_N), \qquad\qquad \dot{q} = (\dot{q}_1, \ldots, \dot{q}_N), \tag{2.64b}$$

to indicate the entire list of the corresponding variables.

The choice of the generalized coordinates must be such that the constraint conditions remain true for *any* values of q_i, i.e., that the functions

$$h_{kj} = h_{kj}\Big(x\big(q(t), t\big), t\Big) \tag{2.65}$$

will satisfy [cf. Eq. (2.17)]

$$\sum_{j=1}^{3n} h_{kj}\, \delta x_j = \sum_{j=1}^{3n} h_{kj} \left(\sum_{r=1}^{N} \frac{\partial x_j}{\partial q_r} \delta q_r \right) = \sum_{r=1}^{N} \hat{h}_{kr} \delta q_r = 0, \tag{2.66}$$

$$k = 1, \ldots, l,$$

for any set of values q_i, where the notation

$$\hat{h}_{kr} = \sum_{j=1}^{3n} h_{kj} \frac{\partial x_j}{\partial q_r} \tag{2.67}$$

was introduced here for convenience. And for real displacements, one has

$$\sum_{j=1}^{3n} h_{kj}\, dx_j + h_{k0}\, dt = \sum_{j=1}^{3n} h_{kj} \left(\sum_{r=1}^{N} \frac{\partial x_j}{\partial q_r} dq_r + \frac{\partial x_j}{\partial t} dt \right) + h_{k0}\, dt$$

$$= \sum_{r=1}^{N} \hat{h}_{kr}\, dq_r + \hat{h}_{k0}\, dt$$

$$= 0, \qquad k = 1, \ldots, l, \tag{2.68}$$

where

$$\hat{h}_{k0} = \sum_{j=1}^{3n} h_{kj} \frac{\partial x_j}{\partial t} + h_{k0}. \tag{2.69}$$

For holonomic conditions, in particular, this means

$$\hat{h}_{kr} = \sum_{j=1}^{3n} \frac{\partial g_k}{\partial x_j} \frac{\partial x_j}{\partial q_r} = \frac{dg_k}{dq_r}, \qquad r = 1, \ldots, N. \tag{2.70}$$

In view of

$$g_k\big(x(q,t),t\big) \equiv 0, \tag{2.71}$$

for any set of q_r, this means that when seen as a function of the q_r, i.e.,

$$0 = g_k\big(x(q,t),t\big) = g_k^*(q,t), \tag{2.72}$$

g_k^* is the function that vanishes identically, and hence neither depends on q_r nor on t, i.e.,

$$\frac{\partial g_k^*}{\partial q_r} = \frac{dg_k}{dq_r} = \sum_{j=1}^{3n} \frac{\partial g_k}{\partial x_j} \frac{\partial x_j}{\partial q_r} = 0, \tag{2.73}$$

$$\frac{\partial g_k^*}{\partial t} = \frac{dg_k}{dt} = \sum_{j=1}^{3n} \frac{\partial g_k}{\partial x_j} \frac{\partial x_j}{\partial t} + \frac{\partial g_k}{\partial t} = 0. \tag{2.74}$$

Note that g_k on the right-hand side here is considered as a function of x and t only; therefore $\partial x_j / \partial t = dx_j / dt = \dot{x}_j$.

In view of Eq. (2.73), we have

$$\hat{h}_{kr} = \frac{dg_k}{dq_r} = 0, \qquad \text{for holonomic constraints,} \tag{2.75}$$

and we conclude that for holonomic conditions, the virtual displacements δq_r of the generalized coordinates in Eq. (2.66) can be chosen completely independently of each other.

Properties of nonholonomic constraints

We can now state the conditions for nonholonomic constraints more precisely: In the nonholonomic case,[58] the virtual displacements δq_r *cannot* be chosen independently, i.e., we cannot conclude that

$$\sum_{r=1}^{N} \hat{h}_{kr}\delta q_r = 0, \qquad k = 1, \ldots, l, \tag{2.76}$$

implies $\hat{h}_{kr} = 0$. Since the N variables q_i describe the degrees of freedom of *macroscopic* motion of the system, this means that nonholonomic constraints are such that the degrees of freedom for infinitesimal displacements are *smaller* than for finite displacements: There are N degrees of freedom for the latter, but only $N - l$ for infinitesimal displacements. The constraint equations for real infinitesimal displacements,

$$\sum_{r=1}^{N} \hat{h}_{kr}dq_r + \hat{h}_{k0}dt = 0, \qquad k = 1, \ldots, l, \tag{2.77}$$

or equivalently

$$\sum_{r=1}^{N} \hat{h}_{kr}\dot{q}_r + \hat{h}_{k0} = 0, \qquad k = 1, \ldots, l, \tag{2.78}$$

then cannot be integrated, i.e., the usual *integrability conditions*

$$\boxed{\frac{\partial \hat{h}_{ki}}{\partial q_j} = \frac{\partial \hat{h}_{kj}}{\partial q_i}, \qquad \frac{\partial \hat{h}_{ki}}{\partial t} = \frac{\partial \hat{h}_{k0}}{\partial q_i}} \tag{2.79}$$

that would ensure that the left-hand side of (2.77) is a total differential (see footnote 19), are *not* satisfied. We will give some examples below to illustrate this behavior of nonholonomic constraints.

2.2.2   *Lagrange function and equations of motion*

Multiplying the Lagrange equations (2.22) by $\partial x_i/\partial q_r$ and summing over i yields

$$\sum_{i=1}^{3n} m_i \ddot{x}_i \frac{\partial x_i}{\partial q_r} = Q_r + \hat{Z}_r, \qquad r = 1, \ldots, N, \tag{2.80}$$

---

[58] A reminder: Nonholonomic constraints in the form of inequalities are excluded from the present discussion; see Sec. 2.2.2.9 for a treatment of such cases.

where

$$Q_r = \sum_{i=1}^{3n} F_i \frac{\partial x_i}{\partial q_r} \tag{2.81}$$

is referred to as a *generalized force component* and, cf. Eq. (2.67),

$$\hat{Z}_r = \sum_{k=1}^{l} \lambda_k \hat{h}_{kr} \tag{2.82}$$

now contains the *components of the generalized forces of constraint* (i.e., $\hat{Z}_r^{(k)} = \lambda_k \hat{h}_{kr}$ is the r-th component of the k-th force).

For holonomic constraints, the components of the generalized constraining forces are actually zero,

$$\boxed{\text{Holonomic:} \qquad \hat{Z}_r = 0} , \tag{2.83}$$

according to (2.75). We have thus succeeded in eliminating the forces of constraint from the Lagrange equations for this case.

Before we comment further on this, let us rewrite the left-hand side of Eq. (2.80) as

$$\sum_{i=1}^{3n} m_i \ddot{x}_i \frac{\partial x_i}{\partial q_r} = \underbrace{\sum_{i=1}^{3n} \frac{d}{dt}\left(m_i \dot{x}_i \frac{\partial x_i}{\partial q_r}\right)}_{\text{Term I}} - \underbrace{\sum_{i=1}^{3n} m_i \dot{x}_i \frac{d}{dt}\frac{\partial x_i}{\partial q_r}}_{\text{Term II}} . \tag{2.84}$$

With

$$\dot{x}_i = \sum_{r=1}^{N} \frac{\partial x_i}{\partial q_r} \dot{q}_r + \frac{\partial x_i}{\partial t}, \tag{2.85}$$

we find

$$\frac{\partial \dot{x}_i}{\partial \dot{q}_j} = \frac{\partial x_i}{\partial q_j}, \tag{2.86}$$

and therefore for term I

$$\sum_{i=1}^{3n} \frac{d}{dt}\left(m_i \dot{x}_i \frac{\partial x_i}{\partial q_r}\right) = \frac{d}{dt} \sum_{i=1}^{3n} m_i \dot{x}_i \frac{\partial \dot{x}_i}{\partial \dot{q}_r} = \frac{d}{dt}\frac{\partial}{\partial \dot{q}_r} \underbrace{\sum_{i=1}^{3n} \frac{m_i}{2} \dot{x}_i^2}_{=T}, \tag{2.87}$$

where the sum on the right is the total kinetic energy T. For term II, we note that

$$\frac{d}{dt}\frac{\partial x_i}{\partial q_r} = \sum_{k=1}^{N} \frac{\partial^2 x_i}{\partial q_r \partial q_k} \dot{q}_k + \frac{\partial^2 x_i}{\partial q_r \partial t} = \frac{\partial}{\partial q_r}\left[\sum_{k=1}^{N} \frac{\partial x_i}{\partial q_k} \dot{q}_k + \frac{\partial x_i}{\partial t}\right], \tag{2.88}$$

and therefore

$$\frac{d}{dt}\frac{\partial x_i}{\partial q_r} = \frac{\partial \dot{x}_i}{\partial q_r}, \tag{2.89}$$

hence

$$\sum_{i=1}^{3n} m_i \dot{x}_i \frac{d}{dt}\frac{\partial x_i}{\partial q_r} = \frac{\partial}{\partial q_r} \underbrace{\sum_{i=1}^{3n} \frac{m_i}{2} \dot{x}_i^2}_{=T} \tag{2.90}$$

is just the partial derivative of the total kinetic energy.

Equation (2.80) now reads

$$\boxed{\frac{d}{dt}\frac{\partial T}{\partial \dot{q}_r} - \frac{\partial T}{\partial q_r} = Q_r + \hat{Z}_r, \qquad r = 1,\ldots,N} \tag{2.91}$$

These equations, sometimes called *Lagrange equations of the mixed type*, comprise one of the most powerful tools in mechanics, applicable to a wide range of problems. They are among the most important equations in classical mechanics. Although the Lagrange equations of the second kind, Eq. (2.104) below, are more 'famous' since they apply to the standard case of conservative forces, it is the present set of equations that is by far more versatile since it is valid for *all* cases, as will be shown in the subsequent sections.

Before we proceed to discuss some special cases, it should be noted that T, Q_r, and $\hat{Z}_r$ here are to be considered as functions of q and $\dot{q}$ and t,

$$T = T(q, \dot{q}, t), \tag{2.92}$$

$$Q_r = Q_r(q, \dot{q}, t), \tag{2.93}$$

$$\hat{Z}_r = \hat{Z}_r(q, \dot{q}, t). \tag{2.94}$$

Since x_i depends only on q_r and t and $\dot{x}_i$ is linear in $\dot{q}_r$ [see Eq. (2.85)], the kinetic energy is at most a quadratic function in the generalized velocities $\dot{q}_r$,

$$T = \sum_{i=1}^{3n} \frac{m_i}{2} \dot{x}_i^2$$

$$= a(q, t) + \sum_{j=1}^{N} b_j(q, t)\, \dot{q}_j + \sum_{j,k=1}^{N} c_{jk}(q, t)\, \dot{q}_j \dot{q}_k, \tag{2.95}$$

with coefficient functions

$$a(q,t) = \sum_{i=1}^{3n} \frac{m_i}{2} \left(\frac{\partial x_i(q,t)}{\partial t} \right)^2, \tag{2.96a}$$

$$b_j(q,t) = \sum_{i=1}^{3n} m_i \frac{\partial x_i(q,t)}{\partial t} \frac{\partial x_i(q,t)}{\partial q_j}, \tag{2.96b}$$

$$c_{jk}(q,t) = \sum_{i=1}^{3n} \frac{m_i}{2} \frac{\partial x_i(q,t)}{\partial q_j} \frac{\partial x_i(q,t)}{\partial q_k}. \tag{2.96c}$$

For scleronomous conditions, the transformation equations (2.62) do not depend explicitly on time and the kinetic energy assumes the particularly simple form

$$\boxed{\text{Scleronomous:} \qquad T = T(q, \dot{q}) = \sum_{j,k=1}^{N} c_{jk}(q)\, \dot{q}_j \dot{q}_k} \tag{2.97}$$

Since, in general, one has

$$\frac{\partial T}{\partial \dot{q}_r} = b_r(q,t) + 2 \sum_{j=1}^{N} c_{rj}(q,t)\, \dot{q}_j, \tag{2.98}$$

the equations (2.91) comprise a set of N second-order differential equations for the generalized coordinates q_r. If the constraints are purely holonomic, $\hat{Z}_r = 0$, and this set then completely determines the q_r. For nonholonomic constraints, one needs in addition the l differential constraint conditions (2.77) to allow determination of the l Lagrange multipliers λ_k contained in $\hat{Z}_r$.

Conservative forces

For conservative forces, there exists a potential $V = V(x)$,

$$F_i = -\frac{\partial V(x)}{\partial x_i}, \tag{2.99}$$

and the generalized force components then become

$$Q_r = \sum_{i=1}^{3n} F_i \frac{\partial x_i}{\partial q_r} = -\sum_{i=1}^{3n} \frac{\partial V}{\partial x_i} \frac{\partial x_i}{\partial q_r} = -\frac{\partial V(q,t)}{\partial q_r}, \tag{2.100}$$

and therefore are also given as partial derivatives of the potential V, which is now to be taken as a function of q and t.[59] The time dependence in $V = V(q,t)$ appears if the variable transformations $x = x(q,t)$ explicitly depend on the time. Since

$$\frac{\partial V(q,t)}{\partial \dot{q}_r} = 0, \tag{2.101}$$

we then obtain a variant of Eq. (2.91), the *Lagrange equations of the second kind for conservative forces*,[60]

$$\boxed{\frac{d}{dt}\frac{\partial L}{\partial \dot{q}_r} - \frac{\partial L}{\partial q_r} = \hat{Z}_r, \qquad r = 1, \ldots, N}, \tag{2.102}$$

where the *Lagrange function*,

$$\boxed{L = L(q,\dot{q},t) = T(q,\dot{q},t) - V(q,t)}, \tag{2.103}$$

measures the excess of the kinetic energy over the potential energy.

2.2.2.1 Conservative holonomic systems

For holonomic constraints, in particular, we have $\hat{Z}_r = 0$, and thus

$$\boxed{\frac{d}{dt}\frac{\partial L}{\partial \dot{q}_r} - \frac{\partial L}{\partial q_r} = 0, \qquad r = 1, \ldots, N}, \tag{2.104}$$

called the *Lagrange equations of the second kind for conservative holonomic systems*. Oftentimes, this set of equations is simply referred to as 'Lagrange equations'.

For conservative holonomic systems, the present Lagrange equations provide the most direct way of arriving at the equations of motion. They only depend on the actual variables describing the degrees of freedom of the system; forces of constraint are completely absent from the formulation.[61] Moreover, deriving the equations of motion requires finding only *one scalar* function, the Lagrange function. It is usually much easier to find this function in terms of the independent variables than writing down the equations of motion directly.

---

[59]Strictly speaking, one should use a different notation depending on whether one considers V as a function of x or of q, t [via $x = x(q,t)$]. However, since the actual physical potential energy does not change, it is customary in physics to use the same notation for the functions describing it irrespective of what variables one chooses. It is usually obvious from the context what the appropriate variables are. The same applies also to the kinetic energy — see e.g., Eq. (2.95) — and to various other quantities in physics.

[60]Note that the same equations obtain even if the force components are given in terms of a time-dependent potential, i.e., if $F_i = -\partial V(x,t)/\partial x_i$.

[61]There may be situations where one would like to know the forces of constraint explicitly. This can be done by deliberately introducing coordinates that are not minimal, with $\hat{Z}_r \neq 0$, as discussed in Sec. 2.2.2.7 below.

Form invariance

One of the most important properties of the Lagrange equations is the fact that they are *form invariant*, or *covariant*, which means they do not change their appearance under a coordinate transformation. This is obvious from the fact that in their derivation we did not specify any particular set of generalized coordinates; in other words, any set will provide equations of the same form.

To show this directly for Eq. (2.104), consider a point transformation[62]

$$q_r = q_r(q', t), \qquad \dot{q}_r = \dot{q}_r(q', \dot{q}', t) \tag{2.105}$$

from variables q to new variables q'. This provides a new Lagrange function given by

$$L'(q', \dot{q}', t) = L\left(q(q', t), \dot{q}(q', \dot{q}', t), t\right). \tag{2.106}$$

Its derivatives are

$$\frac{\partial L'}{\partial q_i'} = \sum_{r=1}^{N} \left(\frac{\partial L}{\partial q_r} \frac{\partial q_r}{\partial q_i'} + \frac{\partial L}{\partial \dot{q}_r} \frac{\partial \dot{q}_r}{\partial q_i'} \right), \tag{2.107a}$$

$$\frac{\partial L'}{\partial \dot{q}_i'} = \sum_{r=1}^{N} \frac{\partial L}{\partial \dot{q}_r} \frac{\partial \dot{q}_r}{\partial \dot{q}_i'} = \sum_{r=1}^{N} \frac{\partial L}{\partial \dot{q}_r} \frac{\partial q_r}{\partial q_i'}, \tag{2.107b}$$

and hence

$$\frac{\partial L'}{\partial q_i'} - \frac{d}{dt} \frac{\partial L'}{\partial \dot{q}_i'} = \sum_{r=1}^{N} \left(\frac{\partial L}{\partial q_r} \frac{\partial q_r}{\partial q_i'} + \frac{\partial L}{\partial \dot{q}_r} \frac{\partial \dot{q}_r}{\partial q_i'} - \frac{d}{dt} \frac{\partial L}{\partial \dot{q}_r} \frac{\partial q_r}{\partial q_i'} \right)$$

$$= \sum_{r=1}^{N} \left(\underbrace{\left[\frac{\partial L}{\partial q_r} - \frac{d}{dt} \frac{\partial L}{\partial \dot{q}_r} \right]}_{=0} \frac{\partial q_r}{\partial q_i'} + \frac{\partial L}{\partial \dot{q}_r} \underbrace{\left[\frac{\partial \dot{q}_r}{\partial q_i'} - \frac{d}{dt} \frac{\partial q_r}{\partial q_i'} \right]}_{=0} \right)$$

$$= 0, \tag{2.108}$$

which are the (form-invariant) Lagrange equations for L' in the new variables.

It is important to realize that *the form invariance of the Lagrange equations holds true even if the equations are transformed to a non-inertial frame*. This is in marked contrast to Newton's equations which require an inertial frame. The only restriction is that both the kinetic and potential energies that go into the Lagrange function $L = T - V$ must be determined in an inertial frame first before being converted to whatever new coordinates one chooses.

---

[62]This transformation is a set of invertible functions that maps a differentiable manifold isomorphically into another, i.e., it is a *diffeomorphism* of smooth manifolds.

Free-fall motion in an accelerated frame

Let us illustrate this by a simple example. Consider a ball of mass m thrown upward with velocity v_0 in an elevator moving upward with velocity

$$u_{\text{elevator}} = u_0 + at, \tag{2.109}$$

as seen from an inertial frame, where u_0 is the elevator's initial velocity and a its acceleration. Clearly, the elevator represents a non-inertial frame. An observer in the inertial frame will see the ball at height z, with velocity $\dot{z}$ and write the Lagrange function as

$$L(z, \dot{z}) = \frac{m}{2}\dot{z}^2 - mgz, \tag{2.110}$$

resulting in the equation of motion

$$\frac{d}{dt}\frac{\partial L}{\partial \dot{z}} - \frac{\partial L}{\partial z} = 0 \quad \Longrightarrow \quad m\ddot{z} = -mg \tag{2.111}$$

describing free-fall motion in an inertial frame due to the downward force of gravity $F_g = -mg$. This is solved by

$$z(t) = z_0 + v_0 t - \frac{1}{2}gt^2 \tag{2.112}$$

where z_0 is the height of the elevator at $t = 0$ when the ball is thrown upward with velocity v_0. Let us now choose a new coordinate

$$z' = z - z_0 - u_0 t - \frac{1}{2}at^2 \tag{2.113}$$

which is the vertical distance in the non-inertial elevator frame. In a Newtonian approach, this gives rise to a pseudoforce $F' = -ma$ and the corresponding equation of motion is

$$m\ddot{z}' = F_g + F' \quad \Longrightarrow \quad m\ddot{z}' = -m(g + a). \tag{2.114}$$

Written in the new variables, the kinetic and potential energies read

$$T = \frac{m}{2}\dot{z}^2 = \frac{m}{2}\left(\dot{z}' + u_0 + at\right)^2, \tag{2.115a}$$

$$V = mgz = mg\left(z' + z_0 + u_0 t + \frac{1}{2}at^2\right), \tag{2.115b}$$

respectively, and the corresponding transformed Lagrange function becomes[63]

$$L'(z', \dot{z}', t) = \frac{m}{2}\left(\dot{z}' + u_0 + at\right)^2 - mg\left(z' + z_0 + u_0 t + \frac{1}{2}at^2\right). \tag{2.116}$$

---

[63] As we shall see later, in Sec. 3.2.1 on gauge transformations, some of the time-dependent terms here can be omitted since they have no bearing on the physics of the situation.

The Lagrange equation for this case retains its usual form,

$$\frac{d}{dt}\frac{\partial L'}{\partial \dot{z}'} - \frac{\partial L'}{\partial z'} = 0 \qquad \Longrightarrow \qquad m(\ddot{z}' + a) + mg = 0, \qquad (2.117)$$

and it immediately provides the correct equation (2.114) in the transformed frame. (Clearly, it would be wrong here to write the kinetic and potential energies as $T' = \frac{m}{2}\dot{z}'^2$ and $V' = mgz'$, respectively.)

Ignorable coordinates

The *generalized*, or *canonical*, *momentum* p_r associated with the generalized coordinate q_r is defined by

$$p_r = \frac{\partial L}{\partial \dot{q}_r}, \qquad r = 1, \ldots, N. \qquad (2.118)$$

If the Lagrange function is given in terms of Cartesian coordinates, this momentum is equal to the usual (linear) momentum components. In general, however, p_r depends on the choice of coordinates q_i and need not have any relationship to what would ordinarily be called a 'momentum'.

If a particular coordinate q_k does not appear in the Lagrange function, i.e.,

$$L = L(q_1, \ldots, q_{k-1}, q_{k+1}, \ldots, q_N, \dot{q}_1, \ldots, \dot{q}_N, t), \qquad (2.119)$$

it is called *ignorable* or *cyclic*. It follows then from (2.104) that

$$\boxed{q_k \text{ ignorable:} \qquad \frac{\partial L}{\partial q_k} = 0 \qquad \Longrightarrow \qquad \frac{\partial L}{\partial \dot{q}_k} \equiv p_k = const}, \qquad (2.120)$$

i.e., the corresponding generalized momentum is conserved.

A simple example

Let us consider the equations of motion of a point mass in spherical coordinates, subject to a potential $V = V(r, \theta, \varphi)$, but without any further constraints. With the Lagrange function

$$L = L(r, \varphi, \theta, \dot{r}, \dot{\varphi}, \dot{\theta}) = \tfrac{1}{2}m(\dot{r}^2 + r^2\dot{\theta}^2 + r^2\sin^2\theta\,\dot{\varphi}^2) - V(r, \theta, \varphi) \qquad (2.121)$$

that describes three degrees of freedom, the equations of motion are immediately found as

$$\frac{d}{dt}(m\dot{r}) - mr(\dot{\theta}^2 + \sin^2\theta\,\dot{\varphi}^2) = -\frac{\partial V}{\partial r}, \qquad (2.122a)$$

$$\frac{d}{dt}(mr^2\dot{\theta}) - mr^2\sin\theta\cos\theta\,\dot{\varphi}^2 = -\frac{\partial V}{\partial \theta}, \qquad (2.122b)$$

$$\frac{d}{dt}(mr^2 \sin^2\theta \, \dot\varphi) = -\frac{\partial V}{\partial \varphi}. \tag{2.122c}$$

For a spherical pendulum of length $r = \ell = const$, with the gravitational potential $V = mgz = mg\ell\cos\theta$, L simplifies to

$$L = L(\theta, \dot\varphi, \dot\theta) = m\ell^2 \left(\frac{1}{2}\dot\theta^2 + \frac{1}{2}\sin^2\theta\,\dot\varphi^2 - \frac{g}{\ell}\cos\theta\right), \tag{2.123}$$

i.e., it has only two degrees of freedom, and the equations of motion are now

$$\ddot\theta - \sin\theta\,\cos\theta\,\dot\varphi^2 - \frac{g}{\ell}\sin\theta = 0, \tag{2.124a}$$

$$\frac{d}{dt}(m\ell^2 \sin^2\theta\,\dot\varphi) = 0. \tag{2.124b}$$

The second equation here is an illustration of a conserved generalized momentum due to a cyclic variable: The azimuthal angle φ does not appear in L and hence the associated canonical momentum

$$p_\varphi = m\ell^2 \sin^2\theta\,\dot\varphi \tag{2.125}$$

is conserved. In this case, p_φ is the angular momentum for the azimuthal motion.

For an ordinary planar pendulum, for which $\dot\varphi = 0$, with oscillations $\vartheta = \pi - \theta$ around the vertical equilibrium at $\vartheta = 0$, the Lagrange function reduces to

$$L = L(\vartheta, \dot\vartheta) = m\ell^2 \left(\frac{1}{2}\dot\vartheta^2 + \frac{g}{\ell}\cos\vartheta\right), \tag{2.126}$$

which has only one degree of freedom, producing the well-known single equation of motion,

$$\ddot\vartheta + \frac{g}{\ell}\sin\vartheta = 0. \tag{2.127}$$

For small amplitudes, using $\cos\vartheta \approx 1 - \vartheta^2/2$, this is reduced further to

$$L = \frac{m\ell^2}{2}\left(\dot\vartheta^2 - \frac{g}{\ell}\vartheta^2\right) \qquad \Longrightarrow \qquad \ddot\vartheta + \frac{g}{\ell}\vartheta = 0, \tag{2.128}$$

which is the well-known equation of motion of a harmonic oscillator with angular frequency $\omega = \sqrt{g/\ell}$.

2.2.2.2   *Vector form of Lagrange equations and Newton's second law*

As an example for how the Lagrange equations (2.104) can be linked back to Newton's equations of motion in a straightforward manner, consider a single particle subject to a conservative force described in Cartesian coordinates. Multiplying the corresponding Lagrange equations,

$$\frac{d}{dt}\frac{\partial L}{\partial \dot{x}_i} - \frac{\partial L}{\partial x_i} = 0, \qquad i = 1, 2, 3, \tag{2.129}$$

by the unit vectors $\mathbf{e}_i$ and summing over i, we obtain

$$\frac{d}{dt}\boldsymbol{\nabla}_{\dot{\mathbf{r}}}L - \boldsymbol{\nabla}_{\mathbf{r}}L = 0 \tag{2.130}$$

which is the vector form of the Lagrange equations. With

$$L = T - V = \frac{m}{2}\dot{\mathbf{r}}^2 - V(\mathbf{r}), \tag{2.131}$$

this may be written as

$$\left(\frac{d}{dt}\boldsymbol{\nabla}_{\dot{\mathbf{r}}} - \boldsymbol{\nabla}_{\mathbf{r}}\right)T = \left(\frac{d}{dt}\boldsymbol{\nabla}_{\dot{\mathbf{r}}} - \boldsymbol{\nabla}_{\mathbf{r}}\right)V, \tag{2.132}$$

thus immediately resulting in Newton's second law

$$m\ddot{\mathbf{r}} = \mathbf{F} \tag{2.133}$$

for the conservative force $\mathbf{F} = -\boldsymbol{\nabla}_{\mathbf{r}}V$.

2.2.2.3   *Generalized potential due to rotating frame*

We had seen in Sec. 1.7 that transforming Newton's equations to non-inertial frames produces pseudoforces. To illustrate how such forces emerge in the Lagrange formalism, let us consider a frame that rotates with an angular velocity $\boldsymbol{\omega}$ about the origin of the inertial frame. This case had been treated previously, in Sec. 1.7.2, where we had found that the position and velocity, respectively, transform as

$$\mathbf{r} = \mathbf{r}' \qquad \text{and} \qquad \dot{\mathbf{r}} = \dot{\mathbf{r}}' + \boldsymbol{\omega} \times \mathbf{r}', \tag{2.134}$$

where the primed quantities pertain to the non-inertial frame.[64] To derive the equations of motion, the form invariance of the Lagrange equations (see p. 116) stipulates that we only need to express the kinetic and potential energies of the inertial frame in the new variables, i.e.,

$$T = \frac{m}{2}\left(\dot{\mathbf{r}}' + \boldsymbol{\omega} \times \mathbf{r}'\right)^2 \qquad \text{and} \qquad V = V(\mathbf{r}'), \tag{2.135}$$

---

[64]We recall here, in particular, that the notation $\dot{\mathbf{r}}'$ denotes the velocity seen in the rotating frame; see Sec. 1.7.2.

and then simply employ the Lagrange equations in the transformed variables,

$$\frac{d'}{dt}\boldsymbol{\nabla}_{\dot{\mathbf{r}}'}(T-V) - \boldsymbol{\nabla}_{\mathbf{r}'}(T-V) = 0, \tag{2.136}$$

where d'/dt here is the time derivative in the rotating frame. We immediately find

$$\left(\frac{d'}{dt}\boldsymbol{\nabla}_{\dot{\mathbf{r}}'} - \boldsymbol{\nabla}_{\mathbf{r}'}\right)T = m\ddot{\mathbf{r}}' + 2m\left(\boldsymbol{\omega} \times \dot{\mathbf{r}}'\right) + m\,\boldsymbol{\omega} \times \left(\boldsymbol{\omega} \times \mathbf{r}'\right) \tag{2.137}$$

and

$$\left(\frac{d'}{dt}\boldsymbol{\nabla}_{\dot{\mathbf{r}}'} - \boldsymbol{\nabla}_{\mathbf{r}'}\right)V = \mathbf{F}, \tag{2.138}$$

and therefore

$$m\ddot{\mathbf{r}}' = \mathbf{F} - 2m\left(\boldsymbol{\omega} \times \dot{\mathbf{r}}'\right) - m\,\boldsymbol{\omega} \times \left(\boldsymbol{\omega} \times \mathbf{r}'\right), \tag{2.139}$$

which indeed is the familiar equation containing on the right-hand side both pseudoforce contributions — the Coriolis and centrifugal terms — in addition to the applied force $\mathbf{F}$ [see Eq. (1.269)].

In this particular case, we may even go one step further. From

$$\begin{aligned}
T - V &= \frac{m}{2}\left(\dot{\mathbf{r}}' + \boldsymbol{\omega} \times \mathbf{r}'\right)^2 - V \\
&= \frac{m}{2}\dot{\mathbf{r}}'^2 + \underbrace{m\dot{\mathbf{r}}' \cdot \left(\boldsymbol{\omega} \times \mathbf{r}'\right) + \frac{m}{2}\left(\boldsymbol{\omega} \times \mathbf{r}'\right)^2 - V}_{=-U},
\end{aligned} \tag{2.140}$$

we see that the *generalized potential*

$$U = U(\mathbf{r}', \dot{\mathbf{r}}') = V(\mathbf{r}') - m\dot{\mathbf{r}}' \cdot \left(\boldsymbol{\omega} \times \mathbf{r}'\right) - \frac{m}{2}\left(\boldsymbol{\omega} \times \mathbf{r}'\right)^2 \tag{2.141}$$

produces the entire force term on the right-hand side of (2.139) according to

$$\left(\frac{d'}{dt}\boldsymbol{\nabla}_{\dot{\mathbf{r}}'} - \boldsymbol{\nabla}_{\mathbf{r}'}\right)U = \mathbf{F} - 2m\left(\boldsymbol{\omega} \times \dot{\mathbf{r}}'\right) - m\,\boldsymbol{\omega} \times \left(\boldsymbol{\omega} \times \mathbf{r}'\right), \tag{2.142}$$

similar to what the right-hand side of (2.132) does for Eq. (2.133). We may thus write the Lagrange equation as

$$\left(\frac{d'}{dt}\boldsymbol{\nabla}_{\dot{\mathbf{r}}'} - \boldsymbol{\nabla}_{\mathbf{r}'}\right)(T' - U) = 0, \tag{2.143}$$

where

$$T' = \frac{m}{2}\dot{\mathbf{r}}'^2 \tag{2.144}$$

is a 'kinetic energy' that has the form familiar from inertial frames. The transformed Lagrange function L' itself has *not* changed, i.e., $L' = T - V = T' - U$.

2.2.2.4   *Generalized potentials defined*

The preceding procedure may now easily be generalized to arbitrary generalized coordinates even if the force components F_i are not conservative. For if there exists a function $U(q,\dot{q},t)$ such that the generalized force components can be expressed as

$$Q_k = \frac{d}{dt}\frac{\partial U}{\partial \dot{q}_k} - \frac{\partial U}{\partial q_k}, \tag{2.145}$$

we may define the Lagrange function as

$$L = L(q,\dot{q},t) = T(q,\dot{q},t) - U(q,\dot{q},t), \tag{2.146}$$

and Eq. (2.104) will remain valid. As in the example above, the function U is called the *generalized potential*.

2.2.2.5   *Generalized potential for electromagnetic forces*

The generalized potential allows us to incorporate electromagnetic forces into the framework of the Lagrange equations. To see this, take the generalized coordinates to be the usual Cartesian position components and consider the potential of a particle of charge[65] e in an electromagnetic field,

$$U(\mathbf{r},\dot{\mathbf{r}},t) = e\Phi(\mathbf{r},t) - \frac{e}{c}\mathbf{A}(\mathbf{r},t)\cdot\dot{\mathbf{r}}, \tag{2.147}$$

where $\Phi(\mathbf{r},t)$ and $\mathbf{A}(\mathbf{r},t)$ are the scalar and vector potentials, respectively, that determine the electric and magnetic fields according to

$$\mathbf{E}(\mathbf{r},t) = -\boldsymbol{\nabla}\Phi(\mathbf{r},t) - \frac{1}{c}\frac{\partial\mathbf{A}(\mathbf{r},t)}{\partial t}, \tag{2.148}$$

$$\mathbf{B}(\mathbf{r},t) = \boldsymbol{\nabla}\times\mathbf{A}(\mathbf{r},t) \tag{2.149}$$

in Gaussian units. The generalized force components then are

$$Q_k = -e\frac{\partial\Phi}{\partial x_k} + \frac{e}{c}\frac{\partial\mathbf{A}}{\partial x_k}\cdot\dot{\mathbf{r}} - \frac{e}{c}\frac{dA_k}{dt}$$

$$= -e\frac{\partial\Phi}{\partial x_k} + \frac{e}{c}\frac{\partial\mathbf{A}}{\partial x_k}\cdot\dot{\mathbf{r}} - \frac{e}{c}\sum_{i=1}^{3}\frac{\partial A_k}{\partial x_i}\dot{x}_i - \frac{e}{c}\frac{\partial A_k}{\partial t}$$

$$= -e\frac{\partial\Phi}{\partial x_k} - \frac{e}{c}\frac{\partial A_k}{\partial t} + \frac{e}{c}\sum_{i=1}^{3}\left(\frac{\partial A_i}{\partial x_k}\dot{x}_i - \frac{\partial A_k}{\partial x_i}\dot{x}_i\right)$$

---

[65]The letter e here denotes *any* charge, i.e., e here is *not* necessarily the elementary unit charge.

$$= -e\frac{\partial \Phi}{\partial x_k} - \frac{e}{c}\frac{\partial A_k}{\partial t} + \frac{e}{c}\sum_{i,m,n=1}^{3}(\delta_{km}\,\delta_{in} - \delta_{kn}\,\delta_{im})\frac{\partial A_n}{\partial x_m}\dot{x}_i$$

$$= -e\frac{\partial \Phi}{\partial x_k} - \frac{e}{c}\frac{\partial A_k}{\partial t} + \frac{e}{c}\sum_{i,m,n,r=1}^{3}\varepsilon_{kir}\,\varepsilon_{rmn}\frac{\partial A_n}{\partial x_m}\dot{x}_i$$

$$= e\left[-\boldsymbol{\nabla}\Phi - \frac{1}{c}\frac{\partial \mathbf{A}}{\partial t}\right]_k + \frac{e}{c}\left[\dot{\mathbf{r}} \times (\boldsymbol{\nabla} \times \mathbf{A})\right]_k, \tag{2.150}$$

where

$$\sum_{r=1}^{3}\varepsilon_{kir}\,\varepsilon_{rmn} = \delta_{km}\,\delta_{in} - \delta_{kn}\,\delta_{im} \tag{2.151}$$

was used. This leads to

$$\mathbf{F} = \mathbf{F}(\mathbf{r}, \dot{\mathbf{r}}, t) = \sum_{k=1}^{3}Q_k\mathbf{e}_k = e\mathbf{E}(\mathbf{r}, t) + \frac{e}{c}\dot{\mathbf{r}} \times \mathbf{B}(\mathbf{r}, t), \tag{2.152}$$

which is the *Lorentz force* well-known from electrodynamics. The (non-relativistic) Lagrange function for a particle in the electromagnetic field is therefore given as

$$L(\mathbf{r}, \dot{\mathbf{r}}, t) = \frac{m}{2}\dot{\mathbf{r}}^2 - e\Phi(\mathbf{r}, t) + \frac{e}{c}\mathbf{A}(\mathbf{r}, t) \cdot \dot{\mathbf{r}}. \tag{2.153}$$

The incorporation of electromagnetic forces will be revisited in Sec. 8.2.8.2 in the context of a relativistic treatment.

2.2.2.6 *Dissipative forces: Rayleigh's function*

In many situations only a part of the forces will be conservative. This can be accommodated by splitting the force components according to

$$F_i = -\frac{\partial V}{\partial x_i} + F_{\text{diss},i}, \tag{2.154}$$

where the first part is conservative and the second part comprises the non-conservative dissipative forces. Defining as before $L = T - V$, this immediately leads to

$$\frac{d}{dt}\frac{\partial L}{\partial \dot{q}_r} - \frac{\partial L}{\partial q_r} = Q_r^*, \qquad r = 1,\ldots,N \tag{2.155}$$

where

$$Q_r^* = \sum_{i=1}^{3n}F_{\text{diss},i}\frac{\partial x_i}{\partial q_r} \tag{2.156}$$

are the generalized force components of the dissipative forces.

Specifically, to be able to describe the friction contributions by a scalar function, one introduces *Rayleigh's dissipation function,*

$$K(\dot{x}) = \frac{1}{2}\sum_{i=1}^{3n} k_i \dot{x}_i^2 \quad\longrightarrow\quad K(q,\dot{q},t) = \frac{1}{2}\sum_{i=1}^{3n} k_i [\dot{x}_i(q,\dot{q},t)]^2, \qquad (2.157)$$

where the k_i are some constants, and whose derivative according to

$$F_{\mathrm{diss},i} = -\frac{\partial K}{\partial \dot{x}_i} = -k_i \dot{x}_i \qquad (2.158)$$

gives rise to friction terms proportional to the velocities. One then finds

$$Q_r^* = -\sum_{i=1}^{3n} \frac{\partial K}{\partial \dot{x}_i}\frac{\partial x_i}{\partial q_r} = -\sum_{i=1}^{3n} \frac{\partial K}{\partial \dot{x}_i}\frac{\partial \dot{x}_i}{\partial \dot{q}_r} = -\frac{\partial K(q,\dot{q},t)}{\partial \dot{q}_r}, \qquad (2.159)$$

which in turn leads to the modified Lagrange equations

$$\frac{d}{dt}\frac{\partial L}{\partial \dot{q}_r} - \frac{\partial L}{\partial q_r} + \frac{\partial K}{\partial \dot{q}_r} = 0, \qquad r = 1,\dots,N, \qquad (2.160)$$

in which the dissipative friction contributions are now incorporated by the additional *scalar* function K.

2.2.2.7   *Determining the forces of constraint*

For holonomic constraints we have eliminated the forces of constraint entirely from the explicit consideration. However, there are situations when one needs to know what those forces are. Examples may be found in many areas of physics and the engineering sciences where knowledge of the constraining forces is important to address questions of structural integrity of buildings, machinery, and other equipment, etc.

The Lagrange equations of the mixed type (2.91) allow one to calculate these forces by deliberately choosing a set of coordinates that is not minimal, i.e., that is larger than the number of degrees of freedom of the system. The constraints then must be reintroduced by appropriate Lagrange multipliers. From (2.91), one immediately has

$$\frac{d}{dt}\frac{\partial L}{\partial \dot{q}_r} - \frac{\partial L}{\partial q_r} = \sum_{k=1}^{l} \lambda_k \frac{\partial g_k}{\partial q_r}, \qquad r = 1,\dots,N, \qquad (2.161)$$

where for simplicity it was assumed that the forces are conservative. The N variables q_r now are not all independent and one needs the supplementary constraint conditions

$$g_k(q,\dot{q},t) = 0, \qquad k = 1,\dots,l, \qquad (2.162)$$

to determine all unknown q_r's and λ_k's.

As an example we are going to determine the constraining force of a planar pendulum. We have treated this case already in Sec. 2.1.6 with the Lagrange equations of the first kind. Using the same variables as there (see Fig. 2.1), the Lagrange function is

$$L = \frac{m}{2}(\dot{r}^2 + r^2\dot{\vartheta}^2) + mgr\cos\vartheta, \tag{2.163}$$

where we deliberately choose not to implement the constraint

$$g(r) = r - l = 0 \tag{2.164}$$

right away. The Lagrange function then yields the equations of motion,

$$\frac{d}{dt}(m\dot{r}) - mr\dot{\vartheta}^2 - mg\cos\vartheta = \lambda\frac{\partial g}{\partial r} = \lambda, \tag{2.165}$$

$$\frac{d}{dt}(mr^2\dot{\vartheta}) + mgr\sin\vartheta = \lambda\frac{\partial g}{\partial\vartheta} = 0. \tag{2.166}$$

To obtain the solution, one needs to consider the last three equations producing

$$\ddot{\vartheta} + \frac{g}{l}\sin\vartheta = 0 \tag{2.167}$$

as the equation for the angular motion; the radial constraining force,

$$Q_r = \lambda = -mg\cos\vartheta - mr\dot{\vartheta}^2, \tag{2.168}$$

is of course the same as the one found previously, in Eq. (2.45). This force compensates the radial component of gravity and centrifugal force. Obviously, as far as structural integrity is concerned, the suspension point of the pendulum in turn must be strong enough to compensate this entire constraining force.

2.2.2.8 *Comparison of methods: Two examples*

Let us now consider two problems — Atwood's machine and a drawbridge with balancing counterweight — to compare the three solution methods of the Newtonian approach, the d'Alembert method, and Lagrange's equations.

Atwood's machine

Atwood's machine, shown in Fig. 2.3, consists of two masses m_1 and m_2 in the field of gravity connected over a massless pulley by a rope of fixed length. The constraint thus is

$$z_1 + z_2 + \ell = 0, \tag{2.169}$$

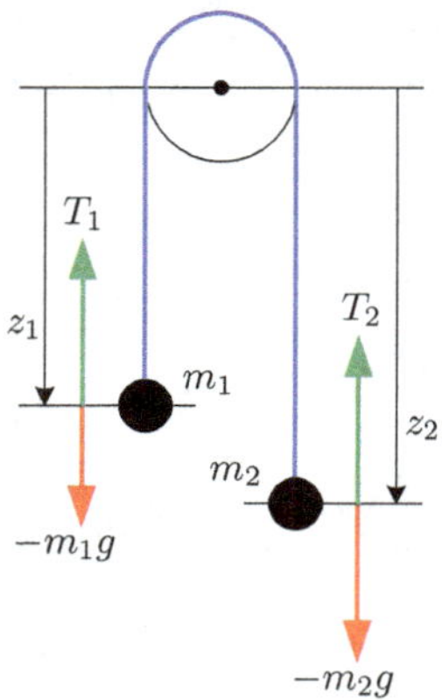

Fig. 2.3   Atwood's machine consists of two masses m_1 and m_2 subject to gravity connected over a massless pulley by a rope of fixed length.

where the (negative) z_i are the vertical positions of the masses as measured from the pulley level and the constant ℓ is the free length of the rope. The string tensions T_i exerting upward forces on the masses are the same in magnitude, of course. Real-world applications of this idealized system are, for example, funicular railways operating on steep inclines where one train moves up (or down) the track while a second train (or a counterweight) moves down (or up) a second parallel track.

Newton: Consulting the diagram, the Newtonian equations of motion are

$$m_1\ddot{z}_1 = -m_1g + T_1 \quad \text{and} \quad m_2\ddot{z}_2 = -m_2g + T_2. \tag{2.170}$$

Using $T_1 = T_2$ and subtracting the equations eliminates the string tension and provides

$$m_1\ddot{z}_1 - m_2\ddot{z}_2 = -(m_1 - m_2)g. \tag{2.171}$$

With the constraint (2.169), we then find the Newtonian equations of motion

$$\ddot{z}_1 = -\frac{m_1 - m_2}{m_1 + m_2}g \quad \text{and} \quad \ddot{z}_2 = -\ddot{z}_1. \tag{2.172}$$

If the masses are the same, there is no acceleration, the gravitational forces are balanced, and all motion is uniform due to initial velocities.[66]

The expression for the string tension,

$$T_1 = T_2 = \frac{2m_1m_2}{m_1 + m_2}g, \tag{2.173}$$

follows now from inserting (2.172) into (2.170). For equal masses, it is just given by the equal weight of the masses.

---

[66] Real-world funicular railways, therefore, require only minimal engine power for their operation.

d'Alembert: The applied forces, $F_i = -m_i g$, due to gravity do virtual work on the masses according to

$$\delta W_1 = F_1 \, \delta z_1 = -m_1 g \, \delta z_1, \tag{2.174a}$$

$$\delta W_2 = F_2 \, \delta z_2 = -m_2 g \, \delta z_2, \tag{2.174b}$$

where the δz_i are the respective virtual displacements. The d'Alembert principle says

$$\sum_{i=1}^{2} (F_i - m_i \ddot{z}_i)\delta z_i = -m_1(g + \ddot{z}_1)\delta z_1 - m_2(g + \ddot{z}_2)\delta z_2 = 0. \tag{2.175}$$

Using the constraint (2.169), we find

$$\left[-(m_1 - m_2)g - (m_1 + m_2)\ddot{z}_1 \right]\delta z_1 = 0 \tag{2.176}$$

and thus again (2.172) since δz_1 is arbitrary.

Note that the 'lost forces' for both masses,

$$F_i - m_i \ddot{z}_1 = -\frac{2m_1 m_2}{m_1 + m_2}, \qquad i = 1, 2, \tag{2.177}$$

are identical and equal to the negative of the string tension (2.173). These forces are lost as cause for acceleration because they need to compensate the string tension.

Lagrange: For the Lagrange formalism, we set up the kinetic and potential energies,

$$T = \frac{m_1}{2}\dot{z}_1^2 + \frac{m_2}{2}\dot{z}_2^2 = \frac{m_1 + m_2}{2}\dot{z}_1^2, \tag{2.178a}$$

$$V = m_1 g z_1 + m_2 g z_2 = (m_1 - m_2)g z_1 + m_2 g \ell, \tag{2.178b}$$

respectively, where the constraint equation (2.169) was used to eliminate z_2 leaving z_1 as the only independent variable. The equation of motion for $L = T - V$ then yields

$$\frac{d}{dt}\frac{\partial L}{\partial \dot{z}_1} - \frac{\partial L}{\partial z_1} = 0 \quad \Rightarrow \quad (m_1 + m_2)\ddot{z}_1 + (m_1 - m_2)g = 0, \tag{2.179}$$

which, once more, leads to (2.172).

In comparing the three approaches, note in particular that to set up the equations of motion (2.170) in the Newtonian approach it was necessary to explicitly consider the string tension as a constraining force, even though we did not actually need to calculate it to determine the equation of motion. By contrast, the string tension does not appear at all explicitly in d'Alembert or Lagrange since constraints are taken care of as a matter of course as integral parts of the respective methods. For d'Alembert, it appears implicitly as the lost force, but for Lagrange it is completely absent. To determine

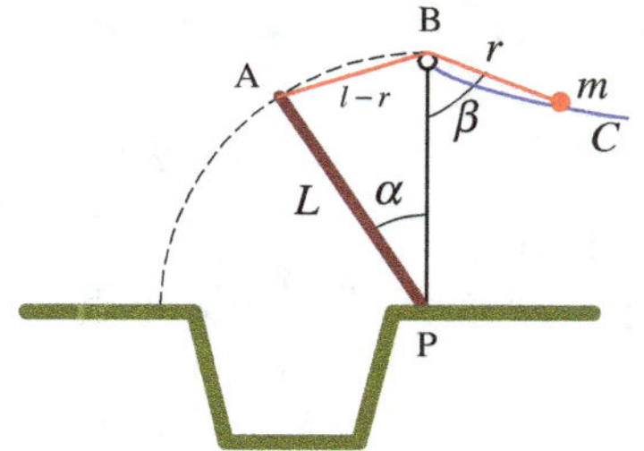

Fig. 2.4   Drawbridge with balancing counterweight of mass m moving frictionlessly along a curve C determined so that the weight of the bridge is exactly balanced for any opening angle α.

the string tension for Lagrange, one would need to explicitly include the constraint (2.169) in terms of Lagrange multipliers.

Drawbridge with balancing counterweight

The task here is to construct a drawbridge with a counterweight of mass m running along a curve C such that the bridge's mass M is balanced for any opening angle α, as shown in Fig. 2.4. (Friction is to be ignored here.)

For a fixed pivot point P of the drawbridge (which is a trivial constraint), the length r on the curve C, the associated angle β, and the opening angle α of the drawbridge are related by two constraints: First, the total length l of the rope is related to r and the opening angle α by

$$(l - r)^2 = 2L^2(1 - \cos\alpha) = 4L^2 \sin^2\frac{\alpha}{2}, \qquad (2.180)$$

which follows from the geometry of the setup. If the bridge is fully open, $\alpha = \pi/2$ and $r = 0$. Hence,

$$l^2 = 2L^2, \qquad (2.181)$$

which shows that

$$\cos\alpha = \frac{2lr - r^2}{l^2} \quad \text{or} \quad \sin\frac{\alpha}{2} = \frac{l - r}{\sqrt{2}\,l}. \qquad (2.182)$$

This provides

$$r = l\left(1 - \sqrt{2}\sin\frac{\alpha}{2}\right), \qquad (2.183)$$

but the remaining constraint that relates the angle β to r is to be determined. The system, therefore, has only one degree of freedom.

Newton: We omit elaborating on applying Newton's equations to this problem because the constraints depend here on the solution $r(\beta)$ in a nontrivial way, making this a rather tedious exercise. We only mention that one may readily guess that the solution $r(\beta)$ must be of the form

$$r(\beta) = al\left(\cos\beta_0 - \cos\beta\right), \tag{2.184}$$

where a is a dimensionless constant and β_0 is the minimal angle β when $r = 0$, with $\beta_0 \leq \beta \leq \beta_l$, with the upper bound given by $r(\beta_l) = l$. The linear $\cos\beta$ dependence is due to the fact that this factor determines the *effective* action of gravity along the rope. This does *not* mean that the string tension is given by $mg\cos\beta$ because the rope is *not* parallel to the tangent of the curve at the location $r(\beta)$ of the mass m leading to other nontrivial factors relating the angles α and β corresponding to equating Eqs. (2.183) and (2.184).

d'Alembert: The only applied forces are the forces of gravity on the mass m of the counterweight and the mass M of the bridge. The virtual work of these forces must vanish, hence

$$M\mathbf{g} \cdot \delta\mathbf{r}_{\mathrm{CM}} + m\mathbf{g} \cdot \delta\mathbf{r} = -Mg\,\delta y_{\mathrm{CM}} - mg\,\delta y = 0, \tag{2.185}$$

where $\mathbf{r}_{\mathrm{CM}} = (x_{\mathrm{CM}}, y_{\mathrm{CM}})$ and $\mathbf{r} = (x, y)$ are the coordinates of the bridge's CM and the mass m with the pivot point P taken as the origin. With (see figure)

$$y_{\mathrm{CM}} = \frac{L}{2}\cos\alpha, \qquad y = L - r\cos\beta, \tag{2.186}$$

one has

$$\frac{ML}{2}g\,\delta(\cos\alpha) - mg\,\delta(r\cos\beta) = 0, \tag{2.187}$$

or upon integration,

$$\frac{ML}{2}g\,\cos\alpha - mgr\,\cos\beta = const. \tag{2.188}$$

If the bridge is completely closed ($r = 0$, $\alpha = \pi/2$), one finds $const = 0$. With Eq. (2.182), this results in

$$\frac{ML}{2}\frac{2l - r}{l^2} - m\cos\beta = 0, \tag{2.189}$$

and finally

$$r = r(\beta) = 2l\left(1 - \sqrt{2}\frac{m}{M}\cos\beta\right). \tag{2.190}$$

This is the equation of a *cartioid*. It is indeed of the form (2.184), with an initial angle determined at $r = 0$ as

$$\cos\beta_0 = \frac{M}{\sqrt{2}m}. \tag{2.191}$$

This also shows that the counterweight must satisfy $m \geq M/\sqrt{2}$ to do its job.

Lagrange: The potential energy of the system is

$$V = Mgy_{\text{CM}} + mgy = Mg\frac{L}{2}\cos\alpha + mg\left(L - r\cos\beta\right) \tag{2.192}$$

or

$$V(r,\xi) = Mg\frac{2lr - r^2}{2\sqrt{2}\,l} - mgr\xi, \quad \text{where} \quad \xi = \cos\beta\ ; \tag{2.193}$$

the irrelevant constant term mgL was dropped here. The holonomic constraint is given by

$$r = f(\beta) \quad \Rightarrow \quad g(r,\xi) = r - f(\xi) = 0, \tag{2.194}$$

where f is the parameterization of C we seek to determine. For the static case, the kinetic energy vanishes and the Lagrange equations are simply

$$\frac{\partial V}{\partial r} = \lambda\frac{\partial g}{\partial r} = \lambda, \quad \frac{\partial V}{\partial \xi} = \lambda\frac{\partial g}{\partial \xi} = -\lambda\frac{df}{d\xi} = -\lambda\frac{dr}{d\xi}, \tag{2.195}$$

and eliminating the Lagrange multiplier λ from the second equation produces

$$\frac{\partial V}{\partial r}\frac{dr}{d\xi} + \frac{\partial V}{\partial \xi} \equiv \frac{dV}{d\xi} = 0. \tag{2.196}$$

This provides a first-order differential equation for r,

$$\left[\frac{M}{\sqrt{2l}}(l - r) - m\xi\right]\frac{dr}{d\xi} - mr = 0, \tag{2.197}$$

that can be solved by the linear ansatz

$$r = a + b\xi. \tag{2.198}$$

The differential equation then reduces to an expression of the form $A + B\xi = 0$ that can only be true for all ξ if both coefficients A and B vanish, i.e.,

$$A = \frac{M}{\sqrt{2l}}(l - a)b - ma = 0 \quad \text{and} \quad B = -\left[\frac{M}{\sqrt{2l}}b + 2m\right]b = 0. \tag{2.199}$$

Hence, non-trivial solutions for the coefficients of (2.198) are

$$b = -2l\sqrt{2}\frac{m}{M} \quad \text{and} \quad a = 2l, \tag{2.200}$$

which then produces the cartioid equation (2.190), as before.

This example shows that implementing constraints for Newton's equations may be very difficult if the constraints depend on the solution in a nontrivial way. By contrast, both d'Alembert and Lagrange formalisms provide here answers in a straightforward manner. One may argue that implementing the constraint in the form of Eq. (2.194) is more straightforward and more elegant in the static form of the Lagrange equations (2.195).

2.2.2.9 *Nonholonomic constraints*

Nonholonomic constraints are incorporated by using

$$\frac{d}{dt}\frac{\partial L}{\partial \dot{q}_r} - \frac{\partial L}{\partial q_r} = \sum_{k=1}^{l} \lambda_k \hat{h}_{kr}, \qquad r = 1,\ldots,N, \qquad (2.201)$$

where again it is assumed that the applied forces are conservative [otherwise one must use the general form (2.91)]. The constraint coefficients $\hat{h}_{kr}$ now cannot be expressed as partial derivatives, i.e., the constraint condition

$$\hat{h}_{k1}dq_1 + \hat{h}_{k2}dq_2 + \ldots + \hat{h}_{k0}dt = 0, \qquad k = 1,\ldots,l \qquad (2.202)$$

is not a total differential. As mentioned above, such constraints are characterized by having less degrees of freedom for infinitesimal displacements than for the macroscopic motion. This will be illustrated with some examples.

Blade motion on plane

Let us consider the motion of a blade on a plane. This problem describes the cutting motion of a knife or of ice-skating blades in an idealized manner. The motion is constrained by the fact that the blade cannot move sideways. Choosing the xy-plane as the plane of motion and the angle the contact line element of the blade makes with the x-axis as ϕ (see Fig. 2.5), the constraint is that $dy/dx = \tan\phi$ or

$$\cos\phi\,dy - \sin\phi\,dx = 0. \qquad (2.203)$$

This is a nonholonomic constraint since it cannot be integrated because the integrability conditions (2.79) are not satisfied. Indeed, with $\hat{h}_{1\phi} = 0$, $\hat{h}_{1y} = \cos\phi$, and $\hat{h}_{1x} = -\sin\phi$, one sees that, for example,

$$0 = \frac{\partial \hat{h}_{1\phi}}{\partial x} \neq \frac{\partial \hat{h}_{1x}}{\partial \phi} = -\cos\phi. \qquad (2.204)$$

While macroscopically one is able to change the direction of the motion of a knife when cutting across a plane, giving it the three degrees of freedom described by x, y, and ϕ, infinitesimal displacements dx and dy completely

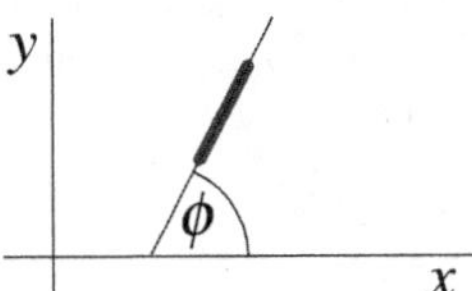

Fig. 2.5   Coordinates used in describing the cutting motion of a blade.

determine the angle ϕ, reducing the 'infinitesimal degrees of freedom' to just two. In view of the non-integrability of the constraining condition, one cannot reduce the number of variables of the macroscopic solutions. Physically, the infinitesimal condition manifests itself in the macroscopic trajectory by the fact that the curve made by the cutting motion in one direction has a well-defined tangent everywhere, i.e., there are no 'kinks' in the curve since the curve is made by a contact line element and not just by a contact point.

This problem cannot be analyzed further without making specific assumptions about the nature of the blade; this will not be done here.

Rolling disk on plane

The frictionless motion of a rolling disk of radius a on the horizontal xy-plane (see Fig. 2.6) is determined by five (macroscopic) degrees of freedom: The position (x, y) of its contact point on the plane, the angle ϕ of its contact-point tangent with the x-axis [this is similar to (x, y, ϕ) from the cutting motion of a blade], the angle ϑ of the disk plane with the xy-plane, and the rolling angle φ between the radial line to the contact point and a given, fixed radial line on the disk. Assuming rolling motion without slippage, a rolling angle increment $d\varphi$ will lead to an advancement by $ds = a\,d\varphi$ along the contact-point tangent, i.e.,

$$dx = -a\cos\phi\,d\varphi, \qquad \text{and} \qquad dy = -a\sin\phi\,d\varphi, \tag{2.205}$$

or

$$\hat{h}_{1x}\,dx + \hat{h}_{1\varphi}\,d\varphi \equiv dx + a\cos\phi\,d\varphi = 0, \tag{2.206a}$$

$$\hat{h}_{2y}\,dy + \hat{h}_{2\varphi}\,d\varphi \equiv dy + a\sin\phi\,d\varphi = 0, \tag{2.206b}$$

with obvious definitions of the coefficient functions $\hat{h}_{kr}$. These two conditions reduce the number of degrees of freedom for infinitesimal motion to

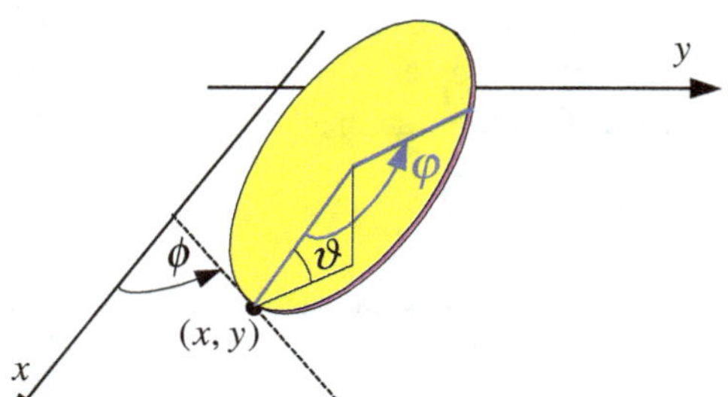

Fig. 2.6 Variables used for the description of a rolling disk on a horizontal plane.

three since the disk cannot move sideways and (as specified) it does not slip while rolling.

Calculating the inertia tensor (see Sec. 5.2) for a thin homogeneous disk (with negligibly small thickness $h \ll a$ compared to the radius a), the rotational kinetic energy is found to be

$$T_{\text{rot}} = \frac{m}{8}a^2 \left[\dot{\vartheta}^2 + \dot{\phi}^2(1 + \cos^2 \vartheta) + 2\dot{\varphi}^2 + 2\dot{\phi}\,\dot{\varphi}\,\cos\vartheta \right]. \tag{2.207}$$

And, using the center-of-mass variables

$$x_c = x - a\cos\vartheta\,\sin\phi, \tag{2.208a}$$

$$y_c = y + a\cos\vartheta\,\cos\phi, \tag{2.208b}$$

$$z_c = a\sin\vartheta, \tag{2.208c}$$

the kinetic energy for translational motion of the center of mass is given by

$$
\begin{aligned}
T_{\text{trans}} &= \frac{m}{2}\left(\dot{x}_c^2 + \dot{y}_c^2 + \dot{z}_c^2 \right) \\
&= \frac{m}{2}\Big[\dot{x}^2 + \dot{y}^2 + a^2(\dot{\vartheta}^2 + \dot{\phi}^2 \cos^2 \vartheta) \\
&\quad - 2a\dot{x}(\dot{\phi}\,\cos\phi\,\cos\vartheta - \dot{\vartheta}\,\sin\phi\,\sin\vartheta) \\
&\quad - 2a\dot{y}(\dot{\phi}\,\sin\phi\,\cos\vartheta + \dot{\vartheta}\,\cos\phi\,\sin\vartheta) \Big].
\end{aligned}
\tag{2.209}
$$

With the potential energy

$$V = mga\sin\vartheta, \tag{2.210}$$

the Lagrange function is then

$$L = T_{\text{rot}} + T_{\text{trans}} - V. \tag{2.211}$$

Setting up the equations of motions is now a simple exercise of applying Eq. (2.201).

2.2.2.10 *Constraints given as inequalities*

Up to now nonholonomic constraints that take the form of inequalities have been excluded from the discussion. To include those, we note that they can always be written in the form

$$h_k(\mathbf{r}_1, \ldots, \mathbf{r}_n, \dot{\mathbf{r}}_1, \ldots, \dot{\mathbf{r}}_n, t) \geq 0, \quad k = 1, 2, \ldots, l. \tag{2.212}$$

We can then break down the solution of this problem into two separate problems. One in which the equalities hold true and one in which they do not. In the latter case, the problem is an unconstrained one. Matching between these two problems must occur for conditions that satisfy the equality. Note also that, if the inequalities do not depend on the velocities, the constrained part of the problem is actually holonomic.

As an illustration of the procedure, let us consider a particle that slides down the surface of a sphere of radius R and calculate the point where it leaves the surface (which marks the point where the trajectory goes over from a constrained to an unconstrained one). Taking the center of the sphere as the origin and using spherical coordinates, the constraint equation here is

$$h(r) = r^2 - R^2 \geq 0, \tag{2.213}$$

i.e., the equality pertains to motion on the sphere whereas the (strict) inequality applies to (unconstrained) motion off the sphere. With initial condition such that the body sits at the top of the sphere, we can ignore the azimuthal angle φ, i.e., the only variables are r and θ. The initial conditions can then be written as

$$\theta(0) = 0, \qquad\qquad\qquad r(0) = R, \tag{2.214a}$$
$$\dot{\theta}(0) = \dot{\theta}_0 \approx 0, \qquad\qquad\qquad \dot{r}(0) = 0, \tag{2.214b}$$

where $\dot{\theta}_0 \approx 0$ means that we consider the case where the particle is nudged out of its (unstable) equilibrium position at the top of the sphere without giving it any initial velocity. The kinetic and potential energies, respectively, are

$$T = \frac{m}{2}\left(\dot{r}^2 + r^2\dot{\theta}^2\right), \tag{2.215}$$

$$V = mgr\cos\theta. \tag{2.216}$$

Taking the constraint first as an equality, this actually corresponds to an holonomic condition since there is no velocity dependence. The Lagrange equations then become

$$\frac{d}{dt}(m\dot{r}) - m(r\dot{\theta}^2 - g\cos\theta) = 2\lambda r, \tag{2.217a}$$

$$\frac{d}{dt}(mr^2\dot{\theta}) - mgr\sin\theta = 0, \tag{2.217b}$$

where the constraining force components appear on the right. As long as the body stays on the surface, $\dot{r} = 0$, and the radial force of constraint,

$$\hat{Z}_r = \lambda\frac{\partial h(r)}{\partial r} = 2\lambda r = mg\cos\theta - mr\dot{\theta}^2, \tag{2.218}$$

will provide the normal force, $mg\cos\theta$, to counteract gravity and the centripetal force, $-mr\dot{\theta}^2$, to keep the body on a circular path. At some critical angle $\theta = \theta_c$, the constraining force will vanish and the motion will become unconstrained. At this point, the mass will leave the surface and the equality of the constraining condition will go over into a strict inequality. The complete solution, therefore, is obtained by first solving (2.217b) with $r = R$. Multiplying both sides by $d\theta$, this equation reads

$$R\frac{d\dot\theta}{dt}d\theta = R\,\dot\theta\,d\dot\theta = g\sin\theta\,d\theta, \qquad (2.219)$$

and integrating once, we obtain

$$R\dot\theta^2 + 2g\cos\theta = 2g, \qquad \text{for} \quad \theta \le \theta_c, \qquad (2.220)$$

where the integration constant, $2g$, on the right is determined by the initial conditions. Using this expression in

$$\hat{Z}_r = 0 = \left[mg\cos\theta - mR\dot\theta^2\right]_{\theta=\theta_c} = mg\left[3\cos\theta - 2\right]_{\theta=\theta_c} \qquad (2.221)$$

provides

$$\theta_c = \arccos\tfrac{2}{3} \approx 48° \qquad (2.222)$$

for the critical angle where the particle leaves the sphere. Finally, to get the complete trajectory, one then solves the unconstrained problem defined by

$$\frac{d}{dt}(m\dot r) - m(r\dot\theta^2 - g\cos\theta) = 0, \qquad (2.223a)$$

$$\frac{d}{dt}(mr^2\dot\theta) - mgr\sin\theta = 0, \qquad (2.223b)$$

with initial conditions determined at $\theta = \theta_c$.

2.2.3 Conservation laws

In this section, we will show how the Lagrangian formulation of mechanics provides energy, momentum, and angular momentum conservation.

2.2.3.1 Energy conservation

For scleronomous conditions, T has the form (2.97) and then $L = L(q, \dot q)$ does not depend explicitly on the time. Considering the general expression

$$\frac{dL}{dt} = \sum_{k=1}^{N}\left[\frac{\partial L}{\partial q_k}\dot q_k + \frac{\partial L}{\partial \dot q_k}\ddot q_k\right] + \frac{\partial L}{\partial t} \qquad (2.224)$$

and

$$\frac{d}{dt}\sum_{k=1}^{N}\frac{\partial L}{\partial \dot{q}_k}\dot{q}_k = \sum_{k=1}^{N}\left[\dot{q}_k\frac{d}{dt}\frac{\partial L}{\partial \dot{q}_k} + \frac{\partial L}{\partial \dot{q}_k}\ddot{q}_k\right] = \sum_{k=1}^{N}\left[\frac{\partial L}{\partial q_k}\dot{q}_k + \frac{\partial L}{\partial \dot{q}_k}\ddot{q}_k\right], \quad (2.225)$$

where the Lagrange equations were used, one finds

$$\frac{d}{dt}\left(\sum_{k=1}^{N}\frac{\partial L}{\partial \dot{q}_k}\dot{q}_k - L\right) = -\frac{\partial L}{\partial t}. \quad (2.226)$$

In other words,

$$\frac{\partial L(q,\dot{q})}{\partial t} = 0 \quad \Rightarrow \quad \sum_{k=1}^{N}\frac{\partial L(q,\dot{q})}{\partial \dot{q}_k}\dot{q}_k - L(q,\dot{q}) = const \quad (2.227)$$

is a constant of motion for $L = L(q,\dot{q})$. With (2.97), i.e.,

$$\sum_{k=1}^{N}\frac{\partial L}{\partial \dot{q}_k}\dot{q}_k = \sum_{k=1}^{N}\frac{\partial T}{\partial \dot{q}_k}\dot{q}_k = 2\sum_{k,j=1}^{N}c_{jk}\dot{q}_j\dot{q}_k = 2T, \quad (2.228)$$

this constant of motion is found to be

$$2T - L = T + V = E = const, \quad (2.229)$$

and this proves, once again, that the total energy is conserved for time-independent conservative systems.

2.2.3.2　Momentum conservation

A system is called *translationally invariant* if

$$L(\mathbf{r}_1,\ldots,\mathbf{r}_n,\dot{\mathbf{r}}_1,\ldots,\dot{\mathbf{r}}_n,t) = L(\mathbf{r}_1 + \mathbf{d},\ldots,\mathbf{r}_n + \mathbf{d},\dot{\mathbf{r}}_1,\ldots,\dot{\mathbf{r}}_n,t), \quad (2.230)$$

for any fixed displacement $\mathbf{d}$. Specifically, taking $\mathbf{d}$ to be infinitesimally small, one then finds

$$0 = L(\mathbf{r}_1 + \mathbf{d},\ldots,\mathbf{r}_n + \mathbf{d},\dot{\mathbf{r}}_1,\ldots,\dot{\mathbf{r}}_n,t) - L(\mathbf{r}_1,\ldots,\mathbf{r}_n,\dot{\mathbf{r}}_1,\ldots,\dot{\mathbf{r}}_n,t)$$

$$= \mathbf{d}\cdot\sum_{i=1}^{n}\boldsymbol{\nabla}_{\mathbf{r}_i}L(\mathbf{r}_1,\ldots,\mathbf{r}_n,\dot{\mathbf{r}}_1,\ldots,\dot{\mathbf{r}}_n,t), \quad (2.231)$$

where a Taylor expansion truncated at terms linear in $\mathbf{d}$ was employed. Since the choice of $\mathbf{d}$ is arbitrary, it follows that translationally invariant Lagrange functions satisfy

$$0 = \sum_{i=1}^{n}\boldsymbol{\nabla}_{\mathbf{r}_i}L(\mathbf{r},\dot{\mathbf{r}},t) = \frac{d}{dt}\sum_{i=1}^{n}\boldsymbol{\nabla}_{\dot{\mathbf{r}}_i}L(\mathbf{r},\dot{\mathbf{r}},t), \quad (2.232)$$

where the second equality follows from in vector form (2.130) of the Lagrange equations. Hence,

$$\frac{d}{dt}\sum_{i=1}^{n}\boldsymbol{\nabla}_{\dot{\mathbf{r}}_i}L(\mathbf{r},\dot{\mathbf{r}},t)=0 \quad\Rightarrow\quad \sum_{i=1}^{n}\boldsymbol{\nabla}_{\dot{\mathbf{r}}_i}L(\mathbf{r},\dot{\mathbf{r}},t)=const. \tag{2.233}$$

Evaluating the left-hand side of the last equation,

$$\sum_{i=1}^{n}\boldsymbol{\nabla}_{\dot{\mathbf{r}}_i}L(\mathbf{r},\dot{\mathbf{r}},t)=\sum_{i=1}^{n}\boldsymbol{\nabla}_{\dot{\mathbf{r}}_i}\left[\sum_{j=1}^{n}\frac{m_j}{2}\dot{\mathbf{r}}_j^2-V(\mathbf{r}_1,\ldots,\mathbf{r}_n)\right]$$

$$=\sum_{i=1}^{n}m_i\dot{\mathbf{r}}_i, \tag{2.234}$$

then provides

$$\sum_{i=1}^{n}m_i\dot{\mathbf{r}}_i=\sum_{i=1}^{n}\mathbf{p}_i=\mathbf{P}=const, \tag{2.235}$$

i.e., the total momentum $\mathbf{P}$ is conserved for a translationally invariant system.

An example of such a system is one in which the n particles interact via central forces, i.e.,

$$L=\sum_{i=1}^{n}\frac{m_i}{2}\dot{\mathbf{r}}_i^2-\frac{1}{2}\sum_{i,j=1}^{n}V(|\mathbf{r}_i-\mathbf{r}_j|), \tag{2.236}$$

as discussed in Sec. 1.6.2.

2.2.3.3 *Angular momentum conservation*

A system is called *rotationally invariant* if

$$L(\mathbf{r}_1,\ldots,\mathbf{r}_n,\dot{\mathbf{r}}_1,\ldots,\dot{\mathbf{r}}_n,t)=L(D\mathbf{r}_1,\ldots,D\mathbf{r}_n,D\dot{\mathbf{r}}_1,\ldots,D\dot{\mathbf{r}}_n,t), \tag{2.237}$$

where $D\mathbf{r}$ symbolically denotes a rotation of the vector $\mathbf{r}$ by a fixed angle about a fixed axis (see Sec. 5.1.1 for more details). For infinitesimally small rotations $d\boldsymbol{\varphi}$, the vectorial rotation increments are described by Eq. (1.252), i.e., $D\mathbf{r}=\mathbf{r}+d\boldsymbol{\varphi}\times\mathbf{r}$, etc.[67] A first-order Taylor expansion in $d\boldsymbol{\varphi}$ then provides

$$0=L(\mathbf{r}+d\boldsymbol{\varphi}\times\mathbf{r},\dot{\mathbf{r}}+d\boldsymbol{\varphi}\times\dot{\mathbf{r}},t)-L(\mathbf{r},\dot{\mathbf{r}},t)$$

$$=\sum_{i=1}^{n}\left[(d\boldsymbol{\varphi}\times\mathbf{r}_i)\cdot\boldsymbol{\nabla}_{\mathbf{r}_i}L(\mathbf{r},\dot{\mathbf{r}},t)+(d\boldsymbol{\varphi}\times\dot{\mathbf{r}})\cdot\boldsymbol{\nabla}_{\dot{\mathbf{r}}_i}L(\mathbf{r},\dot{\mathbf{r}},t)\right]$$

---

[67]For the details of D for this infinitesimal rotation, see Sec. 5.1.1.2.

$$= d\boldsymbol{\varphi} \cdot \sum_{i=1}^{n} \left[\mathbf{r}_i \times \boldsymbol{\nabla}_{\mathbf{r}_i} L(\mathbf{r}, \dot{\mathbf{r}}, t) + \dot{\mathbf{r}}_i \times \boldsymbol{\nabla}_{\dot{\mathbf{r}}_i} L(\mathbf{r}, \dot{\mathbf{r}}, t) \right]$$

$$= d\boldsymbol{\varphi} \cdot \sum_{i=1}^{n} \left[\mathbf{r}_i \times \frac{d}{dt} \boldsymbol{\nabla}_{\dot{\mathbf{r}}_i} L(\mathbf{r}, \dot{\mathbf{r}}, t) + \frac{d\mathbf{r}_i}{dt} \times \boldsymbol{\nabla}_{\dot{\mathbf{r}}_i} L(\mathbf{r}, \dot{\mathbf{r}}, t) \right]$$

$$= d\boldsymbol{\varphi} \cdot \frac{d}{dt} \sum_{i=1}^{n} \left[\mathbf{r}_i \times \underbrace{\boldsymbol{\nabla}_{\dot{\mathbf{r}}_i} L(\mathbf{r}, \dot{\mathbf{r}}, t)}_{=m_i \dot{\mathbf{r}}_i = \mathbf{p}_i} \right]. \tag{2.238}$$

Since $d\boldsymbol{\varphi}$ is arbitrary, this means

$$\frac{d}{dt} \sum_{i=1}^{n} \left[\mathbf{r}_i \times \mathbf{p}_i \right] = 0 \quad \Rightarrow \quad \sum_{i=1}^{n} \left[\mathbf{r}_i \times \mathbf{p}_i \right] = \sum_{i=1}^{n} \mathbf{L}_i = \mathbf{L} = const. \tag{2.239}$$

Thus, the total angular momentum $\mathbf{L}$ is conserved if the Lagrange function is rotationally invariant.

2.2.4   *Summary of Lagrange equations*

For easy reference and because of their central role for finding solutions in practice of problems in classical mechanics, a summary of all relevant Lagrange equations written in terms of generalized coordinates with and without forces of constraint is provided here. This summary will not provide any new equations as such, but it will discuss the relationship between macroscopic and microscopic degrees of freedom, and holonomic and non-holonomic constraints in a unified manner.

Let us first consider only conservative forces. They are incorporated into the scalar Lagrange function L via a potential V in the form

$$L = L(q, \dot{q}, t) = T(q, \dot{q}, t) - V(q, t). \tag{2.240}$$

The general form of the second-order differential Lagrange equations of motion then reads

$$\frac{d}{dt} \frac{\partial L}{\partial \dot{q}_r} - \frac{\partial L}{\partial q_r} = \hat{Z}_r, \qquad r = 1, \ldots, N, \tag{2.241}$$

where N is the number of generalized coordinates and

$$\hat{Z}_r = \hat{Z}_r^{(h)} + \hat{Z}_r^{(n)} \tag{2.242}$$

provides a split of the generalized forces of constraint into holonomic and nonholonomic contributions, with

$$\hat{Z}_r^{(h)} = \sum_{k=1}^{l_h} \lambda_k^{(h)}(q, \dot{q}, t) \frac{\partial g_k(q, t)}{\partial q_r} \qquad \text{(holonomic)} \tag{2.243}$$

and

$$\hat{Z}_r^{(n)} = \sum_{k=1}^{l_n} \lambda_k^{(n)}(q,\dot{q},t)\,\hat{h}_{kr}(q,\dot{q},t) \qquad \text{(nonholonomic).} \qquad (2.244)$$

The functions g_k and $\hat{h}_{kr}$ defining the constraints are given; the Lagrange multipliers $\lambda_k^{(h)}$ and $\lambda_k^{(n)}$ need to be determined by adding constraint conditions in differential form as

$$\sum_{r=1}^{N} \frac{\partial g_k(q,t)}{\partial q_r}\,dq_r + \frac{\partial g_k(q,t)}{\partial t}\,dt = 0, \qquad k = 1,\dots,l_h, \qquad (2.245)$$

for l_h holonomic constraints and

$$\sum_{r=1}^{N} \hat{h}_{kr}(q,\dot{q},t)\,dq_r + \hat{h}_{k0}(q,\dot{q},t)\,dt = 0, \qquad k = 1,\dots,l_n, \qquad (2.246)$$

for l_n nonholonomic constraints. The latter constraints are *non-integrable*, which means that of the mixed partial derivatives,

$$\text{at least one equality} \quad \frac{\partial \hat{h}_{ki}}{\partial q_j} = \frac{\partial \hat{h}_{kj}}{\partial q_i}, \quad \frac{\partial \hat{h}_{ki}}{\partial t} = \frac{\partial \hat{h}_{k0}}{\partial q_i} \quad \text{is } not \text{ true.}$$
$$(2.247)$$

Together, there are

$$l = l_h + l_n \qquad (2.248)$$

differential constraint conditions that reduce the $3n$ Cartesian components to

$$N' = 3n - l \qquad (2.249)$$

microscopic degrees of freedom (DOF).

The holonomic constraints (2.245) provide a perfect differential dg_k and thus are integrable; they can be equivalently written as

$$g_k(q,t) = 0, \qquad k = 1,\dots,l_h, \qquad (2.250)$$

which shows that the functions g_k provide topological constraints that correlate the generalized coordinates q_r, $r = 1,\dots,N$. If all correlations are fully implemented in the choice of generalized coordinates, the number N of Lagrange equations is *minimal* and reflects the actual physical DOF of the system. In this case, the constraint functions (2.250) and the resulting constraint forces vanish identically, $g_k(q,t) \equiv 0$ and $\hat{Z}_r^{(h)} \equiv 0$, leaving a reduced set of Lagrange equations,

$$\frac{d}{dt}\frac{\partial L}{\partial \dot{q}_r} - \frac{\partial L}{\partial q_r} = \hat{Z}_r^{(n)}, \qquad r = 1,\dots,N \qquad (N\text{: physical DOF}), \quad (2.251)$$

where only nonholonomic constraint forces appear explicitly. The *macroscopic* (physical) DOF, therefore, are given by

$$N = N' + l_n = 3n - l_h \qquad (N\text{: physical DOF}). \tag{2.252}$$

Thus, if nonholonomic constraints are present, they are characterized by the fact that the microscopic DOF is *less* that the macroscopic one. If they are absent ($l_n = 0$), we obtain the Lagrange equations in the familiar form

$$\frac{d}{dt}\frac{\partial L}{\partial \dot{q}_r} - \frac{\partial L}{\partial q_r} = 0, \qquad r = 1, \ldots, N \qquad (N\text{: physical DOF}). \tag{2.253}$$

In this case, macroscopic and microscopic degrees of freedom are identical, $N = N'$.

Note, however, that the number of equations, N, in the generic Lagrange equations (2.241) need not be minimal, i.e., it may be chosen larger than the actual physical degrees of freedom if one wishes to explicitly calculate (holonomic) constraint forces according to (2.243) because $\hat{Z}_r^{(h)} \neq 0$ in this case.

If the constraints are given in terms of *inequalities* ($\geq$), the corresponding parts of the trajectories need to be separated, with the equal-sign ($=$) part describing an ordinary constrained motion, either holonomic or nonholonomic, and the unequal-sign ($>$) part corresponding to an unconstrained motion. The two parts of the trajectories need to be matched at the point where the constraining forces vanish as discussed in examples. This procedure can also be generalized for two-sided inequalities (for example, if the motion is confined to the inside of a spherical shell such that $r_1^2 \leq x^2 + y^2 + z^2 \leq r_2^2$, where r_1 and r_2 are inner and outer shell radii, respectively).

Finally, if we have additional *dissipative forces*, they need to be added to the right-hand side of Eq. (2.241). All other considerations apply accordingly to this case as well.

Chapter 3

Variational Principles

The formulation of mechanics given in the previous sections is based on Newton's axioms. In an equivalent axiomatic approach, one may *postulate* that the physical trajectories of a system are given by stationary solutions of a *variational principle* called *Hamilton's Principle*. This approach and its implications will be the subject of the following sections. The great importance of Hamilton's Principle lies in fact that it provides the mathematical basis for quantum-mechanical and field-theoretical formulations of modern physics that go beyond classical mechanics and the applicability of Newton's axioms.

3.1   Basics of Variational Calculus

A function $y = y(x)$ associates a value y with every value of x. Variational principles are formulated in terms of *functionals* $J = J[y]$ which associate a value J with every function $y(x)$. For example,

$$J = J[y] = \int_i^f ds = \int_i^f \sqrt{dx^2 + dy^2} = \int_{x_i}^{x_f} dx \sqrt{1 + \left(\frac{dy}{dx}\right)^2} \qquad (3.1)$$

is the functional that describes the path length along the curve $y = y(x)$ between an initial point $(x_i, y(x_i))$ and a final point $(x_f, y(x_f))$, thus associating a number (the path length) with each curve $y(x)$ that connects the two points.

In the discussion of the behavior of curves the positions of their minima and maxima are frequently of interest. The necessary condition to have a local extremum of $y(x)$ at a position x is that

$$\frac{dy}{dx} = 0, \qquad (3.2)$$

i.e., that the function is stationary. Whether this stationary point is a minimum or a maximum, or perhaps a point of inflection with a horizontal tangent, requires an additional investigation.

By analogy, we may ask the question: For which function will a given functional become stationary? To provide the necessary conditions, analogous to (3.2), for determining the answer to this question is the central task of the calculus of variations.

The relevance of this question can be illustrated by the very problem that stood at the inception of the variational calculus, namely the famous *brachistochrone problem* posed by Johann Bernoulli[68] in 1696 as a challenge to the mathematicians of the time:[69] *Consider a body in the field of gravity sliding down, in the vertical xy-plane, along a curve y(x) without friction from a point A, where it is initially at rest, to a point B. Determine the 'minimal' curve y(x) that requires the least time.* In practical terms: How does one design an optimal package chute? To solve this problem, we seek a minimum for the total time T,

$$T = T[y] = \int_A^B dt = \int_A^B \frac{ds}{v}. \tag{3.3}$$

Employing energy conservation,

$$\tfrac{1}{2}mv^2 + mgy = mgy_A, \tag{3.4}$$

where y_A is the height of the initial position (the positive direction of y is up), this leads to the functional

$$T = T[y] = \int_{x_A}^{x_B} dx \sqrt{\frac{1 + \left(y'(x)\right)^2}{2g\left(y_A - y(x)\right)}}. \tag{3.5}$$

Here and in the following y' denotes the derivative of $y(x)$ with respect to x. The solution to this problem will be given below, on p. 145, once we have developed the variational formalism.

---

[68] The Bernoulli family, of Switzerland, produced some of the most famous mathematicians and physicists of their times. Most notably among them are the brothers Jakob (1655–1705) and Johann (1667–1748), and Johann's son, Daniel (1700–1782). Johann Bernoulli also was the teacher of Leonhard Euler, one of the most influential mathematical physicists of all times (see footnote 72).

[69] The solution was already known to Johann Bernoulli when he posed the problem. Correct solutions were submitted by Jakob Bernoulli (Johann's brother), Newton, Leibniz, and de l'Hôpital. (The latter solution was not published until 1988!)

3.1.1 *Variation without constraints*

The basic problem of variational calculus is the following: Find the function $y(x)$ for which the functional

$$J = J[y] = \int_{x_1}^{x_2} dx\, F(y, y', x) \tag{3.6}$$

is stationary. It is assumed here for simplicity that F depends only on the first derivative, $y' = dy/dx$, of y. Moreover, for the present purpose we may take the values at the end points,

$$y(x_1) = y_1, \qquad y(x_2) = y_2, \tag{3.7}$$

as given — this is referred to as a variational problem with fixed end points. Generalizations of these assumptions are given in Sec. 3.1.1.1.

Let $y(x)$ be the function that makes the functional stationary, and consider a small variation

$$\delta y = \varepsilon \eta(x), \tag{3.8}$$

where ε is infinitesimally small and $\eta(x)$ is an arbitrary function subject to

$$\eta(x_1) = \eta(x_2) = 0 \tag{3.9}$$

so that the end points of $y + \delta y$ remain fixed at the values given in (3.7) even if $\varepsilon \neq 0$. The functional $J[y + \delta y]$ then becomes a function of ε, denoted by

$$J_\eta(\varepsilon) \equiv J[y + \varepsilon\eta] = \int_{x_1}^{x_2} dx\, F(y + \varepsilon\eta, y' + \varepsilon\eta', x). \tag{3.10}$$

In view of the smallness of ε, the Taylor expansion[70] of $J_\eta(\varepsilon)$ around $\varepsilon = 0$ can be truncated at the linear term,

$$J_\eta(\varepsilon) = J_\eta(0) + \varepsilon \left[\frac{dJ_\eta(\varepsilon)}{d\varepsilon} \right]_{\varepsilon=0}. \tag{3.11}$$

In terms of functionals, this expansion reads

$$J[y + \delta y] = J[y] + \delta J, \tag{3.12}$$

where

$$\delta J \equiv \varepsilon \left[\frac{dJ_\eta(\varepsilon)}{d\varepsilon} \right]_{\varepsilon=0} = J_\eta(\varepsilon) - J_\eta(0) \tag{3.13}$$

is the (first-order) variation of J.

---

[70]It is assumed that all functions are reasonably well behaved so that all derivatives exist.

By construction, the functional $J[y + \varepsilon\eta]$ is stationary for $\varepsilon = 0$, i.e., the function $J_\eta(\varepsilon)$ has an extremum for which the usual necessary condition

$$\left[\frac{dJ_\eta(\varepsilon)}{d\varepsilon}\right]_{\varepsilon=0} = 0 \tag{3.14}$$

holds true. With the truncated Taylor expansion

$$F(y + \varepsilon\eta, y' + \varepsilon\eta', x) = F(y, y', x) + \frac{\partial F}{\partial y}\varepsilon\eta + \frac{\partial F}{\partial y'}\varepsilon\eta', \tag{3.15}$$

this can be written as

$$\delta J = \varepsilon\left[\frac{dJ_\eta(\varepsilon)}{d\varepsilon}\right]_{\varepsilon=0} = \int_{x_1}^{x_2} dx\left[\frac{\partial F}{\partial y}\delta y + \frac{\partial F}{\partial y'}\delta y'\right] = 0, \tag{3.16}$$

where[71]

$$\delta y' = \frac{d}{dx}\delta y = \varepsilon\eta'. \tag{3.17}$$

Integration by parts of the second term under the integral then yields

$$\delta J = \left[\frac{\partial F}{\partial y'}\delta y\right]_{x_1}^{x_2} + \int_{x_1}^{x_2} dx\left[\frac{\partial F}{\partial y} - \frac{d}{dx}\frac{\partial F}{\partial y'}\right]\delta y = 0. \tag{3.18}$$

The first term on the right-hand side vanishes because of (3.9), hence the integral must vanish. Since δy is arbitrary, the variation δJ will vanish if and only if the integrand itself vanishes. Hence, we find that for the functional $J[y]$ to be stationary the corresponding function $y(x)$ must satisfy the necessary condition

$$\boxed{\frac{\partial F}{\partial y} - \frac{d}{dx}\frac{\partial F}{\partial y'} = 0} \; . \tag{3.19}$$

---

[71]The result here is a special case for the more general statement that the variation operation δ may always be interchanged in sequence with both differentiation and integration. Specifically for differentiation one immediately finds

$$\delta\frac{df(x)}{dx} = \lim_{\Delta x \to 0}\delta\frac{f(x + \Delta x) - f(x)}{\Delta x} = \lim_{\Delta x \to 0}\frac{\delta f(x + \Delta x) - \delta f(x)}{\Delta x} = \frac{d}{dx}\delta f(x).$$

For integration we write the variation of a function $f(x)$ generically as

$$\delta f(x) = g(x, \varepsilon) - g(x, 0),$$

where $g(x, \varepsilon)$ is a suitably chosen function with $g(x, 0) = f(x)$ and ε is an infinitesimal variation parameter. Hence, considering the definite integral

$$\int \delta f(x)\,dx = \int [g(x, \varepsilon) - g(x, 0)]\,dx = \int g(x, \varepsilon)dx - \int g(x, 0)\,dx = \delta\int f(x)\,dx,$$

we see that integration and variation can be interchanged as well.

This is the *Fundamental Lemma of Variational Calculus*, usually called *Euler's equation*, first derived by Leonhard Euler.[72] It corresponds exactly to the Lagrange equation of the second kind for holonomic conservative systems [see (2.104), with $x \to t$, $y \to q$, and $F \to L$]. Euler's equation is a second-order differential equation, linear in y'', for the desired function $y(x)$.

Shortest distance

We can now determine the shortest distance between two points according to the functional (3.1). For

$$F = F(y') = \sqrt{1 + y'^2} \tag{3.20}$$

the Euler equation is

$$\frac{d}{dx} \frac{y'}{\sqrt{1 + y'^2}} = 0, \tag{3.21}$$

which yields

$$y' = const, \qquad \text{or} \qquad y = ax + b, \tag{3.22}$$

upon integration, where a and b are integration constants determined by the end points. The shortest distance between two points is thus (not surprisingly) found to be a straight line.

Brachistochrone problem

For the brachistochrone problem, one sees that

$$F = F(y, y') = \sqrt{\frac{1 + y'^2}{2g(y_A - y)}} \tag{3.23}$$

---

[72]The Swiss Leonhard Euler (1707–1783) was the most influential and prolific mathematician and physicist of the 18th century. He held prestigious appointments at the St. Petersburg Academy of Science under Czar Peter the Great of Russia, the Berlin Academy of Science under King Frederick the Great of Prussia, and then again at St. Petersburg. His book *Mechanica* (1736-37) was the first description of Newtonian mechanics entirely in the language of mathematical analysis. During his 25 years at Berlin alone, he published 380 articles. Even partial and, later in life, complete blindness did not stop him. With the help of several assistants he continued to publish at an astonishing rate. For example, in a seven-year period around 1772 (when he became totally blind), he published 250 papers. After his death in 1783, the St. Petersburg Academy continued to publish Euler's unpublished work for fifty years! His name is associated with many, many areas of mathematics and physics. Thus simply saying "Euler's equation" without proper context is not sufficient to identify which of the many equations loosely called such one is referring to.

actually does not depend explicitly on the variable x. In view of the correspondence of F to the Lagrange function L, this has the same implication for F as it has for a time-independent L. We may thus follow the reasoning of Sec. 2.2.3.1 and immediately conclude that [cf. Eq. (2.227)]

$$F - y'\frac{\partial F}{\partial y'} = const \equiv a. \tag{3.24}$$

This provides a first-order differential equation for $y(x)$,

$$y' = -\sqrt{\frac{1}{2ga^2(y_A - y)} - 1} = -\sqrt{\frac{2r - y_A + y}{y_A - y}}, \tag{3.25}$$

where $r = 1/4ga^2$ was introduced as a new constant substituting for a. We immediately see here that at the starting point, at $y = y_A$, the derivative is infinite, i.e., the solution will have a cusp at this point — in other words, initially the body will drop straight down. The solution is obtained by a separation of variables via

$$x - x_A = -\int_{y_A}^{y} dy \, \sqrt{\frac{y_A - y}{2r - y_A + y}}, \tag{3.26}$$

where x_A was obtained as the integration constant satisfying the correct boundary conditions. This integral can be found in integral tables providing

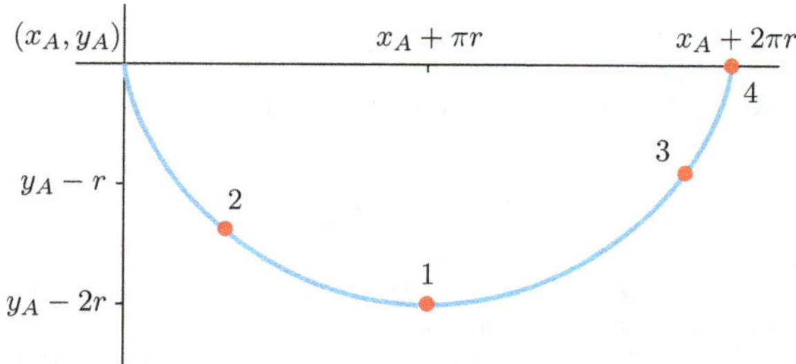

Fig. 3.1   Brachistochrone Problem: Starting at (x_A, y_A), this plot shows one period of the cycloid solution (3.27) with four possible scenarios for end points (x_B, y_B): Point 1, the lowest point of the cycloid, corresponds to the condition $\Delta y/\Delta x \equiv (y_B - y_A)/(x_B - y_A) = -2/\pi$; in that case, $r = (x_B - x_A)/\pi$. Point 2 corresponds to $\Delta y/\Delta x < -2/\pi$ and point 3 to $\Delta y/\Delta x > -2/\pi$. For these cases, r cannot be determined analytically, but one finds from the parametric representation of the cycloid in terms of the rolling angle ϕ that it is given by $1/r = (1 - \cos\phi_B)/(y_A - y_B)$, where ϕ_B is the root of $(\phi_B - \sin\phi_B)/(x_B - x_A) - (1 - \cos\phi_B)/(y_A - y_B) = 0$. The latter equation can be solved by iteration (e.g., by Newton's method). Point 4 corresponds to the case where the end point is at the same height as the initial point; for this case $r = (x_B - x_A)/2\pi$.

the implicit solution

$$\frac{y_A - y}{r} = 1 - \cos\left[\frac{x - x_A}{r} + \sqrt{\frac{y_A - y}{r}\left(2 - \frac{y_A - y}{r}\right)}\right].\qquad (3.27)$$

This equation describes a *cycloid* with period $2\pi r$ (see Fig. 3.1). A cycloid is the curve traced out by a fixed peripheral point on a circle of radius r as the circle rolls on a horizontal line without slipping. For the present problem, the parameter r is determined by the end point (x_B, y_B) (see discussion in Fig. 3.1).

3.1.1.1 Generalizations

Adapting the previous procedure, it is straightforward to find the generalizations listed in the following without any derivations. Further, additional generalizations are possible by combining the features of the following cases.

Several functions: If there are N functions $y_i(x)$, $i = 1, \ldots, N$, with fixed end points, the stationary points of the functional

$$J = J[y_1, \ldots, y_N] = \int_{x_1}^{x_2} dx\, F(y_1, \ldots, y_N, y_1', \ldots, y_N', x)\qquad (3.28)$$

are determined by the N Euler equations,

$$\frac{\partial F}{\partial y_i} - \frac{d}{dx}\frac{\partial F}{\partial y_i'} = 0,\qquad i = 1, \ldots, N,\qquad (3.29)$$

which correspond to the Lagrange equations for N degrees of freedom. These equations, therefore, sometimes are called Euler–Lagrange equations.

Several variables: If there are n independent variables x_i, $i = 1, \ldots, n$, the functional

$$J = J[y] = \int_R dx_1 \ldots \int_R dx_n\, F\left(y, \frac{\partial y}{\partial x_1}, \ldots, \frac{\partial y}{\partial x_n}, x_1, \ldots, x_n\right)\qquad (3.30)$$

becomes stationary if and only if $y = y(x_1, \ldots, x_n)$ satisfies

$$\frac{\partial F}{\partial y} - \sum_{i=1}^{n}\frac{\partial}{\partial x_i}\frac{\partial F}{\partial \frac{\partial y}{\partial x_i}} = 0,\qquad (3.31)$$

where it is assumed that $y(x_1, \ldots, x_n)$ has given (fixed) values at the perimeter of the region R. An example of this case will be treated in Sec. 7.1.

Higher derivatives: If there are higher derivatives, e.g.,

$$J = J[y] = \int_{x_1}^{x_2} dx\, F(y, y', y'', x), \tag{3.32}$$

the Euler equations will also contain higher derivatives. For the present example, taking the end points $y(x_1)$, $y(x_2)$, $y'(x_1)$, $y'(x_2)$ as fixed, one finds

$$\frac{\partial F}{\partial y} - \frac{d}{dx}\frac{\partial F}{\partial y'} + \frac{d^2}{dx^2}\frac{\partial F}{\partial y''} = 0. \tag{3.33}$$

This is a fourth-order differential equation.

Variation of end points: In the previous derivations, it was always assumed that the end points remain fixed [cf. (3.9)], which led to the vanishing of the end-point term in (3.18). If the end points are allowed to vary, this end-point term will vanish nonetheless if the partial derivatives $\partial F/\partial y'$ vanish at the end points, i.e.,

$$\frac{\partial F\left(y(x_1), y'(x_1), x_1\right)}{\partial y'} = 0, \qquad \frac{\partial F\left(y(x_2), y'(x_2), x_2\right)}{\partial y'} = 0. \tag{3.34}$$

This provides two boundary conditions necessary for determining the integration constants. The Euler equations (3.19) themselves remain unchanged.

3.1.2   Variation with constraints

In many practical cases, the function $y(x)$ will need to be determined subject to some constraints. For example, if we suspend a chain between two points in the field of gravity, the equilibrium curve described by the chain will have to be determined by keeping the overall length L of the chain a constant.

For the chain in the field of gravity, the potential energy will have a minimum when the chain is in equilibrium. (If the potential energy were not minimal, the gravity would convert potential energy into kinetic energy, contradicting the assumption that the chain is at its equilibrium configuration.) The functional to be minimized is, therefore,

$$U = U[y] = \int_{1}^{2} dm\, gy = \rho g \int_{x_1}^{x_2} dx\, y\sqrt{1 + y'^2}, \tag{3.35}$$

where 1, 2 enumerate the two suspension points. It is assumed here that the chain has a homogeneous linear mass density $\rho = m/L$, with mass elements $dm = \rho \, ds$. The end points are fixed,

$$y(x_1) = y_1, \qquad y(x_2) = y_2. \tag{3.36}$$

The problem is constrained by the functional

$$C[y] = \int_1^2 ds = \int_{x_1}^{x_2} dx \, \sqrt{1 + y'^2} = L = const. \tag{3.37}$$

Constraints of this form, i.e., $C[y] = const$, are called *isoperimetric*. (The name is derived from the problem of maximizing an area for which the length of its perimeter is given.)

Another type of constraints are *holonomic* constraints. As an example, consider two points P_1 and P_2 in a surface in three-dimensional space parameterized by

$$g(x, y, z) = 0. \tag{3.38}$$

The shortest distance between these two points such that the connecting line is within the surface is called a *geodesic*. It is determined by minimizing

$$J = J[y, z] = \int_1^2 ds = \int_{x_1}^{x_2} dx \, \sqrt{1 + y'^2 + z'^2}, \tag{3.39}$$

where the curves $y = y(x)$ and $z = z(x)$ also must lie within the surface. The variation, therefore, must be taken subject to the holonomic condition (3.38).

Holonomic constraints restrict each point of a function, whereas the less stringent isoperimetric ones only constrain the entire function.

3.1.2.1 *Isoperimetric constraints*

In general, consider seeking the solutions $y(x)$ that make the functional

$$J = J[y] = \int_{x_1}^{x_2} dx \, F(y, y', x) \tag{3.40}$$

minimal under the isoperimetric constraint

$$C = C[y] = \int_{x_1}^{x_2} dx \, G(y, y', x) = const \equiv c, \tag{3.41}$$

with prescribed values

$$y(x_1) = y_1, \qquad y(x_2) = y_2 \tag{3.42}$$

that keep the end points fixed.

It is easy to see that for this problems the variations of the type $\delta y = \varepsilon\eta$ considered previously will not work: Take $y(x)$ as the solution of the problem, then

$$C[y + \varepsilon\eta] = c + O(\varepsilon), \tag{3.43}$$

will not satisfy the constraint unless $\varepsilon \equiv 0$. Clearly, only variations that *always* satisfy the constraint may be used.

This can be achieved by considering variations of the true solution y of the form

$$\delta y = \varepsilon_1\eta_1 + \varepsilon_2\eta_2, \tag{3.44}$$

with two infinitesimal parameters ε_1, ε_2 and two *linearly independent* functions $\eta_1(x)$ and $\eta_2(x)$ taken to vanish at the end points,

$$\eta_1(x_1) = \eta_1(x_2) = 0, \qquad \eta_2(x_1) = \eta_2(x_2) = 0. \tag{3.45}$$

This choice allows us to satisfy

$$C_\eta(\varepsilon_1, \varepsilon_2) \equiv C[y + \varepsilon_1\eta_1 + \varepsilon_2\eta_2] = c, \tag{3.46}$$

which makes $C_\eta(\varepsilon_1, \varepsilon_2)$ a curve in the $(\varepsilon_1, \varepsilon_2)$ parameter space that correlates ε_1 and ε_2 such that the constraint is always satisfied. Moreover, in view of the linear independence of η_1 and η_2, the point $\varepsilon_1 = \varepsilon_2 = 0$ that determines the true solution lies on this curve.

We thus have

$$J_\eta(\varepsilon_1, \varepsilon_2) \equiv J[y + \varepsilon_1\eta_1 + \varepsilon_2\eta_2], \tag{3.47}$$

where, however, ε_1 and ε_2 must satisfy the constraint equation (3.46). Without loss of generality, we can assume that ε_2 is determined in terms of ε_1, i.e., $\varepsilon_2 = \varepsilon_2(\varepsilon_1)$.[73] More precisely, we write

$$J_\eta(\varepsilon_1, \varepsilon_2) \equiv J[y + \varepsilon_1\eta_1 + \varepsilon_2(\varepsilon_1)\eta_2], \tag{3.48}$$

where now the isoperimetric constraint is satisfied for arbitrary functions η_1, η_2. By construction, at $\varepsilon_1 = 0$ we have also $\varepsilon_2 = 0$ (see footnote 73) and the function $J_\eta(\varepsilon_1)$ will be minimal, i.e.,

$$\frac{dJ_\eta(\varepsilon_1, \varepsilon_2)}{d\varepsilon_1} = \frac{\partial J_\eta(\varepsilon_1, \varepsilon_2)}{\partial \varepsilon_1} + \frac{\partial J_\eta(\varepsilon_1, \varepsilon_2)}{\partial \varepsilon_2}\frac{d\varepsilon_2}{d\varepsilon_1} = 0. \tag{3.49}$$

---

[73]In general, ε_2 may be a multi-valued function of ε_1. However, in practice, the relation $\varepsilon_2 = \varepsilon_2(\varepsilon_1)$ will only be needed in the vicinity of $\varepsilon_1 = 0$ where ε_2 is also close to zero. In this vicinity, we may therefore write $\varepsilon_2 = a\varepsilon_1$, where a is some constant, and thus this procedure is well-defined.

To use this equation in practical calculations would require the explicit knowledge of $\varepsilon_2(\varepsilon_1)$.

This complication can be circumvented by considering the fact that around $\varepsilon_1 = \varepsilon_2(\varepsilon_1) = 0$, the constraint may be written as[74]

$$C_\eta(\varepsilon_1, \varepsilon_2) = \varepsilon_2 + h(\varepsilon_1), \tag{3.50}$$

where h is defined as a function of ε_1 as

$$h(\varepsilon_1) = c - \varepsilon_2(\varepsilon_1) \tag{3.51}$$

such that $C_\eta(\varepsilon_1, \varepsilon_2) = c$ in the vicinity of $\varepsilon_1 = \varepsilon_2 = 0$. Hence, we have

$$\frac{\partial C_\eta(\varepsilon_1, \varepsilon_2)}{\partial \varepsilon_2} = 1, \qquad \text{and} \qquad \frac{\partial C_\eta(\varepsilon_1, \varepsilon_2)}{\partial \varepsilon_1} = \frac{dh}{d\varepsilon_1} = -\frac{d\varepsilon_2}{d\varepsilon_1}, \tag{3.52}$$

which makes the two equations

$$\frac{\partial J_\eta(\varepsilon_1, \varepsilon_2)}{\partial \varepsilon_1} - \lambda \frac{\partial C_\eta(\varepsilon_1, \varepsilon_2)}{\partial \varepsilon_1} = 0, \tag{3.53}$$

$$\frac{\partial J_\eta(\varepsilon_1, \varepsilon_2)}{\partial \varepsilon_2} - \lambda \frac{\partial C_\eta(\varepsilon_1, \varepsilon_2)}{\partial \varepsilon_2} = 0, \tag{3.54}$$

for $\varepsilon_1 = \varepsilon_2 = 0$, equivalent to the extremum condition (3.49), as can be seen easily by eliminating λ.[75] Therefore, if we now reinterpret the functions J_η and C_η as being functions of ε_1 alone, i.e.,

$$J_\eta\left(\varepsilon_1, \varepsilon_2(\varepsilon_1)\right) \longrightarrow J_\eta(\varepsilon_1), \qquad C_\eta\left(\varepsilon_1, \varepsilon_2(\varepsilon_1)\right) \longrightarrow C_\eta(\varepsilon_1), \tag{3.55}$$

Eq. (3.53) provides the stationarity condition

$$\frac{d}{d\varepsilon_1}\left[J_\eta(\varepsilon_1) - \lambda C_\eta(\varepsilon_1)\right]_{\varepsilon_1=0} = 0 \tag{3.56}$$

that minimizes the functional

$$\hat{J}[y] = J[y] - \lambda C[y] = \int_{x_1}^{x_2} dx\, \hat{F}(y, y', x), \tag{3.57}$$

---

[74]This expression certainly is not unique. It is chosen here such that the condition (3.56) obtains. If we had chosen to express ε_1 in terms of ε_2, the expression would be $C_\eta(\varepsilon_1, \varepsilon_2) = \varepsilon_1 + \bar{h}(\varepsilon_2)$, with a different function $\bar{h}(\varepsilon_2) = c - \varepsilon_1(\varepsilon_2)$. The stationarity condition (3.56) then would be written as a derivative with respect to ε_2. The final result (3.59) would *not* be affected by this alternative choice; see also following footnote 75.

[75]In other words, using (3.52), we obtain $\lambda = \partial J_\eta/\partial \varepsilon_2$ from (3.54) which produces (3.49) upon insertion into (3.53). Note here that the symmetry of Eqs. (3.53) and (3.54) in ε_1 and ε_2 shows that it is irrelevant whether one considers $\varepsilon_2 = \varepsilon_2(\varepsilon_1)$ or $\varepsilon_1 = \varepsilon_1(\varepsilon_2)$.

with

$$\hat{F} = F(y, y', x) - \lambda G(y, y', x). \tag{3.58}$$

Hence, one may solve the variational problem for this functional *as if no constraint existed.* The differential equation to be solved is then the usual Euler–Lagrange equation for the unconstrained problem,

$$\boxed{\frac{d}{dx}\frac{\partial \hat{F}}{\partial y'} - \frac{\partial \hat{F}}{\partial y} = 0}. \tag{3.59}$$

This will provide a solution $y = y(x, c_1, c_2, \lambda)$ that depends on two integration constants, c_1 and c_2, and the *Lagrange multiplier* λ. These parameters are determined by the boundary conditions (3.42) and the constraint (3.41).

Generalization: If there are several constraints,

$$C_k[y] = \int_{x_1}^{x_2} dx \, G_k(y, y', x) = const \equiv c_k, \qquad k = 1, \ldots, l, \tag{3.60}$$

the Euler equation is written in terms of

$$\hat{F} = F(y, y', x) - \sum_{k=1}^{l} \lambda_k G_k(y, y', x). \tag{3.61}$$

The two integration constants and the l Lagrange multipliers λ_k are completely determined by two boundary conditions and the l constraint equations.

Chain suspended in the field of gravity

This problem was explained on p. 148. According to Eqs. (3.35) and (3.37), the Euler equation needs to be solved for

$$\hat{F}(y, y') = \sqrt{1 + y'^2}\,(y - \lambda). \tag{3.62}$$

Since $\hat{F}$ does not depend explicitly on x, we have

$$\hat{F} - y'\frac{\partial \hat{F}}{\partial y'} = \frac{y - \lambda}{\sqrt{1 + y'^2}} = const \equiv a. \tag{3.63}$$

Solving for y' and integrating provides

$$y(x) = \lambda + a \cosh\left(\frac{x}{a} + b\right), \tag{3.64}$$

with parameters a, b, and λ determined by

$$y(x_1) = y_1, \quad y(x_2) = y_2, \quad \text{and} \quad \int_{x_1}^{x_2} dx\ \sqrt{1 + y'^2} = L. \tag{3.65}$$

The function (3.64) is called a *catenary*, derived from the Latin word for chain.

To be specific, let us consider an example with symmetric suspension points,

$$(x_1, y_1) = (-\ell, 0) \quad \text{and} \quad (x_2, y_2) = (\ell, 0). \tag{3.66}$$

One then finds

$$b = 0 \quad \text{and} \quad \lambda = -a \cosh \frac{\ell}{a}. \tag{3.67}$$

The remaining parameter a in

$$y(x) = a \cosh \frac{x}{a} - a \cosh \frac{\ell}{a} \tag{3.68}$$

is determined by the length, i.e.,

$$L = \int_{-\ell}^{+\ell} dx\ \sqrt{1 + y'^2} = 2a \sinh \frac{\ell}{a}. \tag{3.69}$$

For $L = 2\ell$, the chain is pulled taut between the two suspension points, and $a = 0$ [$\lim_{a \to 0} a \sinh(\ell/a) = 1$]. For $L > 2\ell$, there are two solutions, $a = a_0 > 0$ and $a = -a_0 < 0$. For the suspended chain, one has $a = a_0$.

For $a = -a_0$, the potential energy has a maximum (rather than a minimum). This corresponds to the optimal solution for the construction of an arch made out of incompressible material (like concrete).[76] In an arch of this shape all the forces act along the arch; there are no forces perpendicular to the arch. In other words, the entire weight of the arch is compensated by pressure forces within the arch. This can be seen simply by the fact that for the chain in equilibrium, the weight of each chain mass element is compensated by tangential forces at both ends of each element. There are no lateral forces and all structural forces are transmitted to the base of the inverted catenary.

---

[76]The *Gateway Arch* in St. Louis, Missouri, designed by the Finnish architect Eero Saarinen, is constructed as an inverted *weighted* catenary corresponding to a chain with links becoming progressively lighter towards the center.

3.1.2.2   *Holonomic constraints*

The general problem of the kind describing geodesics is given by seeking the solutions $y_1(x)$ and $y_2(x)$ that make the functional

$$J[y_1, y_2] = \int_{x_1}^{x_2} dx\, F(y_1, y_2, y_1', y_2', x) \tag{3.70}$$

an extremum. Boundary values are taken as fixed. The solution is to satisfy the holonomic constraint

$$g(y_1, y_2, x) = 0. \tag{3.71}$$

Geometrically $\big(x, y_1(x), y_2(x)\big)$ is a curve in the three-dimensional space given by (x, y_1, y_2). The constraint restricts this curve to lie within the surface defined by the constraint equation.

The solution strategy involves turning this problem into a sum of piecewise isoperimetric problems. To this end, the entire interval $[x_1, x_2]$ is divided into p intervals of length $d = (x_2 - x_1)/p$ centered around

$$\xi_i = x_1 - \frac{d}{2} + i\, d, \qquad i = 1, \ldots, p. \tag{3.72}$$

Defining now functions

$$f_i(x) = \begin{cases} 1 & \text{for} \quad \xi_i - d/2 < x < \xi_i + d/2 \ , \\ 0 & \text{else} \ , \end{cases} \tag{3.73}$$

the set of p isoperimetric constraints

$$C_i[y_1, y_2] = \int_{x_1}^{x_2} dx\, g(y_1, y_2, x) f_i(x) = 0, \qquad i = 1, \ldots, p, \tag{3.74}$$

follows directly from (3.71). Also, given the isoperimetric constraints, we recover the holonomic constraint when summing over all i and letting $p \to \infty$, i.e.,

$$\lim_{p \to \infty} \sum_{i=1}^{p} \frac{C_i[y_1, y_2]}{d} = \lim_{p \to \infty} \sum_{i=1}^{p} g(y_1, y_2, \xi_i)\, f_i(x) = g(y_1, y_2, x). \tag{3.75}$$

In other words, in this limit, the isoperimetric constraints (3.74) are equivalent to the holonomic constraint (3.71).

By construction, therefore, solving the isoperimetric variational problem defined by

$$\hat{J}[y_1, y_2] = \int_{x_1}^{x_2} dx\, \hat{F}(y_1, y_2, y_1', y_2', x), \tag{3.76}$$

where

$$\hat{F} = F(y_1, y_2, y_1', y_2', x) - \sum_{i=1}^{p} \lambda_i \, f_i(x) \, g(y_1, y_2, x), \qquad (3.77)$$

provides a solution of the original problem if we let $p \to \infty$. This means, in particular, that the set of p Lagrange multipliers now becomes one continuous function,

$$\lim_{p \to \infty} \sum_{i=1}^{p} \lambda_i \, f_i(x) = \lambda(x). \qquad (3.78)$$

The variational problem then is given in terms of the function

$$\hat{F} = F(y_1, y_2, y_1', y_2', x) - \lambda(x) \, g(y_1, y_2, x), \qquad (3.79)$$

with the Euler–Lagrange equations

$$\boxed{\frac{\partial F}{\partial y_i} - \frac{d}{dx} \frac{\partial F}{\partial y_i'} = \lambda(x) \frac{\partial g}{\partial y_i}, \qquad i = 1, 2} \qquad (3.80)$$

These two differential equations and the constraint equation $g = 0$ determine the three functions $y_1(x)$, $y_2(x)$, and $\lambda(x)$, with the integration constants given by the boundary conditions.

Generalization: For N functions $y_1(x), \ldots, y_N(x)$ and $l < N$ holonomic constraints $g_k(y_1, \ldots, y_N, x) = 0$, there are N Euler–Lagrange equations for

$$\hat{F} = F(y_1, \ldots, y_N, y_1', \ldots, y_N', x) - \sum_{k=1}^{l} \lambda_k(x) \, g_k(y_1, \ldots, y_N, x). \qquad (3.81)$$

Together with the l constraints, the $N + l$ functions $y_i(x)$ and $\lambda_k(x)$ are thus completely determined.

Geodesics

To obtain the geodesics on the surface

$$g(x, y, z) = 0, \qquad (3.82)$$

we write the coordinates in parametric form as a function of t: $x(t)$, $y(t)$, and $z(t)$. For a physical trajectory, the parameter t may be taken as the time, but in general it can be any convenient parametrization. This treatment here is thus more general than what was discussed on p. 149, where x was distinguished from the other coordinates.

Denoting derivatives with respect to t by a dot, the geodesics are then determined by the minimal values of the functional

$$J[x, y, z] = \int_{t_1}^{t_2} dt \, \sqrt{\dot{x}^2 + \dot{y}^2 + \dot{z}^2}. \tag{3.83}$$

The corresponding Euler–Lagrange equations are to be solved for

$$\hat{F} = \sqrt{\dot{x}^2 + \dot{y}^2 + \dot{z}^2} - \lambda(t) \, g(x, y, z). \tag{3.84}$$

The three equations read

$$\frac{d}{dt}\frac{\dot{x}}{v} = -\lambda\frac{\partial g}{\partial x}, \qquad \frac{d}{dt}\frac{\dot{y}}{v} = -\lambda\frac{\partial g}{\partial y}, \qquad \frac{d}{dt}\frac{\dot{z}}{v} = -\lambda\frac{\partial g}{\partial z}, \tag{3.85}$$

where $v = \sqrt{\dot{x}^2 + \dot{y}^2 + \dot{z}^2}$ was used as an abbreviation. In many applications, it will be more convenient to introduce a new parameter s via

$$ds = \sqrt{dx^2 + dy^2 + dz^2} = v \, dt, \tag{3.86}$$

where $s = s(t)$ is called the *path length*. One then finds

$$\frac{d}{dt} = v\frac{d}{ds}, \qquad \dot{x} = \frac{dx(s)}{ds}\frac{ds}{dt} = v \, x'(s), \qquad \frac{d}{dt}\frac{\dot{x}}{v} = v \, x'', \tag{3.87}$$

where the prime now denotes the derivative with respect to s. The corresponding Euler–Lagrange equations are

$$x'' = \mu\frac{\partial g}{\partial x}, \qquad y'' = \mu\frac{\partial g}{\partial y}, \qquad z'' = \mu\frac{\partial g}{\partial z}, \tag{3.88}$$

where the unknown function $\lambda(t)$ was replaced by the, equally unknown, function $\mu(s) = -\lambda/v$.

The preceding equations are valid for arbitrary surfaces. For further discussion, we consider the surface of a sphere of radius R, i.e.,

$$g(x, y, z) = x^2 + y^2 + z^2 - R^2 = 0. \tag{3.89}$$

The Euler–Lagrange equations now are

$$x'' = 2\mu x, \qquad y'' = 2\mu y, \qquad z'' = 2\mu z. \tag{3.90}$$

Differentiating the constraint $g = 0$ twice produces

$$0 = x'^2 + y'^2 + z'^2 + xx'' + yy'' + zz'' = 1 + xx'' + yy'' + zz'', \tag{3.91}$$

where $dx^2 + dy^2 + dz^2 = ds^2$ was used. Employing here the differential equations and the constraint, then yields

$$\mu = -\frac{1}{2R^2} = const, \tag{3.92}$$

which in turn provides

$$x'' = -\frac{x}{R^2}, \qquad y'' = -\frac{y}{R^2}, \qquad z'' = -\frac{z}{R^2}. \tag{3.93}$$

Each solution of these equations has the form $A\sin(s/R) + B\cos(s/R)$, with constants given via the boundary conditions and the constraints. The resulting lines are circles with the circumference $2\pi R$ on the surface of the sphere called *great circles*. The particular solution describing the equator is

$$x(s) = R\cos\frac{s}{R}, \tag{3.94a}$$

$$y(s) = R\sin\frac{s}{R}, \tag{3.94b}$$

$$z(s) = 0, \tag{3.94c}$$

with ratio s/R being the (east) longitude angle.

3.2 Hamilton's Principle

In the preceding section, the equations that determine the stationary solutions of the functionals considered in variational calculus were found to have exactly the same structure as the Lagrange equations that were previously derived from Newton's laws. This equivalence suggests that one may use the variational approach as an equivalent axiomatic foundation for mechanics.

This alternative axiomatic approach is given by *Hamilton's Principle*: *The trajectories $q_i(t)$ of a physical system between times t_1 and t_2 will be such that the action functional*

$$S = S[q] = \int\limits_{t_1}^{t_2} dt\, L\big(q(t), \dot{q}(t), t\big), \tag{3.95}$$

where L is the Lagrange function, will have an extremum, i.e., its variation $\delta S[q]$ will vanish for physically realizable trajectories,

$$\delta S[q] = 0. \tag{3.96}$$

This variation is to be taken for fixed end points,

$$q_i(t_1) = q_{i1}, \qquad q_i(t_2) = q_{i2}, \tag{3.97}$$

for all variables.

This principle is sometimes also called the *principle of stationary action*, or the *principle of least action*. The latter name alludes to the fact that in many practical applications the extremal value of S will be a minimum. This principle was put forward by the Irish mathematician and physicist W. R. Hamilton (1805–1865). The formulation given here is valid for all systems with holonomic and/or differential constraints whose forces can be derived from a potential according to Eqs. (2.100) or (2.145).

The equivalence for differential constraints will be shown in Section 3.2.3 below. For the time being, we will assume in the following that all l constraints are holonomic and that the Lagrange function L depends only on the $N = 3n - l$ independent coordinates, their velocities and the time, i.e.,

$$L(q, \dot{q}, t) = T(q, \dot{q}, t) - V(q, t), \tag{3.98}$$

where q and $\dot{q}$ stand for the list of all N coordinates and velocities. It follows from variational calculus that the corresponding equations of motion are the Euler–Lagrange equations of the second kind,

$$\frac{\partial L}{\partial q_i} - \frac{d}{dt}\frac{\partial L}{\partial \dot{q}_i} = 0, \qquad i = 1, \ldots, N. \tag{3.99}$$

Hamilton's principle is an *integral principle* that is completely equivalent to the *differential principle* on which the previous derivation of the Lagrange equations was based: the N Lagrange equations are valid if and only if Hamilton's principle is satisfied.

One peculiar feature of Hamilton's principle is that the trajectories are not determined in terms of the $2N$ initial values of $q_i(t_1)$ and $\dot{q}_i(t_1)$, as in ordinary Newtonian mechanics, but in terms of the N initial coordinates $q_i(t_1)$ and the N *final* coordinates $q_i(t_2)$ [cf. (3.97)]. This means, in particular, that in the solution the values of the initial velocities depend on the values of the final coordinates. While this may seem as if events in the future determine events in the past, and therefore raise questions about the causality of events, this is not a violation of causality, but merely a reflection of the essentially deterministic nature of this aspect of classical mechanics.

What is the advantage of using Hamilton's Principle?

In practical terms, there is very little difference to the previous formulation since any application to a concrete problem will lead to the Lagrange equations. The major advantage lies on a more formal level: The condition

$$\boxed{\delta \int_{t_1}^{t_2} dt\, L(q, \dot{q}, t) = 0} \tag{3.100}$$

provides a very compact formulation of mechanics in terms of one single equation, as compared to the N Lagrange equations. Moreover, it does not require any particular set of coordinates as long as the chosen set reflects the degrees of freedom of the system.

As we have seen already in many examples, the Lagrange function in general is a particularly simple function of the variables that often is the simplest function that can be found that is compatible with the symmetries of the problem. This can be illustrated by considering the Lagrange function of a free particle described by the position vector $\mathbf{r}(t)$. Restricting the discussion to first-order derivatives, one has

$$L = L(\mathbf{r}, \dot{\mathbf{r}}, t). \tag{3.101}$$

Translational invariance in time and space then implies that L cannot depend either on t or on $\mathbf{r}$. Rotational invariance of space means that L can only depend on a scalar, which leaves only a dependence on $\dot{\mathbf{r}}^2$. The simplest possible ansatz for L then is a monomial in $\dot{\mathbf{r}}^2$,

$$L = A\dot{\mathbf{r}}^2, \tag{3.102}$$

where A is a constant. This ansatz indeed provides the correct equation of motion. The constant A is without consequence for the trajectory; usually it is taken as $m/2$, thus making L equal to the kinetic energy. Note, however, that the present result is only based on symmetry considerations, without any need for introducing the notion of a kinetic energy.

This ease with which symmetries can be introduced into the Lagrange functions is of particularly great importance for many modern developments of physics that go beyond classical mechanics. It is in these applications, in particular, that Hamilton's principle becomes an indispensable tool.

3.2.1 Gauge transformations

The Lagrange function is not unique, i.e., different functions may lead to the same equations of motion. For example, multiplying a given L by a constant, or adding a constant to it, obviously does not change the equations of motion.

A particularly important class of equivalent Lagrange functions is obtained by so-called *gauge transformations*,

$$L(q, \dot{q}, t) \longrightarrow L^*(q, \dot{q}, t) = L(q, \dot{q}, t) + \frac{df(q, t)}{dt}, \tag{3.103}$$

where a total time derivative of an arbitrary scalar function $f(q,t)$ is added to L.[77] The corresponding action functional is

$$S^* = \int\limits_{t_1}^{t_2} dt\, L^* = \int\limits_{t_1}^{t_2} dt\, L + f\big(q(t_2),t_2\big) - f\big(q(t_1),t_1\big). \qquad (3.104)$$

Its first-order variation is

$$\delta S^* = \delta S + \sum_{i=1}^{N} \left[\frac{\partial f}{\partial q_i}\bigg|_{t_2} \delta q_i(t_2) - \frac{\partial f}{\partial q_i}\bigg|_{t_1} \delta q_i(t_1) \right]. \qquad (3.105)$$

Since the end points are kept fixed, one has $\delta q_i(t_2) = \delta q_i(t_1) = 0$, and hence $\delta S^* = \delta S$. The stationarity conditions $\delta S^* = 0$ and $\delta S = 0$, therefore, are identical and thus the resulting equations of motion will be identical.[78]

A change of the Lagrange function under a Galilei transformation is also a gauge transformation, i.e.,

$$L = \frac{m}{2}\dot{\mathbf{r}}^2 \quad \xrightarrow{\ \dot{\mathbf{r}} \to \dot{\mathbf{r}} + \mathbf{v}\ } \quad L^* = \frac{m}{2}\dot{\mathbf{r}}^2 + m\mathbf{v}\cdot\dot{\mathbf{r}} + \frac{m}{2}\mathbf{v}^2$$

$$= L + \frac{d}{dt}\left[m\mathbf{v}\cdot\mathbf{r} + \frac{m}{2}\mathbf{v}^2 t \right], \qquad (3.106)$$

where the additional term generated by the transformation is written as a total time derivative ($\mathbf{v}$ is a constant here).

Gauge transformations derive their name from the gauge transformation of electrodynamics. The Lagrange function of a charged particle in the electromagnetic field was given in Eq. (2.153), i.e.,

$$L(\mathbf{r},\dot{\mathbf{r}},t) = \frac{m}{2}\dot{\mathbf{r}}^2 - e\Phi(\mathbf{r},t) + \frac{e}{c}\mathbf{A}(\mathbf{r},t)\cdot\dot{\mathbf{r}}. \qquad (3.107)$$

The physical $\mathbf{E}$ and $\mathbf{B}$ fields,

$$\mathbf{E} = -\boldsymbol{\nabla}\Phi - \frac{1}{c}\frac{\partial\mathbf{A}}{\partial t}, \qquad \mathbf{B} = \boldsymbol{\nabla}\times\mathbf{A}, \qquad (3.108)$$

are invariant under the following *gauge transformation of the potentials,*

$$\mathbf{A} \longrightarrow \mathbf{A} + \boldsymbol{\nabla}\Lambda(\mathbf{r},t), \qquad \Phi \longrightarrow \Phi - \frac{1}{c}\frac{\partial\Lambda(\mathbf{r},t)}{\partial t}, \qquad (3.109)$$

where $\Lambda(\mathbf{r},t)$ is an arbitrary scalar function. The Lagrange function then changes according to

$$L \longrightarrow L^* = L + \frac{e}{c}\frac{d\Lambda(\mathbf{r},t)}{dt} \qquad (3.110)$$

in the manner considered above.

---

[77]The scalar function f may not depend on $\dot{q}$.

[78]Of course, this also follows directly from the Euler-Lagrange equations themselves.

3.2.2 *Equivalence of Hamilton and d'Alembert*

The d'Alembert principle (2.32) postulates that the virtual work of the constraining forces $\mathbf{Z}_i$ vanishes. To show its equivalence to Hamilton's principle, let us integrate Eq. (2.31) over time,

$$-\int_{t_1}^{t_2} \left(\sum_{i=1}^{N} \mathbf{Z}_i \cdot \delta \mathbf{r}_i \right) dt = \int_{t_1}^{t_2} \left(\sum_{i=1}^{n} (\mathbf{F}_i - m_i \ddot{\mathbf{r}}_i) \cdot \delta \mathbf{r}_i \right) dt, \qquad (3.111)$$

where, as usual, we assume that the virtual variations $\delta \mathbf{r}_i$ vanish at the end points. For conservative forces, we then find

$$\begin{aligned}
\int_{t_1}^{t_2} \left(\sum_{i=1}^{n} \mathbf{F}_i \cdot \delta \mathbf{r}_i \right) dt &= -\int_{t_1}^{t_2} \left(\sum_{i=1}^{n} \boldsymbol{\nabla}_{\mathbf{r}_i} V \cdot \delta \mathbf{r}_i \right) dt \\
&= -\int_{t_1}^{t_2} \delta V \, dt \\
&= -\delta \int_{t_1}^{t_2} V \, dt, \qquad (3.112)
\end{aligned}$$

where V is the potential.[79] Moreover, integration by parts provides

$$\begin{aligned}
\int_{t_1}^{t_2} \left(\sum_{i=1}^{n} m_i \ddot{\mathbf{r}}_i \cdot \delta \mathbf{r}_i \right) dt &= \int_{t_1}^{t_2} \left(\sum_{i=1}^{n} m_i \frac{d\dot{\mathbf{r}}_i}{dt} \cdot \delta \mathbf{r}_i \right) dt \\
&= \underbrace{\left[\sum_{i=1}^{n} m_i \dot{\mathbf{r}} \cdot \delta \mathbf{r}_i \right]_{t_1}^{t_2}}_{=0} - \int_{t_1}^{t_2} \left(\sum_{i=1}^{n} m_i \dot{\mathbf{r}}_i \cdot \frac{d\delta \mathbf{r}_i}{dt} \right) dt \\
&= -\int_{t_1}^{t_2} \left(\sum_{i=1}^{n} m_i \dot{\mathbf{r}}_i \cdot \delta \dot{\mathbf{r}}_i \right) dt \\
&= -\int_{t_1}^{t_2} \left(\sum_{i=1}^{n} \frac{m_i}{2} \delta \dot{\mathbf{r}}_i^2 \right) dt \\
&= -\delta \int_{t_1}^{t_2} \left(\sum_{i=1}^{n} \frac{m_i}{2} \dot{\mathbf{r}}_i^2 \right) dt
\end{aligned}$$

---

[79]For the validity of the last equality here, see footnote 71.

$$= -\delta \int_{t_1}^{t_2} T \, dt, \tag{3.113}$$

where T is the total kinetic energy. Hence,

$$\int_{t_1}^{t_2} \sum_{i=1}^{n} (\mathbf{F}_i - m_i \ddot{\mathbf{r}}_i) \cdot \delta \mathbf{r}_i \, dt = \delta \int_{t_1}^{t_2} (T - V) dt, \tag{3.114}$$

which shows that the time integral of the (negative) virtual work of the constraining forces $\mathbf{Z}_i$ is directly equal to the variation of the action integral $S = \int L \, dt = \int (T - V) dt$, thus establishing the equivalence of Hamilton's principle, $\delta S = 0$, with d'Alembert's principle.

3.2.3 *Equivalence to Lagrange equations with differential constraints*

To show the equivalence of Hamilton's principle to the Lagrange equations obtained for nonholonomic differential constraints, consider L as a function of all $3n$ coordinates. They may be Cartesian coordinates, or any other complete set of coordinates. The variation of the action then yields in the usual manner

$$\delta \int_{t_1}^{t_2} dt \, L = \int_{t_1}^{t_2} dt \sum_{j=1}^{3n} \left[\frac{\partial L}{\partial q_j} - \frac{d}{dt} \frac{\partial L}{\partial \dot{q}_j} \right] \delta q_j = 0. \tag{3.115}$$

The $3n$ virtual displacements δq_j are subject to l constraints and therefore not independent of each other. The relation between the δq_j is given in Eq. (2.17),

$$\sum_{j=1}^{3n} h_{kj} \delta q_j = 0, \tag{3.116}$$

rewritten here explicitly in terms of all $3n$ components. This can be used to introduce Lagrange multipliers according to

$$\int_{t_1}^{t_2} dt \sum_{k=1}^{l} \sum_{j=1}^{3n} \lambda_k h_{kj} \delta q_j = 0. \tag{3.117}$$

The validity of this expression for arbitrary λ's follows directly from (3.116). Adding this to the previous integral then yields

$$\delta \int_{t_1}^{t_2} dt \, L = \int_{t_1}^{t_2} dt \sum_{j=1}^{3n} \left[\frac{\partial L}{\partial q_j} - \frac{d}{dt} \frac{\partial L}{\partial \dot{q}_j} + \sum_{k=1}^{l} \lambda_k h_{kj} \right] \delta q_j = 0. \tag{3.118}$$

The l Lagrange multipliers are now chosen such that l bracketed terms in the sum under the integral vanish individually. This makes the remaining $3n - l$ virtual displacements δq_j independent and hence the remaining bracketed terms must also vanish individually. Hence, *all* terms in the brackets vanish individually, i.e.,

$$\frac{d}{dt}\frac{\partial L}{\partial \dot{q}_j} - \frac{\partial L}{\partial q_j} = \sum_{k=1}^{l} \lambda_k h_{kj}, \qquad j = 1, \ldots, 3n. \tag{3.119}$$

We have thus established the equivalence of Hamilton's principle to the Lagrange equations derived earlier for nonholonomic differential constraints.

3.3 Noether's Theorem

We have seen that symmetries of space-time always lead to conserved quantities: Invariance under time translation leads to energy conservation; invariance under space translation implies (linear) momentum conservation, and rotational invariance provides angular momentum conservation. In this section, it will be shown that any one-parameter transformation that leaves the action functional invariant will lead to a constant of motion. This important result was found by the mathematician Emmy Noether (1882–1935).

Consider infinitesimal transformations of the coordinates $q_i(t)$ and the time t that depend on *one* parameter ε according to

$$q_i \quad \longrightarrow \quad q_i^* = q_i + \varepsilon \psi_i(q, \dot{q}, t), \tag{3.120a}$$

$$t \quad \longrightarrow \quad t^* = t + \varepsilon \theta(q, \dot{q}, t), \tag{3.120b}$$

where ψ_i and θ are some functions that specify the details of the transformation. For example, for time translation, $t \to t^* = t + \varepsilon$, one would have $\theta = 1$ and $\psi_i = 0$. The parameter ε is to be infinitesimal, i.e., in the following, terms that depend on ε^2, or higher powers, can always be neglected and will not be written down.

We may now compare the action functional $S[q(t)]$ with end points t_1 and t_2 with the transformed action functional $S^*[q^*(t^*)]$ with transformed end points t_1^* and t_2^*. If $S[q(t)] = S^*[q^*(t^*)]$, i.e.,

$$\int_{t_1}^{t_2} dt\, L\left(q(t), \frac{dq(t)}{dt}, t\right) = \int_{t_1^*}^{t_2^*} dt^*\, L\left(q^*(t^*), \frac{dq^*(t^*)}{dt^*}, t^*\right), \tag{3.121}$$

the functional is said to be *invariant* under this particular transformation. Expanding the right-hand side around $\varepsilon = 0$,

$$
\begin{aligned}
S^* &= \int_{t_1^*}^{t_2^*} dt^* \, L\left(q^*(t^*), \frac{dq^*(t^*)}{dt^*}, t^*\right) \\[2mm]
&= \int_{t_1}^{t_2} dt \, L\left(q^*(t^*), \frac{dq^*(t^*)}{dt^*}, t^*\right) \frac{dt^*}{dt} \\[2mm]
&= \int_{t_1}^{t_2} dt \left\{ L\left(q(t), \frac{dq(t)}{dt}, t\right) + \varepsilon \frac{d}{d\varepsilon} \left[L\left(q^*(t^*), \frac{dq^*(t^*)}{dt^*}, t^*\right) \frac{dt^*}{dt} \right]_{\varepsilon=0} \right\} \\[2mm]
&= S + \varepsilon \int_{t_1}^{t_2} dt \, \frac{d}{d\varepsilon} \left[L\left(q^*(t^*), \frac{dq^*(t^*)}{dt^*}, t^*\right) \frac{dt^*}{dt} \right]_{\varepsilon=0},
\end{aligned}
\tag{3.122}
$$

this shows that

$$
\frac{d}{d\varepsilon} \left[L\left(q^*(t^*), \frac{dq^*(t^*)}{dt^*}, t^*\right) \frac{dt^*}{dt} \right]_{\varepsilon=0} = 0
\tag{3.123}
$$

is the necessary and sufficient condition for the invariance of S.

To evaluate this further, note that

$$
\frac{dt^*}{dt} = 1 + \varepsilon \frac{d\theta}{dt},
\tag{3.124}
$$

and

$$
\begin{aligned}
\frac{dq_i^*}{dt^*} = \frac{dq_i^*}{dt} \frac{dt}{dt^*} &= \left(\dot{q}_i + \varepsilon \frac{d\psi_i}{dt} \right) \left(1 - \varepsilon \frac{d\theta}{dt} \right) \\[2mm]
&= \dot{q}_i + \varepsilon \frac{d\psi_i}{dt} - \varepsilon \dot{q}_i \frac{d\theta}{dt},
\end{aligned}
\tag{3.125}
$$

where all higher powers of ε were dropped. Inserting these expressions into the invariance condition then yields

$$
\begin{aligned}
0 &= \frac{d}{d\varepsilon} \left[L\left(q_i + \varepsilon \psi_i, \dot{q}_i + \varepsilon \frac{d\psi_i}{dt} - \varepsilon \dot{q}_i \frac{d\theta}{dt}, t + \varepsilon \theta\right) \left(1 + \varepsilon \frac{d\theta}{dt} \right) \right]_{\varepsilon=0} \\[2mm]
&= \sum_{i=1}^{N} \left(\frac{\partial L}{\partial q_i} \psi_i + \frac{\partial L}{\partial \dot{q}_i} \left[\frac{d\psi_i}{dt} - \dot{q}_i \frac{d\theta}{dt} \right] \right) + \frac{\partial L}{\partial t} \theta + L \frac{d\theta}{dt}.
\end{aligned}
\tag{3.126}
$$

Since this is taken at $\varepsilon = 0$, all quantities appearing here are written in terms of untransformed variables. We assume now that the variation of $S[q]$ vanishes, in other words, that the untransformed $q_i(t)$ correspond to the actual physical trajectories of the system. Then the Euler–Lagrange equations

$$\frac{d}{dt}\frac{\partial L}{\partial \dot{q}_i} = \frac{\partial L}{\partial q_i} \tag{3.127}$$

are valid, and in addition we have [cf. (2.226)]

$$\frac{d}{dt}\left(L - \sum_{i=1}^{N}\frac{\partial L}{\partial \dot{q}_i}\dot{q}_i\right) = \frac{\partial L}{\partial t}. \tag{3.128}$$

The previous equation can then immediately be recast in the *invariance condition*

$$\boxed{\frac{d}{dt}\left[\sum_{i=1}^{N}\frac{\partial L}{\partial \dot{q}_i}\psi_i + \left(L - \sum_{i=1}^{N}\frac{\partial L}{\partial \dot{q}_i}\dot{q}_i\right)\theta\right] = 0}, \tag{3.129}$$

providing

$$\boxed{C = C(q,\dot{q},t) \equiv \sum_{i=1}^{N}\frac{\partial L}{\partial \dot{q}_i}\psi_i + \left(L - \sum_{i=1}^{N}\frac{\partial L}{\partial \dot{q}_i}\dot{q}_i\right)\theta = const} \tag{3.130}$$

as the conserved quantity for the physical trajectories. This is a first-order differential equation and thus a first integral of the (second-order) equations of motion.

This is the content of *Noether's Theorem*: *If a functional S is invariant under a one-parameter transformation (3.120), there exists a conserved quantity given by (3.130).* The Noether theorem thus formalizes the relation between a symmetry of the system and the conserved quantity that corresponds to this symmetry.

A simple example is a Lagrange function that does not depend on time, i.e., $L = L(q,\dot{q})$. Considering now the transformation

$$\psi_i = 0: \qquad q_i^* = q_i, \tag{3.131a}$$

$$\theta = 1: \qquad t^* = t + \varepsilon, \tag{3.131b}$$

leading to

$$q_i^*(t^*) = q_i(t), \qquad \frac{dt^*}{dt} = 1, \qquad \frac{dq_i^*}{dt^*} = \frac{dq_i}{dt}, \tag{3.132}$$

the evaluation of the invariance condition yields

$$\frac{d}{d\varepsilon}\left[L\left(q_i^*(t^*),\frac{dq_i^*(t^*)}{dt^*}\right)\frac{dt^*}{dt}\right]_{\varepsilon=0} = \frac{d}{d\varepsilon}L(q_i,\dot{q}_i) = 0. \tag{3.133}$$

The invariance condition, thus, is satisfied and it immediately follows that

$$C = L - \sum_{i=1}^{N} \frac{\partial L}{\partial \dot{q}_i} \dot{q}_i = const \tag{3.134}$$

is a constant of motion. For a conservative system, $-C = E$ is the energy of the system [cf. Eq. (2.227)].

Another example is given by a Lagrange function that does not depend on a particular coordinate q_k, i.e., $\partial L/\partial q_k = 0$. Consider now the transformation

$$\psi_i = \delta_{ik}: \qquad q_i^* = q_i + \varepsilon \delta_{ik}, \tag{3.135a}$$

$$\theta = 0: \qquad t^* = t. \tag{3.135b}$$

This will only affect the k-component of the velocities; however, since

$$\frac{dq_k^*}{dt^*} = \frac{dq_k}{dt}, \tag{3.136}$$

this has no consequence for the Lagrange function and hence the invariance condition $dL/d\varepsilon = 0$ is trivially satisfied. The corresponding conserved quantity,

$$C = \frac{\partial L}{\partial \dot{q}_k} = p_k = const, \tag{3.137}$$

is the generalized momentum (see p. 118, and also the subsequent Sec. 4.1) associated with the coordinate q_k.

Both of these examples are rather trivial. Moreover, the corresponding conserved quantities can also be easily obtained from other considerations. However, the great advantage of Noether's theorem is that it is also applicable in less obvious situations, when it is not immediately clear which quantities would be conserved.

3.3.1   *Generalized Noether Theorem*

The previous investigation presumed the validity of the invariance condition $S^* = S$. The equations of motion, however, are equivalent to $\delta S = 0$. It is, therefore, sufficient to consider the weaker requirement $\delta S^* = \delta S$ instead:

$$\delta \int_{t_1}^{t_2} dt\, L\left(q(t), \frac{dq(t)}{dt}, t \right) = \delta \int_{t_1^*}^{t_2^*} dt^*\, L\left(q^*(t^*), \frac{dq^*(t^*)}{dt^*}, t^* \right). \tag{3.138}$$

The discussion of gauge transformations in Sec. 3.2.1 showed that one can always add the total time derivative of a scalar function to L without changing δS. If in Eq. (3.122), we have

$$
\begin{aligned}
S^* &= \int_{t_1^*}^{t_2^*} dt^* \, L\left(q^*(t^*), \frac{dq^*(t^*)}{dt^*}, t^* \right) \\
&= \int_{t_1}^{t_2} dt \left\{ L\left(q(t), \frac{dq(t)}{dt}, t \right) + \varepsilon \underbrace{\frac{d}{d\varepsilon}\left[L\left(q^*(t^*), \frac{dq^*(t^*)}{dt^*}, t^* \right) \frac{dt^*}{dt} \right]_{\varepsilon=0}}_{=\, \varepsilon \, dh/dt} \right\},
\end{aligned}
$$

$$\tag{3.139}$$

this corresponds to adding a total time derivative $\varepsilon\, dh/dt$ to the original L. One then has $\delta S^* = \delta S$, and the equations of motion remain unchanged. It then follows immediately that the *modified invariance condition* now becomes

$$
\boxed{\; \frac{d}{d\varepsilon}\left[L\left(q^*(t^*), \frac{dq^*(t^*)}{dt^*}, t^* \right) \frac{dt^*}{dt} \right]_{\varepsilon=0} = \frac{d}{dt} h(q,t) \;} \tag{3.140}
$$

leading to a conserved quantity

$$
\boxed{\; C = C(q, \dot{q}, t) \equiv \sum_{i=1}^{N} \frac{\partial L}{\partial \dot{q}_i} \psi_i + \left(L - \sum_{i=1}^{N} \frac{\partial L}{\partial \dot{q}_i} \dot{q}_i \right) \theta - h(q,t) = const \;}
$$

$$\tag{3.141}$$

for the physical trajectories.

As the derivation of this generalization shows, it is closely related to gauge transformations discussed in Sec. 3.2.1.

Galilei invariance

It was shown in Eq. (3.106) that a Galilei transformation is a gauge transformation. As an example of the associated conserved quantity that follows from the generalized Noether Theorem, let us therefore consider a particular Galilei transformation with an infinitesimal relative velocity $\boldsymbol{\varepsilon}$,

$$
\mathbf{r}_i^* = \mathbf{r}_i + \boldsymbol{\varepsilon} t, \qquad t^* = t, \tag{3.142}
$$

for a closed system of n point masses described by

$$
L(\mathbf{r}, \dot{\mathbf{r}}) = \frac{1}{2} \sum_{i=1}^{n} m_i \dot{\mathbf{r}}_i^2 - \frac{1}{2} \sum_{i,j=1}^{n} V_{ij}(|\mathbf{r}_i - \mathbf{r}_j|). \tag{3.143}
$$

Invariance of physical trajectories under the Galilei transformation is referred to simply as *Galilei invariance.*

One has now

$$\frac{d\mathbf{r}_i^*}{dt^*} = \dot{\mathbf{r}}_i + \boldsymbol{\varepsilon}, \qquad \frac{dt^*}{dt} = 1, \qquad |\mathbf{r}_i^* - \mathbf{r}_j^*| = |\mathbf{r}_i - \mathbf{r}_j|, \qquad (3.144)$$

and the invariance condition (3.140) then reads[80]

$$\boldsymbol{\nabla}_{\boldsymbol{\varepsilon}} L(\mathbf{r} + \boldsymbol{\varepsilon}t, \dot{\mathbf{r}} + \boldsymbol{\varepsilon})\big|_{\boldsymbol{\varepsilon}=0} = \sum_{i=1}^{n} \boldsymbol{\nabla}_{\dot{\mathbf{r}}_i} L = \sum_{i=1}^{n} m_i \dot{\mathbf{r}}_i$$

$$= \frac{d}{dt} \sum_{i=1}^{n} m_i \mathbf{r}_i = \frac{d}{dt} M\mathbf{R}, \qquad (3.145)$$

where the center-of-mass vector $\mathbf{R} = \sum_i m_i \mathbf{r}_i / M$ was introduced. In the present vector notation, each of the three components of $M\mathbf{R}$ corresponds to a function h in (3.141) (see footnote 80). The conserved quantity therefore is now

$$\sum_{i=1}^{n} (\boldsymbol{\nabla}_{\dot{\mathbf{r}}_i} L)\, t - M\mathbf{R} = M\left(\dot{\mathbf{R}}t - \mathbf{R}\right) = const. \qquad (3.146)$$

This can be integrated immediately to yield

$$\mathbf{R}(t) = \mathbf{u}t + \mathbf{R}_0, \qquad (3.147)$$

where $\mathbf{u}$ and $\mathbf{R}_0$ are integration constants. This is, of course, the well-known result of uniform center-of-mass motion for a closed system. It is equivalent to having a conserved total momentum for the system (identical to the momentum of the center of mass), which follows from employing the transformation $\mathbf{r}^* = \mathbf{r} + \boldsymbol{\varepsilon}$. Invariance under the particular transformation considered here thus provides six first integrals. Altogether there are ten first integral corresponding to the ten parameters of the Galilei transformations (see Secs. 1.5.4 and 1.6.1). The remaining four integrals are furnished by energy conservation and angular momentum conservation.

---

[80]We employ here a straightforward vector generalization of the invariance condition. Its validity follows immediately by considering first a Galilei transformation in the x direction, then one in the y and then in the z directions, and combining the (independent) results of all three. The Lagrange function is separately invariant under all three, and the resulting conserved quantity is the vector with components made out of the three individual conserved quantities. In other words, the total time derivative that is added to the Lagrange function under this Galilei transformation is $d(M\boldsymbol{\varepsilon} \cdot \mathbf{R})/dt$ [cf. Eq. (3.106)].

Chapter 4

Hamiltonian Mechanics

Yet another alternative approach to classical mechanics, *Hamiltonian Mechanics*, is formulated in terms of a function — the Hamilton function H — that is closely related to the total mechanical energy of the system. Because of the paramount importance of the concept of energy in physics in general, Hamiltonian Mechanics is particularly well suited for going beyond classical mechanics. Indeed, the formulation of quantum mechanics as conceived in the 1920s by Heisenberg and Schrödinger, in particular, was strongly influenced by Hamiltonian mechanics.

4.1 Hamilton's Equations

The generalized momentum,

$$p_r = \frac{\partial L(q, \dot{q}, t)}{\partial \dot{q}_r}, \qquad r = 1, \dots, N. \tag{4.1}$$

was introduced in Eq. (2.118). With its help, the Lagrange equations (2.104) may be simply written as

$$\frac{\partial L(q, \dot{q}, t)}{\partial q_r} = \dot{p}_r, \qquad r = 1, \dots, N. \tag{4.2}$$

Note, however, that this is *not* a differential equation for p_r since L does not depend explicitly on p_r. Obviously, to make it into a differential equation, one needs to eliminate the variable $\dot{q}_r$ in favor of p_r, i.e., one must be able to *invert* Eq. (4.1),

$$\boxed{\; p_r = \frac{\partial L(q, \dot{q}, t)}{\partial \dot{q}_r} \quad \xrightarrow{\text{invert}} \quad \dot{q}_r = \dot{q}_r(q, p, t) \;} \tag{4.3}$$

One then has a *first-order* differential equation for p_r,

$$\frac{\partial L(q, p, t)}{\partial q_r} = \dot{p}_r, \qquad r = 1, \dots, N, \tag{4.4}$$

where $L(q, p, t)$ is now taken as a function of the q's and p's via[81]

$$L(q, p, t) \equiv L\big(q, \dot{q}(q, p, t), t\big). \tag{4.5}$$

The equations (4.4) and $\dot{q}_r = \dot{q}_r(q, p, t)$ from the invertibility requirement (4.3) thus will turn the N second-order differential equations for q_r of the Lagrange approach into $2N$ first-order differential equations for p_r and q_r.

In formalizing the procedure outlined here, as a first step, we express the total differential of $L(q, \dot{q}, t)$ as

$$\begin{aligned}
dL &= \frac{\partial L}{\partial t} dt + \sum_{r=1}^{N} \left(\frac{\partial L}{\partial q_r} dq_r + \frac{\partial L}{\partial \dot{q}_r} d\dot{q}_r \right) \\
&= \frac{\partial L}{\partial t} dt + \sum_{r=1}^{N} \left(\dot{p}_r dq_r + p_r d\dot{q}_r \right) \\
&= \frac{\partial L}{\partial t} dt + \sum_{r=1}^{N} \left[\dot{p}_r dq_r - \dot{q}_r dp_r + d\left(p_r \dot{q}_r \right) \right],
\end{aligned} \tag{4.6}$$

where the product rule was used for $d(p_r \dot{q}_r)$ in the last step. Rearranging this result according to

$$d\left(\sum_{r=1}^{N} p_r \dot{q}_r - L \right) = -\frac{\partial L}{\partial t} dt + \sum_{r=1}^{N} \left(-\dot{p}_r dq_r + \dot{q}_r dp_r \right), \tag{4.7}$$

the right-hand side here is seen to be equal to the total differential of a function on the left called the *Hamilton function*,

$$\boxed{\; H(q, p, t) = \sum_{r=1}^{N} p_r \, \dot{q}_r (q, p, t) - L(q, p, t) \;}, \tag{4.8}$$

where $L(q, p, t)$ is given by (4.5) and $\dot{q}_r(q, p, t)$ by (4.3). Since on the other hand

$$dH = \frac{\partial H}{\partial t} dt + \sum_{r=1}^{N} \left(\frac{\partial H}{\partial q_r} dq_r + \frac{\partial H}{\partial p_r} dp_r \right), \tag{4.9}$$

comparison of the coefficients of the independent differentials dt, dq_r, and dp_r with those of (4.7) then shows

$$\frac{\partial H}{\partial t} = -\frac{\partial L}{\partial t} \tag{4.10}$$

---

[81]We mention without proof that Eq. (4.1) can be inverted according to (4.3) if and only if

$$\det \left[\frac{\partial^2 L}{\partial \dot{q}_i \partial \dot{q}_j} \right] \neq 0.$$

The case of vanishing determinant was first treated by P. A. M. Dirac, Can. J. of Math. **2**, 129 (1950). Dirac (1902–1984) was one of the founding fathers of quantum mechanics.

and

$$\boxed{-\frac{\partial H}{\partial q_r} = \dot{p}_r, \qquad \frac{\partial H}{\partial p_r} = \dot{q}_r, \qquad r = 1, \ldots, N}. \qquad (4.11)$$

These are the desired $2N$ first-order differential equations; they are called *Hamilton's equations of motion*, or *canonical equations of motion*.[82] This set of equations is completely equivalent to the N second-order Lagrange equations (2.104) and, therefore, provide an alternative way of formulating dynamical equations of motion.[83] They are valid for conservative systems, without any dissipative forces. The corresponding equations for dissipative forces are readily found from Eq. (2.155) as

$$-\frac{\partial H}{\partial q_r} + Q_r^* = \dot{p}_r, \qquad \frac{\partial H}{\partial p_r} = \dot{q}_r, \qquad (4.12)$$

where Q_r^* is a component of the generalized dissipative force defined in Eq. (2.156).

For practical applications in mechanics, the Hamilton equations usually do not have any advantage over the Lagrange equations. However, as alluded to in the introductory remarks to the present chapter, the Hamilton formalism is extremely useful for investigating the relationship between classical mechanics and quantum mechanics; it is therefore of great formal importance. In general, systems that are governed by equations of the same structure as the canonical equations are referred to as *Hamiltonian systems* or *canonical systems*. Hamiltonian systems are not restricted to problems in mechanics; they can be found in all areas of physics, in biology, in meteorology, and even in economics.

Some examples

As a first example, we derive the Hamilton function in polar coordinates for the three-dimensional motion of a free point mass. The Lagrange function is given by [cf. (2.121)]

$$L(r, \theta, \dot{r}, \dot{\varphi}, \dot{\theta}) = \frac{m}{2}(\dot{r}^2 + r^2\dot{\theta}^2 + r^2\sin^2\theta\,\dot{\varphi}^2). \qquad (4.13)$$

---

[82]The term 'canonical' for these equations was introduced by Jacobi in 1837. Despite the fact that its meaning is somewhat unclear (Jacobi himself gave no explanation), it has been used ever since. Most likely the term was inspired by the *Canon Law* that regulates the Catholic Church and its members.

[83]Recalling the discussion at the beginning of this section, the first of Hamilton's equations here is equivalent to (4.4) and the second corresponds to $\dot{q}_r = \dot{q}_r(q, p, t)$ from the invertibility requirement (4.3).

The generalized momenta are

$$p_r = \frac{\partial L}{\partial \dot{r}} = m\dot{r} \Rightarrow \dot{r} = \frac{p_r}{m}, \tag{4.14a}$$

$$p_\varphi = \frac{\partial L}{\partial \dot{\varphi}} = mr^2 \sin^2\theta\, \dot{\varphi} \Rightarrow \dot{\varphi} = \frac{p_\varphi}{mr^2 \sin^2\theta}, \tag{4.14b}$$

$$p_\theta = \frac{\partial L}{\partial \dot{\theta}} = mr^2\dot{\theta} \Rightarrow \dot{\theta} = \frac{p_\theta}{mr^2}. \tag{4.14c}$$

The free Hamilton function for three-dimensional motion then becomes

$$H(r, \theta, p_r, p_\varphi, p_\theta) = \dot{r}p_r + \dot{\varphi}p_\varphi + \dot{\theta}p_\theta - L$$

$$= \frac{1}{2m}\left(p_r^2 + \frac{p_\varphi^2}{r^2 \sin^2\theta} + \frac{p_\theta^2}{r^2}\right). \tag{4.15}$$

The Hamilton function for planar motion in the xy-plane is found by putting $\theta = 90°$ and $p_\theta = 0$.

As another example, we consider the motion of a particle in a potential. With

$$L = \frac{m}{2}\dot{\mathbf{r}}^2 - V(\mathbf{r}, t) = \frac{m}{2}\left(\dot{x}^2 + \dot{y}^2 + \dot{z}^2\right) - V(x, y, z, t), \tag{4.16}$$

the generalized momenta,

$$p_x = \frac{\partial L}{\partial \dot{x}} = m\dot{x}, \qquad p_y = \frac{\partial L}{\partial \dot{y}} = m\dot{y}, \qquad p_z = \frac{\partial L}{\partial \dot{z}} = m\dot{z}, \tag{4.17}$$

are just the usual linear momentum components. Using a vector notation, the Hamilton function then is

$$H(\mathbf{r}, \mathbf{p}, t) = \frac{\mathbf{p}^2}{2m} + V(\mathbf{r}, t), \tag{4.18}$$

leading to the canonical equations

$$\dot{\mathbf{p}} = -\boldsymbol{\nabla}V(\mathbf{r}, t), \qquad \dot{\mathbf{r}} = \frac{\mathbf{p}}{m}, \tag{4.19}$$

which obviously is equivalent to the Lagrange equations $m\ddot{\mathbf{r}} = -\boldsymbol{\nabla}V$.

It should be emphasized that the Hamilton function must always be written in terms of the q's and p's only. Working with expressions mixing q's, $\dot{q}$'s, and p's will lead to errors.

4.1.1 Conservation laws

Energy conservation is particularly simple with the Hamilton function. Considering

$$
\begin{aligned}
\frac{dH}{dt} &= \frac{\partial H}{\partial t} + \sum_{r=1}^{N} \left(\frac{\partial H}{\partial q_r} \dot{q}_r + \frac{\partial H}{\partial p_r} \dot{p}_r \right) \\
&= \frac{\partial H}{\partial t} + \underbrace{\sum_{r=1}^{N} \left(\frac{\partial H}{\partial q_r} \frac{\partial H}{\partial p_r} - \frac{\partial H}{\partial p_r} \frac{\partial H}{\partial q_r} \right)}_{=0} \\
&= \frac{\partial H}{\partial t} \, ;
\end{aligned}
\tag{4.20}
$$

in other words

$$
\frac{dH}{dt} = \frac{\partial H}{\partial t} = -\frac{\partial L}{\partial t}.
\tag{4.21}
$$

For scleronomous conditions, we have $\partial L/\partial t = 0$ and hence the conservation law

$$
\boxed{ \frac{\partial H}{\partial t} = 0 \quad \Rightarrow \quad H = const }.
\tag{4.22}
$$

For scleronomous *conservative* systems, this conserved quantity is just the total energy,

$$
H(q, p, t) = T + V.
\tag{4.23}
$$

This follows immediately from the considerations of Section 2.2.3.1 since the right-most equation of (2.227) is a direct statement of $H = const$.

Since the q_r dependence of the Hamilton function H is the same as that of L, a variable that is ignorable in L will also be ignorable in H, and hence

$$
\boxed{ q_k \text{ cyclic:} \quad \frac{\partial H}{\partial q_k} = 0 \quad \Rightarrow \quad p_k = const }.
\tag{4.24}
$$

4.1.2 Hamilton's principle revisited

The Hamilton principle may also be used to derive the canonical equations. Taking the original variational principle for the functional $S[q]$,

$$
\delta S[q] = \delta \int_{t_1}^{t_2} dt \, L(q, \dot{q}, t) = 0,
\tag{4.25}
$$

we may express L in terms of H via (4.8) and reinterpret this as a variational principle for the functional $S[q, p]$,

$$\delta S[q, p] = \delta \int_{t_1}^{t_2} dt \left(\sum_{i=1}^{N} p_i \dot{q}_i - H(q, p, t) \right) = 0, \tag{4.26}$$

with fixed end points

$$\delta q(t_1) = \delta q(t_2) = 0, \qquad \delta p(t_1) = \delta p(t_2) = 0. \tag{4.27}$$

The corresponding Euler–Lagrange equations,

$$\left(\frac{d}{dt} \frac{\partial}{\partial \dot{q}_r} - \frac{\partial}{\partial q_r} \right) \left(\sum_{i=1}^{N} p_i \dot{q}_i - H(q, p, t) \right) = \dot{p}_r + \frac{\partial H}{\partial q_r} = 0 \tag{4.28a}$$

$$\left(\frac{d}{dt} \frac{\partial}{\partial \dot{p}_r} - \frac{\partial}{\partial p_r} \right) \left(\sum_{i=1}^{N} p_i \dot{q}_i - H(q, p, t) \right) = -\dot{q}_r + \frac{\partial H}{\partial p_r} = 0, \tag{4.28b}$$

then immediately provide Hamilton's equations.

4.1.3   *Poisson brackets*

The observables F of physical systems can only depend on the coordinates q_i, the momenta p_i, and the time t, i.e., $F = F(q, p, t)$. Hence, their time derivative is given by

$$\frac{dF}{dt} = \sum_{i=1}^{N} \left(\frac{\partial F}{\partial q_i} \dot{q}_i + \frac{\partial F}{\partial p_i} \dot{p}_i \right) + \frac{\partial F}{\partial t}$$

$$= \sum_{i=1}^{N} \left(\frac{\partial F}{\partial q_i} \frac{\partial H}{\partial p_i} - \frac{\partial F}{\partial p_i} \frac{\partial H}{\partial q_i} \right) + \frac{\partial F}{\partial t}, \tag{4.29}$$

where the canonical equations were employed in the last step. Defining the *Poisson bracket* of two functions $F(q, p, t)$ and $G(q, p, t)$ by

$$[F, G] = \sum_{i=1}^{N} \left(\frac{\partial F}{\partial q_i} \frac{\partial G}{\partial p_i} - \frac{\partial F}{\partial p_i} \frac{\partial G}{\partial q_i} \right), \tag{4.30}$$

this result can be written as

$$\boxed{\frac{dF}{dt} = [F, H] + \frac{\partial F}{\partial t}}. \tag{4.31}$$

This equation determines the time derivative of an arbitrary physical quantity F in terms of that quantity's Poisson bracket with the Hamilton function H.

Special cases are the time derivative of H itself, i.e.,

$$\frac{dH}{dt} = [H, H] + \frac{\partial H}{\partial t} = \frac{\partial H}{\partial t}, \tag{4.32}$$

which shows, once again, that $H = const$ if it does not explicitly depend on time. The time derivatives of q_r and p_r are

$$\dot{p}_r = [p_r, H], \qquad \dot{q}_r = [q_r, H]. \tag{4.33}$$

These are, of course, the canonical equations in their most concise form.

Further applications of the Poisson-bracket notation for the partial derivative of an arbitrary function $F(q, p, t)$ provide

$$\frac{\partial F}{\partial q_j} = [F, p_j], \qquad \text{and} \qquad \frac{\partial F}{\partial p_j} = -[F, q_j]. \tag{4.34}$$

Special cases are immediately obtained by taking F to be either q_i or p_i, viz.

$$[q_i, q_j] = 0, \qquad [p_i, p_j] = 0, \qquad [q_i, p_j] = \delta_{ij}. \tag{4.35}$$

While they allow a very concise formulation of many results, the Poisson brackets do not provide anything that could not be obtained otherwise. Their importance lies on a formal level: They are the classical analogue of the quantum-mechanical commutator. Indeed, the quantum analogues of many classical relations can be simply found by replacing the classical Poisson bracket of the physical quantities of interest by the quantum-mechanical commutator of the corresponding operators, i.e.,

$$\underbrace{[A, B]_{\mathrm{CM}}}_{\text{classical Poisson bracket}} \qquad \longrightarrow \qquad \underbrace{\tfrac{1}{i\hbar}[A_{\mathrm{op}}, B_{\mathrm{op}}]_{\mathrm{QM}}}_{\text{quantum-mechanical commutator}} . \tag{4.36}$$

Applying this to Eqs. (4.35), for example, immediately provides the canonical commutator relations for the quantum-mechanical position and momentum operators which are the basis for Heisenberg's famous uncertainty relation.

Properties of Poisson brackets

For A, B, and C being arbitrary functions of q, p, t, the following properties hold true.

$$[A, B] = -[B, A] \qquad \Longrightarrow \qquad [A, A] = 0, \tag{4.37}$$

$$[A, B + C] = [A, B] + [A, C], \tag{4.38}$$

$$[A, BC] = B[A, C] + [A, B]C, \tag{4.39}$$

$$\text{Jacobi identity:} \qquad [A, [B, C]] + [B, [C, A]] + [C, [A, B]] = 0. \tag{4.40}$$

4.1.4 *Legendre transformations*

The transition from variables q_r and $\dot{q}_r$ to the canonically conjugate variables q_r and p_r that replaced the description in terms of the Lagrange function $L(q, \dot{q}, t)$ by the one in terms of the Hamilton function $H(q, p, t)$ is an example of the general method of *Legendre transformations*.

Generically, Legendre transformations describe the transition of variables x, y of a function $f(x, y)$ to new variables x, u of a new function $g(x, u)$ where u is defined by

$$u = \frac{\partial f}{\partial y}. \tag{4.41}$$

The total differential of f is

$$df = \frac{\partial f}{\partial x} dx + \frac{\partial f}{\partial y} dy = \frac{\partial f}{\partial x} dx + u\, dy. \tag{4.42}$$

Defining a function g by

$$g = f - uy, \tag{4.43}$$

where the additive term uy depends on the old variable y (to be replaced) and the new variable u, it has the total differential

$$dg = df - u\, dy - y\, du = \frac{\partial f}{\partial x} dx - y\, du. \tag{4.44}$$

Since this shows that the total differential of g is completely determined in terms of total differentials dx and du, this means that

$$g = g(x, u), \qquad \text{with} \qquad dg = \frac{\partial g}{\partial x} dx + \frac{\partial g}{\partial u} du. \tag{4.45}$$

Comparing coefficients then yields

$$\frac{\partial f}{\partial x} = \frac{\partial g}{\partial x}, \qquad y = -\frac{\partial g}{\partial u}. \tag{4.46}$$

The Legendre transformation thus provides an equivalent description in terms of the function $g(x, u)$ if the Eqs. (4.46) are satisfied. For the preceding applications, these equations correspond to the Hamilton equations (i.e., the analogy is $f \to -L$, $x \to q$, $y \to \dot{q}$, $u \to -p$, $g \to H$).

4.2 Canonical Transformations

One of the major advantages of the Lagrange formalism is that the Lagrange equations are form invariant (see p. 116) under *point*, or *contact*, *transformations*

$$q_i = q_i(Q_1, \ldots, Q_N, t), \tag{4.47}$$

which replace the old coordinates q by new coordinates Q. We now ask the question whether it is possible to transform the variables q and p of the Hamilton formalism to new variables Q and P, i.e.,

$$q_i = q_i(Q_1, \ldots, Q_N, P_1, \ldots, P_N, t), \qquad (4.48\text{a})$$

$$p_i = p_i(Q_1, \ldots, Q_N, P_1, \ldots, P_N, t), \qquad (4.48\text{b})$$

such that the form of the Hamilton equations remains the same for the new variables. Such a transformation shall be called a *canonical transformation.* In other words, canonical transformations are defined by

$$\left.\begin{array}{c} \text{variables } q,\, p \\[4pt] H(q, p, t) \\[6pt] \dfrac{\partial H}{\partial p_i} = \dot{q}_i, \quad -\dfrac{\partial H}{\partial q_i} = \dot{p}_i \end{array}\right\} \longrightarrow \left\{\begin{array}{c} \text{variables } Q,\, P \\[4pt] K(Q, P, t) \\[6pt] \dfrac{\partial K}{\partial P_i} = \dot{Q}_i, \quad -\dfrac{\partial K}{\partial Q_i} = \dot{P}_i \end{array}\right.$$

where $K(Q, P, t)$ is the Hamilton function in the new variables.

In Section 4.1.2, it was shown that the canonical equations follow from Hamilton's principle. Hence, a transformation will be canonical if

$$\delta \int_{t_1}^{t_2} dt \left(\sum_{i=1}^{N} p_i \dot{q}_i - H(q, p, t) \right) = \delta \int_{t_1}^{t_2} dt \left(\sum_{i=1}^{N} P_i \dot{Q}_i - K(Q, P, t) \right) = 0.$$

$$(4.49)$$

According to Sec. 3.2.1 and the generalized Noether theorem in Sec. 3.3.1, to describe the same physical system (albeit in different variables), the two integrands can at most differ by the total time derivative of a scalar function. We then have the requirement that

$$\sum_{i=1}^{N} p_i \dot{q}_i - H(q, p, t) = \sum_{i=1}^{N} P_i \dot{Q}_i - K(Q, P, t) + \frac{d\Phi}{dt}, \qquad (4.50)$$

where Φ is a function of $2N$ independent variables and of time. Nontrivial functions Φ must depend on both old and new variables (otherwise they would have no bearing on the canonical transformation itself and can be omitted). This leaves four possibilities:

$$\Phi_1(q, Q, t), \quad \Phi_2(q, P, t), \quad \Phi_3(p, Q, t), \quad \Phi_4(p, P, t). \qquad (4.51)$$

Each type of these functions can be shown to lead to a canonical transformation; they are therefore called the *generators of the canonical transformation.*

Choice 1: $\Phi_1(q, Q, t)$

For this choice, the additional term in (4.50) is explicitly given as

$$\frac{d\Phi_1(q, Q, t)}{dt} = \sum_{i=1}^{N} \left(\frac{\partial \Phi_1}{\partial q_i} \dot{q}_i + \frac{\partial \Phi_1}{\partial Q_i} \dot{Q}_i \right) + \frac{\partial \Phi_1}{\partial t}. \tag{4.52}$$

Comparison of the coefficients of $\dot{q}_i$ and $\dot{Q}_i$ in (4.50) then yields

$$p_i = \frac{\partial \Phi_1(q, Q, t)}{\partial q_i} = p_i(q, Q, t), \qquad i = 1, \ldots, N \tag{4.53}$$

$$P_i = -\frac{\partial \Phi_1(q, Q, t)}{\partial Q_i} = P_i(q, Q, t), \qquad i = 1, \ldots, N \tag{4.54}$$

$$K(Q, P, t) = H(q, p, t) + \frac{\partial \Phi_1(q, Q, t)}{\partial t}. \tag{4.55}$$

These equations completely specify the canonical equations: Solving the $2N$ equations (4.53) and (4.54) for $q = q(Q, P, t)$ and $p = p(Q, P, t)$ and replacing q and p on the right-hand side of (4.55) by these functions yields the new Hamilton function $K(Q, P, t)$.

The entire transformation, therefore, is *generated* by the particular choice of Φ. Choosing Φ judiciously, a given problem may become considerably simpler in the new variables.

Note also that the choice of generating function may 'scramble' the meaning of coordinate and momentum variables. For example, the choice

$$\Phi_1(q, Q) = \sum_{i=1}^{N} q_i Q_i \tag{4.56}$$

produces

$$p_i = \frac{\partial \Phi_1}{\partial q_i} = Q_i, \qquad P_i = -\frac{\partial \Phi_1}{\partial Q_i} = -q_i, \tag{4.57}$$

and therefore (up to a sign) completely exchanges the meaning of coordinates and momenta.

As a simple example, let us consider the well-known case of the harmonic oscillator. Its Hamilton function is

$$H(q, p) = \frac{p^2}{2m} + \frac{m\omega^2 q^2}{2} \; ; \tag{4.58}$$

it is conserved and equal to the total energy $H = E$. Taking the generator as

$$\Phi(q, Q) = \frac{m}{2}\omega q^2 \cot Q, \tag{4.59}$$

one has

$$p = \frac{\partial \Phi}{\partial q} = m\omega q \cot Q, \qquad P = -\frac{\partial \Phi}{\partial Q} = \frac{m\omega q^2}{2\sin^2 Q}, \qquad (4.60)$$

which provides

$$q = \sqrt{\frac{2P}{m\omega}} \sin Q, \qquad p = \sqrt{2m\omega P}\cos Q. \qquad (4.61)$$

The new Hamilton function then becomes

$$K(Q,P) = H\big(q(Q,P),p(Q,P)\big) = \omega P \cos^2 Q + \omega P \sin^2 Q = \omega P. \qquad (4.62)$$

The new variable Q, therefore, is ignorable, and the canonical equations are

$$\dot{Q} = \frac{\partial K}{\partial P} = \omega, \qquad \dot{P} = -\frac{\partial K}{\partial Q} = 0, \qquad (4.63)$$

with the solution

$$P = const = \frac{K}{\omega} = \frac{E}{\omega}, \qquad Q(t) = \omega t + \phi, \qquad (4.64)$$

where E and ϕ are integration constants. This then immediately provides the well-known solution for the harmonic oscillator,

$$q(t) = \sqrt{\frac{2E}{m\omega^2}} \sin(\omega t + \phi). \qquad (4.65)$$

Choice 2: $\Phi_2(q, P, t)$

Using a Legendre transformation (see Sec. 4.1.4),

$$\Phi_2(q,P,t) = \Phi_1(q,Q,t) + \sum_{j=1}^{N} P_j Q_j, \qquad (4.66)$$

this choice can be related to the previous treatment of Φ_1. The resulting transformation equations then are immediately found as

$$p_j = \frac{\partial \Phi_2}{\partial q_j}, \qquad Q_j = \frac{\partial \Phi_2}{\partial P_j}, \qquad K = H + \frac{\partial \Phi_2}{\partial t}. \qquad (4.67)$$

Note here that the particular choice

$$\Phi_2(q,P) = \sum_{i=1}^{N} q_i P_i \qquad (4.68)$$

provides

$$p_i = \frac{\partial \Phi_2}{\partial q_i} = P_i, \qquad Q_i = \frac{\partial \Phi_2}{\partial p_i} = q_i, \qquad (4.69)$$

i.e., this is the *identity transformation*.

Choosing the Φ_2 generator as

$$\Phi_2(q, P, t) = \sum_{i=1}^{N} F_i(q_1, \ldots, q_N, t) P_i, \qquad (4.70)$$

where F_i is an arbitrary function of the coordinates, this results in

$$Q_i = F_i(q_1, \ldots, q_N, t). \qquad (4.71)$$

Since this is the most general contact transformation [see (4.47)], this means that *every contact transformation is canonical.*

Choice 3: $\Phi_3(p, Q, t)$

Again, this case is related to the treatment of Φ_1 by the Legendre transformation

$$\Phi_3(p, Q, t) = \Phi_1(q, Q, t) - \sum_{j=1}^{N} q_j p_j, \qquad (4.72)$$

leading to transformation equations

$$q_j = -\frac{\partial \Phi_3}{\partial p_j}, \qquad P_j = -\frac{\partial \Phi_3}{\partial Q_j}, \qquad K = H + \frac{\partial \Phi_3}{\partial t}. \qquad (4.73)$$

Choice 4: $\Phi_4(p, P, t)$

For this choice, the double Legendre transformation

$$\Phi_4(p, P, t) = \Phi_1(q, Q, t) + \sum_{j=1}^{N} P_j Q_j - \sum_{j=1}^{N} q_j p_j \qquad (4.74)$$

relates this to Φ_1, and the resulting transformation equations are

$$q_j = -\frac{\partial \Phi_4}{\partial p_j}, \qquad Q_j = \frac{\partial \Phi_4}{\partial P_j}, \qquad K = H + \frac{\partial \Phi_4}{\partial t}. \qquad (4.75)$$

Canonical invariance

Necessary and sufficient conditions that given transformations $Q = Q(q, p, t)$ and $P = P(q, p, t)$ are canonical are

$$[Q_i, Q_j] = 0, \qquad [P_i, P_j] = 0, \qquad [Q_i, P_j] = \delta_{ij}, \qquad (4.76)$$

for all i, j.

4.2.1 *Infinitesimal canonical transformations*

If the new and the old coordinates only differ by an infinitesimal amount,

$$Q_i = q_i + \delta q_i, \qquad P_i = p_i + \delta p_i, \tag{4.77}$$

where the δ prefix now signifies small changes, the generating function can only differ by an infinitesimal contribution from the identity transformation (4.68) and hence can be written as

$$\Phi_2(q, P) = \sum_{i=1}^{N} q_i P_i + \varepsilon \hat{G}(q, P), \tag{4.78}$$

where ε is an infinitesimally small parameter and $\hat{G}$ is some function that defines the particular infinitesimal transformation. The transformation equations (4.67) then yield

$$P_i - p_i = \delta p_i = -\varepsilon \frac{\partial \hat{G}}{\partial q_i}, \tag{4.79}$$

$$Q_i - q_i = \delta q_i = \varepsilon \frac{\partial \hat{G}}{\partial P_i} = \varepsilon \frac{\partial \hat{G}}{\partial p_i}, \tag{4.80}$$

where terms of order ε^2 are neglected. As is seen here, the infinitesimal changes of the positions and momenta depend directly on the choice of $\hat{G}$ and $\hat{G}$, therefore, is called the *generator of infinitesimal transformations*.

Consider now the particular choices $\hat{G} = H(q, p)$ and $\varepsilon = dt$; then

$$\delta p_i = -dt \frac{\partial H}{\partial q_i} = \dot{p}_i dt = dp_i, \tag{4.81}$$

$$\delta q_i = dt \frac{\partial H}{\partial p_i} = \dot{q}_i dt = dq_i. \tag{4.82}$$

This transformation advances the coordinates q_i and momenta p_i at time t to the coordinates $q_i + dq_i$ and momenta $p_i + dp_i$ at time $t + dt$. The time evolution of the system during the interval dt, therefore, may be described by infinitesimal canonical transformations generated by the Hamilton function. The evolution in a finite time interval can be obtained by a series of infinitesimal ones. Moreover, since two successive canonical transformations are equivalent to a single canonical transformation, one obtains the entire time evolution of the system as a canonical transformation that is a continuous function of the time. Hence, *the Hamilton function is the generator of the system evolution in time.*

The change of an arbitrary physical quantity $F(q,p)$ under an infinitesimal canonical transformations is given by

$$\delta F = \sum_{i=1}^{N} \left(\frac{\partial F}{\partial q_i} \delta q_i + \frac{\partial F}{\partial p_i} \delta p_i \right) = \varepsilon \sum_{i=1}^{N} \left(\frac{\partial F}{\partial q_i} \frac{\partial \hat{G}}{\partial p_i} - \frac{\partial F}{\partial p_i} \frac{\partial \hat{G}}{\partial q_i} \right), \qquad (4.83)$$

or

$$\delta F = \varepsilon [F, \hat{G}]. \qquad (4.84)$$

Specifically, if one takes $F = H$, the infinitesimal change of the Hamilton function is

$$\delta H = \varepsilon [H, \hat{G}] = \varepsilon \left(\frac{\partial \hat{G}}{\partial t} - \frac{d\hat{G}}{dt} \right), \qquad (4.85)$$

where the last equality follows from (4.31). Therefore, if $\hat{G}$ is a constant of motion (i.e., $d\hat{G}/dt = 0$) that does not depend explicitly on time (i.e., $\partial\hat{G}/\partial t = 0$), the Hamilton function will not change (i.e., $\delta H = 0$) under this transformation. Hence, *the constants of motion are the generators of the infinitesimal canonical transformations that leave the Hamilton function invariant.*

4.3   Liouville's Theorem

One may take the N generalized variables q_r as the coordinates of an N-dimensional space, called the *configuration space*. Analogously, the N generalized momenta p_r form the N-dimensional *momentum space*. Together, the $2N$ variables $(q_1, \ldots, q_N, p_1, \ldots, p_N)$ constitute the $2N$-dimensional *phase space*. The spatial position of the system at a time t is described by a point in configuration space. Analogously, the velocity state of the system is given by a point in momentum space. The complete state of the system at a particular time is then given by a point in phase space. For this reason, phase space is also equivalently called *state space*.

4.3.1   *Hamiltonian systems*

For Hamiltonian systems, the time evolution of a phase space point is described by the canonical equations (4.11). For such systems, the trajectory in phase space — also called a *flow* — completely describes the time evolution of the entire system. The flow of phase-space points[84] within a given volume is found to obey

---

[84]We use here the loose language common in this context. The phase-space points, of course, do not move. What is meant is the time evolution of the system from one phase-space point to the next.

> **Liouville's Theorem:**
> In Hamiltonian systems, the size of a given volume in phase space remains constant as time evolves. (4.86)

For its proof, we note that during an infinitesimal time dt, the phase-space points evolve according to

$$q_i \to q_i' = q_i + \dot{q}_i dt = q_i + \frac{\partial H}{\partial p_i} dt, \tag{4.87a}$$

$$p_i \to p_i' = p_i + \dot{p}_i dt = p_i - \frac{\partial H}{\partial q_i} dt, \tag{4.87b}$$

which corresponds to an infinitesimal canonical transformation from the phase-space variables (q, p) to new variables (q', p'). The relationship between the volume element in the old variables,

$$dV = dq_1 \ldots dq_N dp_1 \ldots dp_N = \prod_{i=1}^{N} dq_i\, dp_i, \tag{4.88}$$

and the corresponding element dV' in the new variables,

$$dV' = dq_1' \ldots dq_N' dp_1' \ldots dp_N' = \prod_{i=1}^{N} dq_i'\, dp_i', \tag{4.89}$$

is given by

$$dV' = \det J \, dV, \tag{4.90}$$

where

$$J = \{J_{ij}\} = \begin{pmatrix} \dfrac{\partial q_i'}{\partial q_j} & \dfrac{\partial q_i'}{\partial p_j} \\[2ex] \dfrac{\partial p_i'}{\partial q_j} & \dfrac{\partial p_i'}{\partial p_j} \end{pmatrix}, \tag{4.91}$$

is the $2N \times 2N$ *Jacobian* of the variable transformation. With the explicit form of the transformation given, we have

$$J = \begin{pmatrix} \delta_{ij} + \dfrac{\partial^2 H}{\partial q_j \partial p_i} dt & \dfrac{\partial^2 H}{\partial p_j \partial p_i} dt \\[2ex] -\dfrac{\partial^2 H}{\partial q_j \partial q_i} dt & \delta_{ij} - \dfrac{\partial^2 H}{\partial p_j \partial q_i} dt \end{pmatrix}$$

$$= \mathbb{1} + \begin{pmatrix} \dfrac{\partial^2 H}{\partial q_j \partial p_i} & \dfrac{\partial^2 H}{\partial p_j \partial p_i} \\[2ex] -\dfrac{\partial^2 H}{\partial q_j \partial q_i} & -\dfrac{\partial^2 H}{\partial p_j \partial q_i} \end{pmatrix} dt, \tag{4.92}$$

where $\mathbb{1}$ is the $2N{\times}2N$ unit matrix. Using now the general result that $\det(\mathbb{1}+\varepsilon M)=1+\varepsilon\,\mathrm{Tr}\,M+\mathcal{O}(\varepsilon^2)$ for any matrix that differs from unity by an infinitesimal matrix εM, with Tr denoting the trace operation, we then have

$$\det J = 1 + dt \sum_{i=1}^{N}\left(\frac{\partial^2 H}{\partial q_i \partial p_i} - \frac{\partial^2 H}{\partial p_i \partial q_i}\right) + \mathcal{O}(dt^2) = 1 + \mathcal{O}(dt^2) \qquad (4.93)$$

and thus

$$\frac{d}{dt}\det J = 0, \qquad (4.94)$$

or $dV' = dV$. It follows then that the size of the volume of an arbitrary region of phase space,

$$V = \int_V dV, \qquad (4.95)$$

is an invariant for Hamiltonian systems.[85] This does *not* mean that the shape of the volume need stay the same. (However, the connectedness properties of the volume is preserved by time evolution, thus simply connected volumes will stay simply connected, etc.)

4.3.1.1   *Continuity equation*

In statistical mechanics, in particular, where one deals with systems comprised of a *very* large number of particles[86], it would be hopeless to attempt computing the flow of the phase space points for all particles. Instead, one forms averages over an *ensemble* of identical, non-interacting systems. The probability of having a system given by a particular phase-space point can then be described in terms of a *density function* $\rho = \rho(q,p,t)$ that tells us how many identical states,

$$dN_V = \rho\,dV, \qquad (4.96)$$

can be found within the volume dV at the N-dimensional phase-space point $(q,p) \equiv (q_1,\ldots,q_N,p_1,\ldots,p_N)$ at a particular time t. The invariance of the volume element under time evolution means that

$$\frac{dN_V}{dt} = \rho\frac{dV}{dt} = 0, \qquad (4.97)$$

---

[85]This is one of the so-called *integral invariants of Poincaré*, also known as *canonical invariants*.

[86]We recall that there are 6×10^{23} molecules in a mole of gas.

i.e., the number of states dN_V within the volume stays constant. Since then both dN_V and dV are constant, we have $\rho = const$ providing

$$\frac{d\rho}{dt} = \frac{\partial \rho}{\partial t} + \sum_{i=1}^{N} \left(\frac{\partial \rho}{\partial q_i} \dot{q}_i + \frac{\partial \rho}{\partial p_i} \dot{p}_i \right) = \frac{\partial \rho}{\partial t} + [\rho, H] = 0, \tag{4.98}$$

or

$$\boxed{\frac{\partial \rho}{\partial t} = -[\rho, H]} \tag{4.99}$$

which is known as *Liouville's equation*.

We may rewrite the Poisson bracket in Liouville's equation using

$$\begin{aligned}
\frac{\partial \rho}{\partial q_i} \dot{q}_i + \frac{\partial \rho}{\partial p_i} \dot{p}_i &= \frac{\partial(\rho \dot{q}_i)}{\partial q_i} + \frac{\partial(\rho \dot{p}_i)}{\partial p_i} - \rho \left(\frac{\partial \dot{q}_i}{\partial q_i} + \frac{\partial \dot{p}_i}{\partial p_i} \right) \\
&= \frac{\partial(\rho \dot{q}_i)}{\partial q_i} + \frac{\partial(\rho \dot{p}_i)}{\partial p_i} - \rho \underbrace{\left(\frac{\partial^2 H}{\partial q_i \partial p_i} - \frac{\partial^2 H}{\partial p_i \partial q_i} \right)}_{=0} \\
&= \frac{\partial(\rho \dot{q}_i)}{\partial q_i} + \frac{\partial(\rho \dot{p}_i)}{\partial p_i},
\end{aligned} \tag{4.100}$$

to obtain

$$[\rho, H] = \sum_{i=1}^{N} \left[\frac{\partial(\rho \dot{q}_i)}{\partial q_i} + \frac{\partial(\rho \dot{p}_i)}{\partial p_i} \right]. \tag{4.101}$$

The quantities $\rho \dot{q}_i$ and $\rho \dot{p}_i$ appearing here play the role of 'current-density components' for the flow along the phase-space trajectories that describe the time evolution of the system.[87] Therefore, written in the form

$$\boxed{\frac{\partial \rho}{\partial t} + \sum_{i=1}^{N} \left[\frac{\partial(\rho \dot{q}_i)}{\partial q_i} + \frac{\partial(\rho \dot{p}_i)}{\partial p_i} \right] = 0}, \tag{4.102}$$

we obtain the *continuity equation* that describes the continuous evolution of the system from one phase-space point to the next.

The continuity equation was derived here for Hamiltonian systems. In actual fact, however, it holds true also for dissipative systems as well since, as we shall see, its validity depends only on the fact that physical systems behave smoothly under time evolution. It is thus generally valid, even for cases where (4.99) is not.

---

[87]This is in analogy to the current density $\mathbf{j} = \rho \mathbf{v}$ of continuum mechanics (see Sec. 7.3.1), where ρ is the mass-density distribution and $\mathbf{j}$ describes the flow of mass elements $\rho \, dV$ with a velocity $\mathbf{v}$.

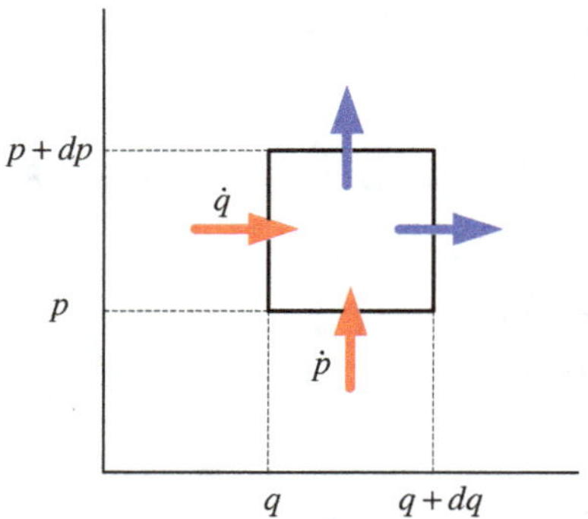

Fig. 4.1 Schematic two-dimensional depiction of phase space. The arrows indicate the trajectory flow into and out of the phase space volume $dV = dq\,dp$ along the q and p axes.

Alternative derivation of the continuity equation

To see this, let us consider the phase-space picture given in Fig. 4.1. The net flow of the number of phase-space points per unit time along the q_i axis is then given by the difference of what flows out of the volume along q_i and what flows in, i.e.,

$$I_{q_i} = \left(\rho + \frac{\partial \rho}{\partial q_i}\dot{q}_i dt\right)\left(\frac{dV}{dt} + \frac{\partial V}{\partial q_i}\frac{dq_i}{dt}\right) - \rho\frac{dV}{dt}, \qquad (4.103)$$

where both density and volume element at $q_i + dq_i$ have been expanded up to first order. Using

$$\frac{\partial V}{\partial q_i}\frac{dq_i}{dt} = \frac{\partial \dot{q}_i}{\partial q_i}dV \qquad (4.104)$$

and retaining only first order terms in the resulting expression, one has

$$I_{q_i} = \left(\frac{\partial \rho}{\partial q_i}\dot{q}_i + \rho\frac{\partial \dot{q}_i}{\partial q_i}\right)dV = \frac{\partial(\rho\dot{q}_i)}{\partial q_i}dV. \qquad (4.105)$$

The analogous result holds true for the flow along the p_i axis; one has

$$I_{p_i} = \frac{\partial(\rho\dot{p}_i)}{\partial p_i}dV. \qquad (4.106)$$

The total flow out of the volume along all phase-space axes is given by summing over all variables, i.e.,

$$I = \sum_{i=1}^{N}\left(I_{q_i} + I_{p_i}\right) = \sum_{i=1}^{N}\left[\frac{\partial(\rho\dot{q}_i)}{\partial q_i} + \frac{\partial(\rho\dot{p}_i)}{\partial p_i}\right]dV \qquad (4.107)$$

This total flow must be due to a change in the density within the volume, i.e.,

$$I = -\frac{\partial \rho}{\partial t} dV, \qquad (4.108)$$

where the minus sign accounts for the fact that a negative derivative $\partial \rho / \partial t$ — i.e., a diminishing density ρ — corresponds to a positive (outgoing) flow. Equating the last two expressions for I and dividing out the volume element dV then indeed reproduces the continuity equation (4.102). The only assumption that went into this derivation is that phase space has a smooth topology and that the density ρ is a differentiable function of its variables.

4.3.1.2 *Statistical equilibrium*

If there is no explicit time dependence, we have

$$\text{at equilibrium:} \qquad [\rho, H] = 0, \qquad (4.109)$$

which, by definition, describes the statistical equilibrium where the system does not change. An important class of time-independent densities can be written as

$$\rho = \rho\big(H(q, p)\big), \qquad (4.110)$$

whose time independence is readily seen from

$$\begin{aligned}
\frac{\partial \rho}{\partial t} &= -\sum_{i=1}^{N} \left(\frac{\partial \rho}{\partial q_i} \frac{\partial H}{\partial p_i} - \frac{\partial \rho}{\partial p_i} \frac{\partial H}{\partial q_i} \right) \\
&= -\frac{\partial \rho}{\partial H} \sum_{i=1}^{N} \left(\frac{\partial H}{\partial q_i} \frac{\partial H}{\partial p_i} - \frac{\partial H}{\partial p_i} \frac{\partial H}{\partial q_i} \right) \\
&= 0. \qquad (4.111)
\end{aligned}$$

The time-independent, and thus conserved, Hamilton function here is the total energy of the system, $H(q, p) = E = const.$ An example for this type of density is the *Boltzmann distribution*

$$\rho = e^{-H(q,p)/kT}, \qquad (4.112)$$

where T is the absolute temperature of the system and k is the Boltzmann constant. For a free particle, in particular, we have $H = \mathbf{p}^2/2m$; the corresponding $\rho = \exp(-\mathbf{p}^2/2mkT) = \exp(-m\mathbf{v}^2/2kT)$ then provides a Gaussian distribution of the velocities $\mathbf{v} = \mathbf{p}/m$.

4.3.2 Dissipative systems

For dissipative systems, the Hamilton equations are [see Eq. (4.12)]

$$\dot{p}_i = -\frac{\partial H}{\partial q_i} + Q_i^*, \qquad \dot{q}_i = \frac{\partial H}{\partial p_i}, \tag{4.113}$$

where Q_i^* is a generalized dissipative force component. The transformations (4.87) read now

$$q_i' = q_i + \dot{q}_i dt = q_i + \frac{\partial H}{\partial p_i} dt, \tag{4.114a}$$

$$p_i' = p_i + \dot{p}_i dt = p_i - \frac{\partial H}{\partial q_i} dt + Q_i^* dt, \tag{4.114b}$$

and the Jacobian is

$$J = \mathbb{1} + \begin{pmatrix} \dfrac{\partial^2 H}{\partial q_j \partial p_i} & \dfrac{\partial^2 H}{\partial p_j \partial p_i} \\[2mm] -\dfrac{\partial^2 H}{\partial q_j \partial q_i} & -\dfrac{\partial^2 H}{\partial p_j \partial q_i} \end{pmatrix} dt + \begin{pmatrix} 0 & 0 \\[2mm] \dfrac{\partial Q_i^*}{\partial q_j} & \dfrac{\partial Q_i^*}{\partial p_j} \end{pmatrix} dt, \tag{4.115}$$

leading to

$$\det J = 1 + dt \sum_{i=1}^{N} \frac{\partial Q_i^*}{\partial p_i} + \mathcal{O}(dt^2). \tag{4.116}$$

Hence, Liouville's theorem does not hold and the phase-space volume occupied by the ensemble is not conserved.

For the total time derivative of ρ we find

$$\begin{aligned} \frac{d\rho}{dt} &= \frac{\partial \rho}{\partial t} + \sum_{i=1}^{N} \left(\frac{\partial \rho}{\partial q_i} \dot{q}_i + \frac{\partial \rho}{\partial p_i} \dot{p}_i \right) \\ &= \frac{\partial \rho}{\partial t} + \underbrace{\sum_{i=1}^{N} \left[\frac{\partial(\rho \dot{q}_i)}{\partial q_i} + \frac{\partial(\rho \dot{p}_i)}{\partial p_i} \right]}_{=0} - \rho \sum_{i=1}^{N} \frac{\partial Q_i^*}{\partial p_i}. \end{aligned} \tag{4.117}$$

The first two terms correspond to the continuity equation (4.102) and thus vanish. Hence, we have

$$\boxed{\frac{d\rho}{dt} = \frac{\partial \rho}{\partial t} + [\rho, H] = -\rho \sum_{i=1}^{N} \frac{\partial Q_i^*}{\partial p_i}} \tag{4.118}$$

for dissipative systems, which replaces Liouville's equation (4.99).

4.4 Hamilton–Jacobi Equations

In the Hamilton–Jacobi approach, one seeks to transform *all* new canonical coordinates Q_i and momenta P_i into constants,

$$Q_i = \alpha_i = const, \qquad P_i = \beta_i = const. \tag{4.119}$$

The corresponding inverse transformations, $q_i = q_i(Q, P, t) = q_i(\alpha, \beta, t)$ and $p_i = p_i(Q, P, t) = p_i(\alpha, \beta, t)$, will then immediately provide the complete solution of the problem since the Q's and P's are constant and the q's and p's, therefore, can only depend on time.

Since the canonical equations for constant P and Q are

$$\frac{\partial K}{\partial Q_i} = -\dot{P}_i = 0, \qquad \frac{\partial K}{\partial P_i} = \dot{Q}_i = 0, \tag{4.120}$$

the new Hamilton function

$$K = H + \frac{\partial \Phi}{\partial t} \tag{4.121}$$

can only depend on time. The problem becomes particularly simple if we choose a generating function Φ that makes K vanish identically. Denoting the generating function that produces $K \equiv 0$ by S, it must satisfy

$$H(q, p, t) + \frac{\partial S}{\partial t} = 0. \tag{4.122}$$

This generating function S is called Hamilton's *action*, or *principal*, *function*. We take the variable dependence of S to be that of a Φ_2 generator, i.e., $S = S(q, P, t) = S(q, \beta, t)$. This provides the transformation relations

$$p_j = \frac{\partial S(q, \beta, t)}{\partial q_j}, \qquad Q_j = \frac{\partial S(q, \beta, t)}{\partial \beta_j} = \alpha_j, \tag{4.123}$$

where (4.119) was used already. Replacing the p's by partial derivatives, Eq. (4.122) now becomes

$$\boxed{H\left(q_1, \ldots, q_N, \frac{\partial S}{\partial q_1}, \ldots, \frac{\partial S}{\partial q_N}, t\right) + \frac{\partial S}{\partial t} = 0} \, . \tag{4.124}$$

This is the *Hamilton–Jacobi equation*; it is a first-order partial differential equation for the function S in $N+1$ variables q_i and t since, by construction, the $S = S(q, \beta, t)$ that goes into this equation is only a function of the $N + 1$ independent variables q_i and t, with $P_i = const$. The solution, therefore, does not automatically provide the constant P_i's. The complete solution, however, yields $N + 1$ integration constants $b_1, \ldots, b_{N+1}$. Since S itself does not enter the Hamilton–Jacobi equation (only its derivatives),

S is only determined up to an additive constant which we may choose to be equal to b_{N+1}. It has no bearing on the desired transformation and can be ignored. We may now identify, without loss of generality, the remaining N integration constants $b_1, \ldots, b_N$ with the constant momenta $\beta_1, \ldots, \beta_N$, i.e., $b_i = \beta_i$.[88] The solution $S = S(q, \beta, t)$ now fully determines the transformation equations (4.123). Inversion of the right-most equation in (4.123), in particular, provides the desired trajectories

$$q_i = q_i(\alpha, \beta, t) \tag{4.125}$$

as functions of time only, with the $2N$ constants α_i and β_i determined by the initial conditions.

Equivalence to Hamilton's principle

By construction, Hamilton's action function S is the generator of a canonical transformation to constant coordinates and momenta. Since the solution of the Hamilton–Jacobi equations yields the complete solution of the mechanical problem, they therefore provide yet *another equivalent basic formulation of classical mechanics*. This equivalence between the Hamilton–Jacobi equation (a first-order partial differential equation) and the $2N$ canonical equations of motion (a set of ordinary first-order differential equations) can be seen immediately by considering their relation to Hamilton's variational principle. For $S = S(q, \beta, t)$, one has

$$\frac{dS}{dt} = \sum_{i=1}^{N} \frac{\partial S}{\partial q_i} \dot{q}_i + \frac{\partial S}{\partial t} = \sum_{i=1}^{N} p_j \dot{q}_i - H = L, \tag{4.126}$$

where Eqs. (4.123), (4.122), and (4.8) were used. The action function S, therefore, differs only by an additive constant from the action integral of Hamilton's principle, i.e.,

$$S = \int dt\, L + const. \tag{4.127}$$

Since S is only determined up to an additive constant by the Hamilton–Jacobi equations, the constant here may be chosen as zero without loss of generality.

---

[88]More generally, we may also choose the β_i as arbitrary independent functions of the N integration constants, i.e., $\beta_i = \beta_i(b_1, \ldots, b_N)$. This does not alter the physical solutions since, whatever the choice, they will be determined by the initial conditions.

4.4.1   *Time-independent Hamilton function*

If the Hamilton function does not explicitly depend on time, the Hamilton–Jacobi equation reads

$$H\left(q, \frac{\partial S}{\partial q}\right) + \frac{\partial S}{\partial t} = 0 \tag{4.128}$$

and then S may be at most linear in time, of the form

$$S(q, \beta, t) = W(q, \beta) - Et, \tag{4.129}$$

where E is one of the N (non-trivial) integration constants of (4.124), i.e., one of the β_j is equal to E. The *time-independent Hamilton–Jacobi equation* then becomes

$$H\left(q_1, \ldots, q_N, \frac{\partial W}{\partial q_1}, \ldots, \frac{\partial W}{\partial q_N}\right) = E, \tag{4.130}$$

which is a partial differential equation for W. The constant E here is the energy, of course. Apart from the time-independence, the solution strategy proceeds essentially as for the time-dependent case.

4.4.2   *Separation of variables*

Partial differential equations are notoriously difficult to solve. In general, therefore, the Hamilton–Jacobi equations offer little practical advantage for the solution of a problem *unless* the principal function S can be split into a sum of functions with each depending on separate sets of variables. A variable q_j is called *separable* if the principal function has the form

$$S(q, \beta, t) = S_j(q_j, \beta, t) + S'(q_1, \ldots, q_{j-1}, q_{j+1}, \ldots, q_N, \beta, t) \tag{4.131}$$

and the Hamilton–Jacobi equations can be split into two equations, one involving only S_j and the other only S'. The Hamilton–Jacobi equations become *completely separable* for

$$S(q, \beta, t) = \sum_{j=1}^{N} S_j(q_j, \beta, t) \; ; \tag{4.132}$$

one then has

$$H_j\left(q_j, \frac{\partial S_j}{\partial q_j}, t\right) + \frac{\partial S_j}{\partial t} = 0, \qquad j = 1, \ldots, N. \tag{4.133}$$

If the Hamilton function does not depend on the time, the S_j can be written as

$$S_j(q_j, \beta, t) = W_j(q_j, \beta) - \beta_j t, \tag{4.134}$$

according to the results of the preceding section, and the Hamilton–Jacobi equations assume then the simple form

$$H_j\left(q_j, \frac{dW_j(q,\beta)}{dq_j}\right) = \beta_j, \qquad j = 1,\ldots,N. \tag{4.135}$$

These equations constitute now a set of *ordinary* first-order differential equations that can be solved completely by quadrature. For this completely separable case, the *separation constants* β_j need not be identifiable with energies, of course.

It is easy to show that *cyclic variables are always separable*. Taking H to be time-independent, the corresponding Hamilton–Jacobi equation is

$$H\left(q_2,\ldots,q_N,\gamma_1,\frac{\partial W}{\partial q_2},\ldots,\frac{\partial W}{\partial q_N}\right) = \beta_1, \tag{4.136}$$

where we have assumed q_1 to be cyclic, with the associated momentum $p_1 = \gamma_1 = const$. The separation ansatz

$$W = W_1(q_1,\beta) + W'(q_2,\ldots,q_N,\beta) \tag{4.137}$$

then immediately leads to

$$H\left(q_2,\ldots,q_N,\gamma_1,\frac{\partial W'}{\partial q_2},\ldots,\frac{\partial W'}{\partial q_N}\right) = \beta_1, \tag{4.138}$$

and

$$\frac{\partial W}{\partial q_1} = \frac{dW_1}{dq_1} = p_1 = \gamma_1, \tag{4.139}$$

with the solution

$$W_1 = \gamma_1 q_1 \tag{4.140}$$

(up to irrelevant constants). For a cyclic variable q_1, we therefore find

$$W(q,\beta) = W'(q_2,\ldots,q_N,\beta) + \gamma_1 q_1. \tag{4.141}$$

Note that this implies that the separability of the Hamilton–Jacobi equations depends on the choice of variables. For example, the one-body central-force problem is separable in polar coordinates (see subsequent section), but not in Cartesian coordinates. Unfortunately, there are no general criteria that allow one to determine how to choose the variables that make a given problem separable (see [1] for more details).

4.4.3 Examples

Free fall

Considering only motion in the vertical xz-plane, the free fall in the field of gravity $\mathbf{g} = -g\mathbf{e}_z$ is governed by the time-independent Hamilton function

$$H = \frac{1}{2m}\left(p_x^2 + p_z^2\right) + mgz. \tag{4.142}$$

The corresponding Hamilton–Jacobi equation is

$$\frac{1}{2m}\left(\frac{\partial W(x,z,E,\beta)}{\partial x}\right)^2 + \frac{1}{2m}\left(\frac{\partial W(x,z,E,\beta)}{\partial z}\right)^2 + mgz = E, \tag{4.143}$$

where we have used the fact that, in addition to E, W can only depend on one more constant β. Making the separation ansatz

$$W = W_1(x,E,\beta) + W_2(z,E,\beta), \tag{4.144}$$

this then leads to

$$\left(\frac{\partial W_1}{\partial x}\right)^2 + \left(\frac{\partial W_2}{\partial z}\right)^2 + 2m^2 gz = 2mE. \tag{4.145}$$

Since the only term that depends on x is $(\partial W_1/\partial x)^2$, it must be equal to a constant, which we choose such that

$$\frac{dW_1}{dx} = \sqrt{\beta}\ ; \tag{4.146}$$

hence,

$$\frac{dW_2}{dz} = \sqrt{2mE - \beta - 2m^2 gz}. \tag{4.147}$$

Up to irrelevant additive constants, this integrates to

$$W_1 = \sqrt{\beta}x, \qquad W_2 = -\frac{1}{3m^2 g}\left(2mE - \beta - 2m^2 gz\right)^{3/2}, \tag{4.148}$$

and therefore

$$W(x,z,E,\beta) = \sqrt{\beta}x - \frac{1}{3m^2 g}\left(2mE - \beta - 2m^2 gz\right)^{3/2}. \tag{4.149}$$

The transformation equations then follow from (4.123) with $S = W - Et$; they are

$$p_x = \frac{\partial W}{\partial x} = \sqrt{\beta}, \tag{4.150}$$

$$p_z = \frac{\partial W}{\partial z} = \sqrt{2mE - \beta - 2m^2 gz}, \tag{4.151}$$

$$\frac{\partial W}{\partial E} - t = -\frac{1}{mg}\sqrt{2mE - \beta - 2m^2 gz} - t = \alpha_1. \tag{4.152}$$

$$\frac{\partial W}{\partial \beta} = \frac{x}{2\sqrt{\beta}} + \frac{1}{2m^2 g}\sqrt{2mE - \beta - 2m^2 gz} = \alpha_2. \tag{4.153}$$

These equations provide $x(t)$, $z(t)$, $p_x(t)$, and $p_z(t)$. Solved for z, Eq. (4.152) yields

$$z(t) = -\frac{g}{2}t^2 + c_1 t + c_2, \tag{4.154}$$

with some constants c_1, c_2. Similarly, we find the usual parabola in the xz-plane from (4.153). This fully determines $x(t)$ and $z(t)$. Equations (4.150) and (4.151) then provide $p_x = m\dot{x}$ and $p_z = m\dot{z}$.

Central-force problem

We know from the treatment of the central-force problem in Sec. 1.4.2.1 that the corresponding motion is confined to a plane. Nevertheless, to see how this simplification is obtained within the Hamilton–Jacobi formalism, we will treat the problem here in three dimensions. In spherical coordinates, the Hamilton function of the central-force problem then is [cf. (4.15)]

$$H(r, \theta, p_r, p_\varphi, p_\theta) = \frac{1}{2m}\left(p_r^2 + \frac{p_\varphi^2}{r^2 \sin^2\theta} + \frac{p_\theta^2}{r^2}\right) + V(r), \tag{4.155}$$

where the azimuthal angle φ is cyclic. Making a separation ansatz

$$W = W_r(r) + W_\theta(\theta) + W_\varphi(\varphi), \tag{4.156}$$

we can then deduce immediately

$$W_\varphi = \gamma_\varphi \varphi, \tag{4.157}$$

where

$$\gamma_\varphi \equiv p_\varphi = \frac{\partial W}{\partial \varphi} = \frac{dW_\varphi}{d\varphi} \tag{4.158}$$

is the constant canonical momentum associated with φ. Identifying the constant Hamilton function with the conserved total energy E, the Hamilton–Jacobi equation $H = E$ then can be written as

$$-r^2\left[\left(\frac{dW_r}{dr}\right)^2 + 2mV(r) - 2mE\right] = \left(\frac{dW_\theta}{d\theta}\right)^2 + \frac{\gamma_\varphi^2}{\sin^2\theta}. \tag{4.159}$$

The left hand-side here depends only on r whereas the right-hand side is only a function of θ. In view of the independence of the two variables, both sides must be equal to a common constant. Denoting this (positive) constant by ℓ^2, the Hamilton–Jacobi equations reduce to two separate equations,

$$\left(\frac{dW_\theta}{d\theta}\right)^2 + \frac{\gamma_\varphi^2}{\sin^2\theta} = \ell^2, \qquad (4.160)$$

and

$$\left(\frac{dW_r}{dr}\right)^2 + \frac{\ell^2}{r^2} + 2mV(r) = 2mE, \qquad (4.161)$$

thus providing a set of single-variable ordinary differential equations that can be solved by quadrature. The last equation, in particular, corresponds to the (time-independent) Hamilton function of the equivalent one-dimensional problem [cf. (1.310)]

$$H = \frac{p_r^2}{2m} + \frac{\ell^2}{2mr^2} + V(r) = E. \qquad (4.162)$$

With $p_r = m\dot{r}$, this immediately leads to the familiar differential equation (1.311), i.e.,

$$dt = \frac{dr}{\sqrt{\frac{2}{m}\left[E - V(r) - \frac{\ell^2}{2mr^2}\right]}}. \qquad (4.163)$$

Obviously, the separation constant ℓ is the magnitude of the conserved total angular momentum.[89] Writing Eq. (4.160) in the form

$$p_\theta^2 + \frac{p_\varphi^2}{\sin^2\theta} = \ell^2 \quad \longrightarrow \quad d\psi \equiv \sqrt{d\theta^2 + \sin^2\theta\, d\varphi^2} = \frac{\ell}{mr^2}dt, \qquad (4.164)$$

this provides the angular orbit variable ψ in the fixed orbital plane of the motion. Usually this plane is chosen as the xy-plane (with $\theta = 90°$ and $d\theta = 0$ producing $d\psi \equiv d\varphi$).

---

[89]The angular momentum vector $\mathbf{L}$ is given by $\mathbf{L} = \mathbf{r} \times \mathbf{p} = p_\theta \mathbf{e}_\varphi - p_\varphi \mathbf{e}_\theta / \sin\theta$.

Chapter 5

Mechanics of Rigid Bodies

A *rigid body* is a system of n point masses m_i at positions $\mathbf{r}_i$ such that all pairwise distances

$$|\mathbf{r}_i - \mathbf{r}_j| = c_{ij} = const$$

are kept fixed at constant values. Rigid bodies, thus, are (idealized) models of solid objects whose internal motion (vibration, torsion, etc.) can be neglected.

For $n = 3$, there are $l = 3$ constraints. Of the $3n = 9$ coordinates, therefore, only $N = 3n - l = 6$ remain independent. The three components of the position of each additional point mass are fixed by three distances to the three original mass points; in other words, no additional degrees of freedom will arise when adding more particles. Hence, for $n \geq 3$:

> A rigid body possesses $N = 6$ degrees of freedom.

Kinematically, these degrees of freedom are divided into three degrees of freedom for translational and three for rotational motion. (For $n = 2$, there are only five — three translational and two rotational — degrees of freedom; and for $n = 1$, only the three translational degrees of freedom remain.)

5.1 Kinematics

The description of the location and orientation of a rigid body in space is facilitated by introducing a coordinate system B that is fixed within the body and then consider its kinematical relation to an inertial coordinate system S fixed in space (see Fig. 5.1). The three translational degrees of freedom correspond to the three independent components of the displacement vector $\mathbf{r}_0$ that provides the location O_B of B's origin with respect to the origin O_S of S. The remaining three degrees of freedom of rotation

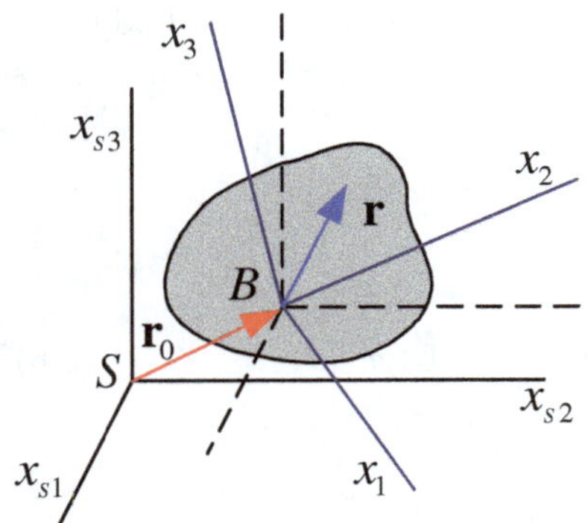

Fig. 5.1 Relation between external inertial coordinate system S, with coordinates x_{s1}, x_{s2}, x_{s3}, and a coordinate system B fixed in the rigid body, with coordinates x_1, x_2, x_3. The origin of B is displaced from the origin of S by a vector $\mathbf{r}_0$. The axes of B are rotated with respect to the (dashed) axes obtained by parallel displacement of S. Also shown is an arbitrary vector $\mathbf{r}$; the *same* vector can be represented as $\vec{r}_B$ within B or as $\vec{r}_S$ within S [see Eq. (5.7)].

pertain to the orientation of the axes of B with respect to the axes obtained by parallel translation of S so that O_S and O_B coincide. The treatment of the translation motion follows the procedures established previously. The major new feature in understanding the mechanics of a rigid body involves the kinematics and dynamics of its rotation.

To develop a description for this rotation, we use the notation

	coordinates	unit vectors	
inertial space system S:	x_{s1}, x_{s2}, x_{s3}	$\mathbf{u}_1$, $\mathbf{u}_2$, $\mathbf{u}_3$	(5.1)
body system B:	x_1, x_2, x_3	$\mathbf{e}_1$, $\mathbf{e}_2$, $\mathbf{e}_3$	

for the coordinates and unit vectors of S and B. Sometimes, the notation $(x, y, z) \equiv (x_{s1}, x_{s2}, x_{s3})$ will also be employed. Note that, in general, the unit vectors $\mathbf{e}_i = \mathbf{e}_i(t)$ of the body axes are time-dependent, i.e., in general, the body system B is not an inertial frame.

5.1.1 Rotations: orthogonal transformations

The unit vectors $\mathbf{e}_i$ of B are related to the unit vectors $\mathbf{u}_j$ of S by [cf. Eq. (1.9)]

$$\mathbf{e}_i = \mathbf{e}_i \cdot \mathbf{1} = \sum_{j=1}^{3} (\mathbf{e}_i \cdot \mathbf{u}_j)\mathbf{u}_j = \sum_{j=1}^{3} D_{ij}\mathbf{u}_j. \tag{5.2}$$

where

$$D_{ij} = \mathbf{e}_i \cdot \mathbf{u}_j = \cos(\mathbf{e}_i, \mathbf{u}_j) \tag{5.3}$$

is the *direction cosine* between the unit vectors $\mathbf{e}_i$ and $\mathbf{u}_j$. Conversely, we have

$$\mathbf{u}_i = \mathbf{u}_i \cdot \mathbf{1} = \sum_{j=1}^{3} (\mathbf{u}_i \cdot \mathbf{e}_j)\mathbf{e}_j = \sum_{j=1}^{3} D_{ji}\mathbf{e}_j. \tag{5.4}$$

The nine direction cosines D_{ij} thus describe how the axes of the body system B are rotated with respect to the axes of S. The respective components x_i and x_{sj} in B and S of a given vector $\mathbf{r}$,

$$x_i = \mathbf{r} \cdot \mathbf{e}_i, \qquad \text{and} \qquad x_{sj} = \mathbf{r} \cdot \mathbf{u}_j, \tag{5.5}$$

are related by

$$x_i = \mathbf{r} \cdot \mathbf{e}_i = \sum_{j=1}^{3} D_{ij}\mathbf{r} \cdot \mathbf{u}_j = \sum_{j=1}^{3} D_{ij}x_{sj}. \tag{5.6}$$

Using the column-vector representations (see Sec. 1.1.1)

$$\mathbf{r} \to \vec{r}_B = \begin{pmatrix} x_1 \\ x_2 \\ x_3 \end{pmatrix} \qquad \text{and} \qquad \mathbf{r} \to \vec{r}_S = \begin{pmatrix} x_{s1} \\ x_{s2} \\ x_{s3} \end{pmatrix} \tag{5.7}$$

for the vector $\mathbf{r}$ in both systems, Eq. (5.6) can be written as

$$\vec{r}_B = D\vec{r}_S \quad \longleftrightarrow \quad \begin{pmatrix} x_1 \\ x_2 \\ x_3 \end{pmatrix} = \begin{pmatrix} D_{11} & D_{12} & D_{13} \\ D_{21} & D_{22} & D_{23} \\ D_{31} & D_{32} & D_{33} \end{pmatrix} \begin{pmatrix} x_{s1} \\ x_{s2} \\ x_{s3} \end{pmatrix}, \tag{5.8}$$

where the elements of the matrix D are given by the direction cosines. D is called the *rotation matrix* that relates the representation $\vec{r}_S$ to the representation $\vec{r}_B$. This particular rotation is called *passive* since it applies to the description of the *same vector* in two *different systems* that are rotated with respect to each other.

The nine elements D_{ij} of the rotation matrix are not independent. Indeed, we find from the orthogonality of the unit vectors,

$$\delta_{ij} = \mathbf{e}_i \cdot \mathbf{e}_j = \left(\sum_{k=1}^{3} D_{ik}\mathbf{u}_k \right) \cdot \left(\sum_{l=1}^{3} D_{jl}\mathbf{u}_l \right)$$

$$= \sum_{k,l=1}^{3} D_{ik}D_{jl}\mathbf{u}_k \cdot \mathbf{u}_l = \sum_{k,l=1}^{3} D_{ik}D_{jl}\delta_{kl}$$

$$= \sum_{k=1}^{3} D_{ik}D_{jk}. \tag{5.9}$$

This result can also be written in matrix form,

$$\boxed{DD^{\mathrm{T}} = D^{\mathrm{T}}D = \mathbb{1} \quad \longleftrightarrow \quad \sum_{k=1}^{3} D_{ik}D_{jk} = \sum_{k=1}^{3} D_{ki}D_{kj} = \delta_{ij}}\,,$$

$$\tag{5.10}$$

where the corresponding transposed relation is included here as well. This *orthogonality condition*, when written out in detail, provides six conditions for the nine real elements of D, leaving *three independent parameters* that completely determine D, and, hence, there are three degrees of freedom for rotations of the rigid body. Geometrically, the orthogonality condition is a reflection of the fact that the length of a vector does not change under a rotation,

$$\mathbf{r} \cdot \mathbf{r} = r^2 = \sum_{i=1}^{3} x_i^2 = \sum_{i=1}^{3} \left(\sum_{j=1}^{3} D_{ij} x_{sj} \right) \left(\sum_{k=1}^{3} D_{ik} x_{sk} \right)$$
$$= \sum_{j,k=1}^{3} \left(\sum_{i=1}^{3} D_{ij} D_{ik} \right) x_{sj} x_{sk}$$
$$= \sum_{j=1}^{3} x_{sj}^2, \tag{5.11}$$

where the last equality is valid if and only if the orthogonality condition holds true.

Orthogonal transformations, in general, are characterized by the fact that their inverse is equal to the transposed,

$$D^{-1} = D^{\mathrm{T}}. \tag{5.12}$$

For $C = BA$, where A, B are orthogonal transformations, one finds

$$CC^{\mathrm{T}} = (BA)(BA)^{\mathrm{T}} = BAA^{\mathrm{T}}B^{\mathrm{T}} = BB^{\mathrm{T}} = \mathbb{1}, \tag{5.13}$$

in other words, successive orthogonal transformations are orthogonal themselves, and one easily verifies that orthogonal transformations form a *group*.[90] For rotations in three dimensions, this orthogonal group is called $O(3)$. The group properties, in particular, imply that *successive rotations can be expressed as a single rotation*. The real 3×3 rotation matrices are a representation of this group, with the usual matrix multiplication being the group composition rule. The orthogonality condition shows that

$$1 = \det \mathbb{1} = \det (DD^{\mathrm{T}}) = (\det D)(\det D^{\mathrm{T}}) = (\det D)^2, \tag{5.14}$$

---

[90] A group $\mathcal{G} = (G, \circ)$ is defined as a set of elements $G = (a, b, c, \ldots)$ with a composition rule "$\circ$" that specifies how two group elements are to be 'multiplied' with each other. The basic rules (axioms) for a group are: **Closure:** If a and b are in the group, then $c = a \circ b$ is also in the group. **Associativity:** If a, b, and c are in the group, then $(a \circ b) \circ c = a \circ (b \circ c)$. **Identity:** There is an element e of the group such that for any element a of the group, $a \circ e = e \circ a = a$. **Inverse:** For any element a of the group, there is an element a^{-1} such that $a \circ a^{-1} = a^{-1} \circ a = e$. — A simple example for a group is the set of all integer numbers with the composition rule of the usual addition, "$+$". Closure and associativity are obvious, the identity element is 0, and the negative integers are the inverses of the respective positive ones and *vice versa*.

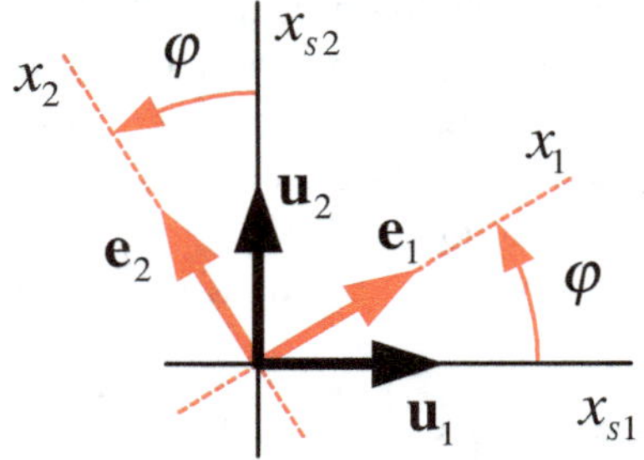

Fig. 5.2   The body system B, with unit vectors $\mathbf{e}_1$, $\mathbf{e}_2$, is obtained here from the inertial system S, with unit vectors $\mathbf{u}_1$, $\mathbf{u}_2$, by a planar rotation by an angle φ about the x_{S3}-axis, i.e., the z-axes of both B and S coincide and point out of the plane of the page.

i.e.,

$$\det D = \pm 1. \tag{5.15}$$

Proper rotations evolve continuously from the identity rotation, $\mathbb{1}$. Since the latter has $\det \mathbb{1} = +1$, proper rotations always have a determinant $+1$. They form the subgroup $SO(3)$ of $O(3)$, written as $SO(3) \subset O(3)$. Elements with determinant -1 do not form a group (e.g., they lack the identity element, $\mathbb{1}$); they can be related to the elements of $SO(3)$ by multiplication by $\mathsf{P} = -\mathbb{1}$. The *parity operator* P has $\det \mathsf{P} = -1$ and it changes a right-handed into a left-handed coordinate system, and *vice versa*. Hence, the 'rotations' with determinant -1 involve a discontinuous mirror operation.

Example: Rotation about the z-axis

For a planar rotation about the z-axis of S (see Fig. 5.2) by an angle φ, the rotation matrix is given by

$$D_z = \begin{pmatrix} \mathbf{e}_1 \cdot \mathbf{u}_1 & \mathbf{e}_1 \cdot \mathbf{u}_2 & \mathbf{e}_1 \cdot \mathbf{u}_3 \\ \mathbf{e}_2 \cdot \mathbf{u}_1 & \mathbf{e}_2 \cdot \mathbf{u}_2 & \mathbf{e}_2 \cdot \mathbf{u}_3 \\ \mathbf{e}_3 \cdot \mathbf{u}_1 & \mathbf{e}_3 \cdot \mathbf{u}_2 & \mathbf{e}_3 \cdot \mathbf{u}_3 \end{pmatrix} = \begin{pmatrix} \cos\varphi & \sin\varphi & 0 \\ -\sin\varphi & \cos\varphi & 0 \\ 0 & 0 & 1 \end{pmatrix}, \tag{5.16}$$

i.e., it depends on a single angle. The remaining two parameters of the three-dimensional rotation are determined by the fact that the rotation axis is given by $\mathbf{u}_3 = \mathbf{e}_3$ (since a unit vector is fully specified by two parameters).

5.1.1.1   Euler angles

One of the standard ways of describing the three rotational degrees of freedom about an arbitrary axis is by means of the three *Euler angles*. They specify how the system B is obtained from S by the three successive rotations shown in Fig. 5.3. The Euler angles may be used as generalized

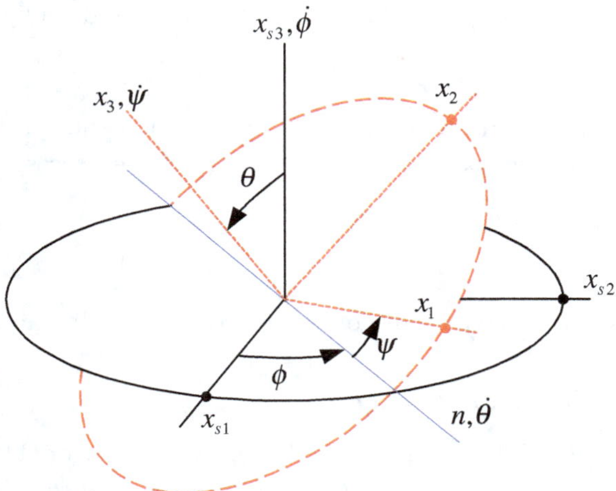

Fig. 5.3   The body system B (dashed and dotted lines) is obtained from the inertial system S (solid lines) by three successive rotations by the Euler angles ϕ, θ, and ψ: First, a rotation by ϕ about the x_{s3}-axis brings the intermediate x_1-axis into the node line n. Second, a tilt of the x_{s3}-axis around the node line by θ brings x_{s3} into the new x_3-axis. And third, a rotation about the new x_3-axis by an angle ψ brings the intermediate x_1-axis from the node line to its final direction. This completely determines the orientation of the body's (x_1, x_2, x_3) axes with respect to the external (x_{s1}, x_{s2}, x_{s3}) system. The angular velocities $\dot{\phi}$, $\dot{\theta}$, and $\dot{\psi}$ associated with each rotation are indicated at the respective rotation axis.

coordinates in the Lagrange and Hamilton functions to describe the degrees of freedom of rotation.

Each step involves a rotation about a particular axis and, therefore, corresponds to a planar rotation similar in form to Eq. (5.16). Consulting Fig. 5.3, the first rotation is one about the z-axis by an angle ϕ,

$$D_z(\phi) = \begin{pmatrix} \cos\phi & \sin\phi & 0 \\ -\sin\phi & \cos\phi & 0 \\ 0 & 0 & 1 \end{pmatrix}. \tag{5.17}$$

The subsequent tilt of the xy-plane by an angle θ around the node line corresponds to a rotation

$$D_x(\theta) = \begin{pmatrix} 1 & 0 & 0 \\ 0 & \cos\theta & \sin\theta \\ 0 & -\sin\theta & \cos\theta \end{pmatrix} \tag{5.18}$$

around the new x-axis. The final rotation by ψ,

$$D_z(\psi) = \begin{pmatrix} \cos\psi & \sin\psi & 0 \\ -\sin\psi & \cos\psi & 0 \\ 0 & 0 & 1 \end{pmatrix}, \tag{5.19}$$

is about the new x_3-axis. An arbitrary rotation thus is given by the product

$$D_{BS}(\phi, \theta, \psi) = D_z(\psi)\, D_x(\theta)\, D_z(\phi). \tag{5.20}$$

It should be emphasized that this matrix specifies how the body coordinates are obtained from the space coordinates. The reverse process,

$$D_{SB}(\phi, \theta, \psi) = D_{BS}^{\mathrm{T}}(\phi, \theta, \psi) = D_z^{\mathrm{T}}(\phi)\, D_x^{\mathrm{T}}(\theta)\, D_z^{\mathrm{T}}(\psi)$$
$$= D_z(-\phi)\, D_x(-\theta)\, D_z(-\psi), \tag{5.21}$$

describes how the space axes are obtained from the body axes.

5.1.1.2 *Infinitesimal rotations*

The preceding description of rotations in terms of orthogonal 3×3 rotation matrices shows that, in general, the end result of several rotations depends on the particular sequence of rotations. Mathematically, this property of rotations is reflected in the fact that matrix multiplication is not commutative, i.e, for two rotations D_1 and D_2, in general $D_1 D_2 \neq D_2 D_1$.

The situation becomes much simpler if we consider *infinitesimal rotations*. We shall show presently that they can be represented by *vectors*, instead of matrices. Successive infinitesimal rotations will then be given by adding the corresponding vectors. The resulting vector does not depend on the particular sequence of performing this addition, and, hence, *infinitesimal rotations commute*.

For infinitesimal rotations, the resulting vector $\mathbf{r}'$ after the rotation differs from the original vector $\mathbf{r}$ only by an infinitesimal amount. Therefore, we can write

$$\mathbf{r}' = D_\epsilon \mathbf{r} = (\mathbb{1} + \epsilon)\,\mathbf{r}, \tag{5.22}$$

i.e., the infinitesimal rotation matrix

$$D_\epsilon = \mathbb{1} + \epsilon \tag{5.23}$$

differs from unity only by an infinitesimal matrix ϵ, such that $\epsilon^2 \approx 0$. Since for two such infinitesimal rotations we have

$$D_{\epsilon_1} D_{\epsilon_2} = (\mathbb{1} + \epsilon_1)(\mathbb{1} + \epsilon_2) = \mathbb{1} + \epsilon_1 + \epsilon_2 + \epsilon_1\epsilon_2$$
$$= \mathbb{1} + \epsilon_1 + \epsilon_2, \tag{5.24a}$$
$$D_{\epsilon_2} D_{\epsilon_1} = (\mathbb{1} + \epsilon_2)(\mathbb{1} + \epsilon_1) = \mathbb{1} + \epsilon_1 + \epsilon_2 + \epsilon_2\epsilon_1$$
$$= \mathbb{1} + \epsilon_1 + \epsilon_2, \tag{5.24b}$$

such rotations indeed commute. The inverse of D_ϵ is then immediately found as

$$D_\epsilon^{-1} = \mathbb{1} - \epsilon. \tag{5.25}$$

On the other hand, the orthogonality condition provides

$$D_\epsilon^{-1} = D_\epsilon^{\mathrm{T}} = (\mathbb{1} + \epsilon)^{\mathrm{T}} = \mathbb{1} + \epsilon^{\mathrm{T}}, \tag{5.26}$$

which means that ϵ must satisfy the condition

$$\epsilon = -\epsilon^{\mathrm{T}}. \tag{5.27}$$

This is the definition of a *skew-symmetric*, or *anti-symmetric*, matrix. Skew-symmetric matrices have vanishing diagonal elements and, in three dimensions, only three independent elements. Without loss of generality, we may then write

$$\epsilon = \begin{pmatrix} 0 & -d\varphi_3 & d\varphi_2 \\ d\varphi_3 & 0 & -d\varphi_1 \\ -d\varphi_2 & d\varphi_1 & 0 \end{pmatrix}, \tag{5.28}$$

where $d\varphi_1$, $d\varphi_2$, and $d\varphi_3$ correspond to the three independent parameters that determine a rotation.

The change of the vector $\mathbf{r}$ due to the rotation is then

$$d\mathbf{r} = \mathbf{r}' - \mathbf{r} = \epsilon\,\mathbf{r} = \begin{pmatrix} x_3\,d\varphi_2 - x_2\,d\varphi_3 \\ x_1\,d\varphi_3 - x_3\,d\varphi_1 \\ x_2\,d\varphi_1 - x_1\,d\varphi_2 \end{pmatrix}. \tag{5.29}$$

The right-hand side here is a vector cross product, i.e.,

$$\boxed{\text{infinitesimal rotation:} \qquad d\mathbf{r} = d\boldsymbol{\varphi} \times \mathbf{r}}, \tag{5.30}$$

where the components of the vector

$$d\boldsymbol{\varphi} = \begin{pmatrix} d\varphi_1 \\ d\varphi_2 \\ d\varphi_3 \end{pmatrix} \tag{5.31}$$

are the three infinitesimal parameters of ϵ.

Equation (5.30) reproduces the result found already in Eq. (1.252) of Sec. 1.7.2. It describes the infinitesimal *active* rotation of a vector $\mathbf{r}$ about an axis $d\boldsymbol{\varphi}$ by an amount $d\varphi = |d\boldsymbol{\varphi}|$ (see also Fig. 1.13). We consider here active rotations since in the following, we will describe the changes of the position vectors of the point masses of the rotating rigid body *as seen from the inertial frame S*.

5.1.1.3 *Lie-algebra interlude*

The rotation group $SO(3)$ of 3×3 rotation matrices $D(\phi, \theta, \psi)$ is an example of a *continuous group* whose members depend on continuous parameters and therefore can be differentiated. They, in particular, evolve continuously from the identity and thus can be studied by investigating the corresponding infinitesimal transformation of the generic structure $T(\varepsilon) = 1 + \varepsilon G$, where ε is an infinitesimal parameter and $G = dT/d\varepsilon|_{\varepsilon=0}$ (up to some constant factor) is called a generator of the group. Such groups are called *Lie groups*.[91] Many of the important transformations in physics can indeed be described as continuous deformations of the identity operation, which makes the study of Lie groups very relevant for physics. Understanding Lie groups is greatly facilitated by investigating the relationship between the generators of the group in question. We will show here some of the pertinent findings for the example of $SO(3)$. However, what follows is by no means meant to be a comprehensive treatment and we refer to the mathematical-physics literature for more details on Lie groups and Lie algebras (see, for example, [8] or [9]).

By inspection of Eq. (5.28), we see that introducing the three *Hermitian* matrices[92]

$$
\mathsf{J}_x = i\begin{pmatrix} 0 & 0 & 0 \\ 0 & 0 & -1 \\ 0 & 1 & 0 \end{pmatrix}, \qquad
\mathsf{J}_y = i\begin{pmatrix} 0 & 0 & 1 \\ 0 & 0 & 0 \\ -1 & 0 & 0 \end{pmatrix},
$$

$$
\mathsf{J}_z = i\begin{pmatrix} 0 & -1 & 0 \\ 1 & 0 & 0 \\ 0 & 0 & 0 \end{pmatrix},
\tag{5.32}
$$

this equation can be written as

$$
\epsilon = -i\, d\boldsymbol{\varphi} \cdot \mathbf{J},
\tag{5.33}
$$

where $\mathbf{J}$ is the vector-like structure

$$
\mathbf{J} = (\mathsf{J}_x, \mathsf{J}_y, \mathsf{J}_z).
\tag{5.34}
$$

Each element of this 'vector' here is a matrix that by itself can 'operate' on the true three-dimensional vectors of the three-dimensional space, and

---

[91]Sophus Lie (1842–1899), Norwegian mathematician.

[92]The elements of the matrix J_k are given by $(\mathsf{J}_k)_{lm} = -i\varepsilon_{klm}$. The factors i here are not necessary; they are introduced to make the resulting expressions compatible with what one usually uses in quantum mechanics.

hence a structure like this is called a *vector operator*.[93] The infinitesimal rotation (5.23) is now written as

$$D_\epsilon = D_\varepsilon(\hat{\mathbf{n}}, \varphi) = \mathbb{1} - i\,\varepsilon\,\varphi\hat{\mathbf{n}} \cdot \mathbf{J}, \tag{5.35}$$

where φ is some fixed finite angle and ε is an infinitesimal parameter given by

$$\varepsilon\varphi = |d\boldsymbol{\varphi}| \tag{5.36}$$

and

$$\hat{\mathbf{n}} = \frac{d\boldsymbol{\varphi}}{|d\boldsymbol{\varphi}|} \tag{5.37}$$

is the unit vector along the rotation axis.

The significance of this procedure is seen immediately by writing $\varepsilon = 1/N$, where N is a large integer, and considering N repeated applications of the same infinitesimal rotation, i.e.,

$$[D_\varepsilon(\hat{\mathbf{n}}, \varphi)]^N = \left(\mathbb{1} - i\frac{\varphi\hat{\mathbf{n}} \cdot \mathbf{J}}{N}\right)^N. \tag{5.38}$$

Clearly, in the limit of $N \to \infty$, this corresponds to a rotation by a finite angle φ about the fixed axis $\hat{\mathbf{n}}$ since for successive rotations about the same axis the rotation angles simply add up. One finds[94]

$$\lim_{N\to\infty} \left(\mathbb{1} - i\frac{\varphi\,\hat{\mathbf{n}} \cdot \mathbf{J}}{N}\right)^N = \lim_{N\to\infty} \sum_{k=0}^{N} \binom{N}{k} \left(-i\frac{\varphi\,\hat{\mathbf{n}} \cdot \mathbf{J}}{N}\right)^k$$

$$= \sum_{k=0}^{\infty} \frac{(-i\varphi\,\hat{\mathbf{n}} \cdot \mathbf{J})^k}{k!}$$

$$= e^{-i\varphi\,\hat{\mathbf{n}}\cdot\mathbf{J}}, \tag{5.39}$$

where the last step is a formal expression for the preceding Taylor series. This result,

$$\boxed{D(\hat{\mathbf{n}}, \varphi) = e^{-i\varphi\,\hat{\mathbf{n}}\cdot\mathbf{J}}}, \tag{5.40}$$

is the *unitary representation* of the rotation operation for an *active* finite rotation about an axis $\hat{\mathbf{n}}$ by an angle φ. Obviously, for rotations about any

---

[93]See also the paragraph surrounding Eq. (5.43).

[94]Use

$$\lim_{N\to\infty} \binom{N}{k}\frac{1}{N^k} = \frac{1}{k!} \qquad \text{(for any finite k).}$$

one of the Cartesian axis — say the i-th axis, i.e., $\hat{\mathbf{n}} \cdot \mathbf{J} = \mathsf{J}_i$ — only the corresponding matrix J_i would appear in the exponent. Hence, the J_i are called the *generators* of rotation about the i-th axis since they completely determine the structure of the resulting rotation operation.

Note that (5.40) is completely equivalent to expressing the rotation matrix in terms of the three Euler angles. One immediately finds that one can write

$$D(\phi, \theta, \psi) = \mathrm{e}^{-i\psi \mathsf{J}_z}\, \mathrm{e}^{-i\theta \mathsf{J}_x}\, \mathrm{e}^{-i\phi \mathsf{J}_z}, \tag{5.41}$$

where ϕ, θ, and ψ are the Euler angles as defined previously.

In the last relation in terms of the Euler angles, the order of the exponential expressions is significant since, in general, the generators J_i do not commute. One finds

$$\boxed{[\mathsf{J}_k, \mathsf{J}_l] = i \sum_{m=1}^{3} \varepsilon_{klm} \mathsf{J}_m, \quad \text{for} \quad k, l = 1, 2, 3,} \tag{5.42}$$

where $[A, B] = AB - BA$ denotes the *commutator* of the two matrices A and B. This equation defines the *commutator Lie algebra*[95] *of rotations* (or, more generically, *of angular momenta*). Its importance is due to the fact that this equation alone completely defines rotations since in three dimensions, up to trivial similarity transformations, only the matrices given in (5.32) solve this commutator relation.

Even more generally, taken as a *defining* relation between three *abstract* operators J_i, this algebra provides the larger group of *special unitary transformations in two dimensions*, $SU(2)$. One can find finite-dimensional realizations — called *representations* — of this algebra in terms of matrices of any dimension N, and one can also find infinite-dimensional realizations in terms of differential operators on function spaces. The $SO(3)$ matrices are but the three-dimensional representation of such a realization. In quantum mechanics, to name other examples, the N-dimensional representations allow one to describe particles of spin $S = (N - 1)/2$.

From a practical point of view, one of the most important aspects of working with generators is that we can now immediately determine whether three operators A_1, A_2, and A_3 — *any* three operators that act on the elements of the space in question — can be regarded as the components of a vector operator $\mathbf{A} = (A_1, A_2, A_3)$ since any vector must behave in analogy

---

[95]Loosely speaking, an *algebra* is a binary multiplicative structure $\mathscr{V} \times \mathscr{V} \to \mathscr{V}$ on the elements of a vector space $\mathscr{V}$ that forms a vector space itself. See the literature on mathematical physics for more details.

to (5.29) under rotations. We mention without proof that $\mathbf{A} = (A_1, A_2, A_3)$ is a vector operator if and only if the three A_i satisfy

$$\left[\mathsf{J}_k, A_l \right] = i \sum_{m=1}^{3} \varepsilon_{klm} A_m, \quad \text{for} \quad k, l = 1, 2, 3. \tag{5.43}$$

This result remains true whatever the space in question is ($\mathbb{R}^n$, function space, etc.) and whatever the detailed nature of the corresponding operators are (matrices, differential operators, etc.). Equation (5.42) shows then that $\mathbf{J}$ of Eq. (5.42) is indeed a vector operator for the present three-dimensional case.

5.1.2   *Angular velocities*

Next, let us look at the change of a generic position vector $\mathbf{r}$ of one mass point as seen from S. Denoting the same position in S by $\mathbf{r}_S$, we have (cf. Fig. 5.1)

$$\mathbf{r}_S = \mathbf{r}_0 + \mathbf{r}. \tag{5.44}$$

Its velocity in S is

$$\mathbf{v}_S = \frac{d\mathbf{r}_S}{dt} = \frac{d\mathbf{r}_0}{dt} + \frac{d\mathbf{r}}{dt}. \tag{5.45}$$

All time derivatives here describe the position changes as seen in the space-fixed inertial frame S. The time derivative

$$\mathbf{v}_0 = \frac{d\mathbf{r}_0}{dt} \tag{5.46}$$

is the velocity of the origin of B. Relative to B, the vector $\mathbf{r}$ is a constant since its position is fixed by the rigid-body constraints. Its change in S can only be due to a possible rotation of B relative to S. According to (5.30), it is given by

$$\frac{d\mathbf{r}}{dt} = \frac{d\boldsymbol{\varphi}}{dt} \times \mathbf{r} = \boldsymbol{\omega} \times \mathbf{r}, \tag{5.47}$$

where

$$\boldsymbol{\omega}(t) = \frac{d\boldsymbol{\varphi}}{dt} \tag{5.48}$$

is the angular velocity with which the system B rotates with respect to S. The direction of $\boldsymbol{\omega}$ is the rotation axis.[96] We thus have

$$\mathbf{v}_S = \mathbf{v}_0 + \boldsymbol{\omega} \times \mathbf{r}. \tag{5.49}$$

---

[96]The sense of rotation is given by the right-hand rule: Pointing in the direction of $\boldsymbol{\omega}$ with the thumb of the right hand, the fingers of the right hand curl in the direction of the rotation.

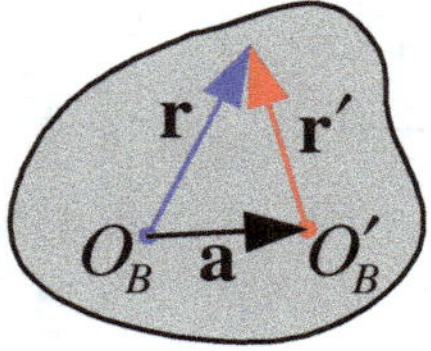

Fig. 5.4   The origin of the coordinate system fixed within the rigid body is changed from O_B to O'_B by a constant vector $\mathbf{a}$. The position vector changes from $\mathbf{r}$ to $\mathbf{r}' = \mathbf{r} - \mathbf{a}$.

Since the choice of the origin of B was arbitrary, it is natural to ask how this result would change under a change of origin. Changing O_B to O'_B by a constant vector $\mathbf{a}$ (see Fig. 5.4), the position vector of the mass point in question is now

$$\mathbf{r}' = \mathbf{r} - \mathbf{a}, \tag{5.50}$$

and the new origin is now seen at

$$\mathbf{r}'_0 = \mathbf{r}_0 + \mathbf{a}. \tag{5.51}$$

Hence,

$$\mathbf{r}'_S = \mathbf{r}'_0 + \mathbf{r}' = \mathbf{r}_0 + \mathbf{r} = \mathbf{r}_S, \tag{5.52}$$

which is the obvious result that the position of the mass point as seen from S does not depend on how one chooses the origin of B. Its velocity thus remains $\mathbf{v}_S$, but it is now expressed as

$$\mathbf{v}_S = \mathbf{v}'_0 + \boldsymbol{\omega}' \times \mathbf{r}' = \mathbf{v}'_0 - \boldsymbol{\omega}' \times \mathbf{a} + \boldsymbol{\omega}' \times \mathbf{r}, \tag{5.53}$$

which is derived exactly the same way as (5.49). Direct comparison with the latter equation produces

$$\mathbf{v}_0 + \boldsymbol{\omega}' \times \mathbf{a} - \mathbf{v}'_0 = (\boldsymbol{\omega}' - \boldsymbol{\omega}) \times \mathbf{r}. \tag{5.54}$$

The right hand-side here depends on $\mathbf{r}$, the left-hand side does not. For this equality to be true for arbitrary choices of $\mathbf{r}$, the right-hand side must vanish. This requires

$$\boldsymbol{\omega}' = \boldsymbol{\omega}, \tag{5.55}$$

i.e., *the angular velocity of the rotation is independent of the choice of the origin of B.* Moreover, as a vector, it does not depend on the orientation of the axes of B (only the components of a particular representation will depend on the axes orientation). The angular velocity $\boldsymbol{\omega}$, therefore, is a quantity that characterizes the rotation of the rigid body, without any reference to a particular coordinate system. By contrast, the translational velocity of the origin,

$$\mathbf{v}'_0 = \mathbf{v}_0 + \boldsymbol{\omega} \times \mathbf{a}, \tag{5.56}$$

does depend on the position of origin. How one chooses this position is a matter of convenience.

Note that, in view of the properties of infinitesimal rotations discussed in the previous section, angular velocities can be added: For two angular velocities $\boldsymbol{\omega}_1$ and $\boldsymbol{\omega}_2$, the associated infinitesimal rotations are $d\boldsymbol{\varphi}_1 = \boldsymbol{\omega}_1 dt$ and $d\boldsymbol{\varphi}_2 = \boldsymbol{\omega}_2 dt$. The resulting change of $\mathbf{r}$ then is

$$dr = d\mathbf{r}_1 + d\mathbf{r}_2 = (d\boldsymbol{\varphi}_1 + d\boldsymbol{\varphi}_2) \times \mathbf{r} = (\boldsymbol{\omega}_1 + \boldsymbol{\omega}_2) \times \mathbf{r}\, dt, \tag{5.57}$$

i.e., the angular velocities add up,

$$\boldsymbol{\omega} = \boldsymbol{\omega}_1 + \boldsymbol{\omega}_2. \tag{5.58}$$

The angular velocity can now be described in terms of the Euler angles. Each of the three rotations that enter the construction of an arbitrary rotation is associated with an angular velocity. Consulting Fig. 5.3, one finds easily

$$\boldsymbol{\omega}_\phi = \dot{\phi}\,\mathbf{u}_3, \qquad \boldsymbol{\omega}_\theta = \dot{\theta}\,\mathbf{n}, \qquad \boldsymbol{\omega}_\psi = \dot{\psi}\,\mathbf{e}_3, \tag{5.59}$$

where

$$\mathbf{n} = \cos\psi\,\mathbf{e}_1 - \sin\psi\,\mathbf{e}_2 \tag{5.60}$$

is the unit vector along the node line. Since angular velocities can be added, the resulting total angular velocity is simply

$$\boldsymbol{\omega} = \boldsymbol{\omega}_\phi + \boldsymbol{\omega}_\theta + \boldsymbol{\omega}_\psi = \dot{\phi}\,\mathbf{u}_3 + \dot{\theta}\,\mathbf{n} + \dot{\psi}\,\mathbf{e}_3. \tag{5.61}$$

This vector can be expressed either in the space frame S or the body frame B by projecting out components according

$$S: \quad \omega_{si} = \boldsymbol{\omega} \cdot \mathbf{u}_i, \qquad\qquad B: \quad \omega_i = \boldsymbol{\omega} \cdot \mathbf{e}_i\ ; \tag{5.62}$$

for example,

$$\boxed{\boldsymbol{\omega} \longrightarrow \vec{\omega} = \begin{pmatrix} \omega_1 \\ \omega_2 \\ \omega_3 \end{pmatrix} = \begin{pmatrix} \boldsymbol{\omega} \cdot \mathbf{e}_1 \\ \boldsymbol{\omega} \cdot \mathbf{e}_2 \\ \boldsymbol{\omega} \cdot \mathbf{e}_3 \end{pmatrix} = \begin{pmatrix} \dot{\phi}\sin\theta\sin\psi + \dot{\theta}\cos\psi \\ \dot{\phi}\sin\theta\cos\psi - \dot{\theta}\sin\psi \\ \dot{\phi}\cos\theta + \dot{\psi} \end{pmatrix}} . \tag{5.63}$$

This represents $\boldsymbol{\omega}$ in the coordinates of the body system B as a function of the Euler angles and their time derivatives.

5.2   Inertia Tensor

5.2.1   *Kinetic energy*

The total kinetic energy of the n point masses forming the rigid body is given by

$$T = \frac{1}{2} \sum_{k=1}^{n} m_k \, \mathbf{v}_k^2, \tag{5.64}$$

where $\mathbf{v}_k$ is the velocity of the k-th body as seen in S according to Eq. (5.49); the index S was dropped here.

Before we proceed, let us allow for the possibility that the mass distribution is continuous. We must then make the replacement

$$\sum_{k=1}^{n} m_k \longrightarrow \int dm \equiv \int_V d^3r \, \rho(\mathbf{r}), \tag{5.65}$$

where ρ describes the mass density and the integration extends over the volume of the body. Note, however, that the discrete case is recovered by utilizing the particular density function

$$\rho(\mathbf{r}) = \sum_{k=1}^{n} m_k \, \delta(\mathbf{r} - \mathbf{r}_k), \tag{5.66}$$

where $\delta(\mathbf{r} - \mathbf{r}_k)$ is the Dirac delta distribution (see Appendix B). In view of this, the symbol $\int dm$ is going to be used in the following to cover either continuous or discrete mass distributions.

The kinetic energy of the rigid body then reads[97]

$$T = \frac{1}{2} \int dm \, \mathbf{v}^2. \tag{5.67}$$

Using (5.49), this becomes

$$T = \frac{1}{2} \int dm \left(\mathbf{v}_0^2 + 2(\boldsymbol{\omega} \times \mathbf{r}) \cdot \mathbf{v}_0 + (\boldsymbol{\omega} \times \mathbf{r})^2 \right). \tag{5.68}$$

The second term can be rewritten as

$$\int dm \, (\boldsymbol{\omega} \times \mathbf{r}) \cdot \mathbf{v}_0 = (\mathbf{v}_0 \times \boldsymbol{\omega}) \cdot \int dm \, \mathbf{r} = M(\mathbf{v}_0 \times \boldsymbol{\omega}) \cdot \mathbf{R}_B, \tag{5.69}$$

---

[97]More explicitly this should be written as

$$T = \frac{1}{2} \int dm(\mathbf{r}) \, [\mathbf{v}(\mathbf{r})]^2,$$

which sums up the squared velocities of the mass elements dm at $\mathbf{r}$ over the entire volume.

where $\mathbf{R}_B = \int dm\,\mathbf{r}/M$ is the center-of-mass position of the rigid body within B and $M = \int dm$ its total mass. If we choose the origin of B either to be fixed (i.e., $\mathbf{v}_0 = 0$) or to coincide with the center of mass then

$$(\mathbf{v}_0 \times \boldsymbol{\omega}) \cdot \mathbf{R}_B = \begin{cases} 0 & \text{for} \quad \mathbf{v}_0 = 0, \\ 0 & \text{for} \quad \mathbf{R}_B = 0. \end{cases} \tag{5.70}$$

The first case describes pure rotation, without any translation. For example, in treating a spinning top whose contact point with the supporting surface is kept fixed, it would be natural to choose the latter point as the origin of the body system. *In the following, we will always consider one of these two cases.* The kinetic energy, therefore, can be simplified as

$$T = \frac{M}{2}\mathbf{v}_0^2 + \frac{1}{2} \int dm\,(\boldsymbol{\omega} \times \mathbf{r})^2 = T_{\text{trans}} + T_{\text{rot}}, \tag{5.71}$$

where T_{trans} and T_{rot} describe the translational and rotational contributions.

Noting that

$$(\boldsymbol{\omega} \times \mathbf{r})^2 = \omega^2 r^2 - (\boldsymbol{\omega} \cdot \mathbf{r})^2 = \sum_{i=1}^{3} \omega_i^2 r^2 - \sum_{i,j=1}^{3} \omega_i x_i \omega_j x_j$$

$$= \sum_{i,j=1}^{3} \left(r^2 \delta_{ij} - x_i x_j\right) \omega_i\,\omega_j, \tag{5.72}$$

where the ω_i are the components of $\boldsymbol{\omega}$ in B [cf. Eq. (5.63)] and the x_i are the components of the position $\mathbf{r}$, the rotational kinetic energy can be written as

$$T_{\text{rot}} = \frac{1}{2} \int dm\,(\boldsymbol{\omega} \times \mathbf{r})^2 = \frac{1}{2} \sum_{i,j=1}^{3} \Theta_{ij}\,\omega_i\,\omega_j, \tag{5.73}$$

where

$$\boxed{\text{Inertia tensor:} \qquad \Theta_{ij} = \int dm\,\left(r^2\,\delta_{ij} - x_i x_j\right)} \tag{5.74}$$

provides the elements of the *inertia tensor*. All quantities in this definition pertain to the body system B; the inertia tensor in this system, therefore, is *time-independent*. It is a quantity that characterizes the mass distribution in the rigid body *for a given choice of body system B*.

The term 'tensor' refers to the properties of Θ_{ij} under rotations. Its essential aspect is that the (Cartesian) indices of Θ behave like the Cartesian components of a vector under such transformations (see Sec. 5.2.2).[98]

---

[98]Specifically, a tensor of this kind is called *Cartesian*. There are also *spherical* tensors.

For the present applications, depending on circumstances, we can treat the Θ_{ij} as the components of a 3×3 matrix,

$$\Theta = \begin{pmatrix} \Theta_{11} & \Theta_{12} & \Theta_{13} \\ \Theta_{21} & \Theta_{22} & \Theta_{23} \\ \Theta_{31} & \Theta_{32} & \Theta_{33} \end{pmatrix}, \tag{5.75}$$

or we can form a *dyad operator*,[99]

$$\boldsymbol{\Theta} = \sum_{i,j=1}^{3} \mathbf{e}_i \, \Theta_{ij} \, \mathbf{e}_j, \tag{5.76}$$

whose left- and right-handed action on vectors is defined by the scalar product according to

$$\mathbf{a} \cdot \boldsymbol{\Theta} = \sum_{i,j=1}^{3} (\mathbf{a} \cdot \mathbf{e}_i) \, \Theta_{ij} \mathbf{e}_j, \qquad \boldsymbol{\Theta} \cdot \mathbf{a} = \sum_{i,j=1}^{3} \mathbf{e}_i \Theta_{ij} \, (\mathbf{e}_j \cdot \mathbf{a}). \tag{5.77}$$

Note that in view of the symmetry of the inertia tensor, $\Theta_{ij} = \Theta_{ji}$, these two expressions are actually identical. A dyad acting on one vector produces another vector; in general, the resulting vector is not aligned with the original vector.

The matrix notation is useful if we have Θ act on vectors that are given in a representation with components in the body system B. For example, with the column-vector representation $\vec{\omega}$ of the angular velocity $\boldsymbol{\omega}$ given in (5.63), the rotational kinetic energy is written as

$$\boxed{ T_{\text{rot}} = \frac{1}{2} \sum_{i,j=1}^{3} \Theta_{ij} \, \omega_i \, \omega_j = \frac{\vec{\omega}^{\mathrm{T}} \Theta \vec{\omega}}{2} = \frac{\boldsymbol{\omega} \cdot \boldsymbol{\Theta} \cdot \boldsymbol{\omega}}{2} } . \tag{5.78}$$

The last equality employs the dyad $\boldsymbol{\Theta}$. The dyad operator acts directly on the vector, without any need for a particular representation.

5.2.2   *Similarity transformation*

It follows from the definition that the tensor Θ_{ij} itself is recovered from the dyad by

$$\Theta_{ij} = \mathbf{e}_i \cdot \boldsymbol{\Theta} \cdot \mathbf{e}_j. \tag{5.79}$$

In fact, the dyad operator is independent of any particular orientation of the body system B [but it does depend on the choice of origin, see (5.89)].

---

[99]We have encountered a dyad operator already in Eq. (1.9), where $\mathbf{1}$ is the dyad operator of the unit tensor, i.e., $\mathbf{1} = \sum_{i,j} \mathbf{e}_i \delta_{ij} \mathbf{e}_j$. See also footnote 3.

To see this, let us consider another body system B', with unit vectors $\mathbf{e}'_i$, that is rotated with respect to B, but shares the same origin. One then finds

$$
\begin{aligned}
\mathbf{e}'_k \cdot \mathbf{\Theta} \cdot \mathbf{e}'_l &= \sum_{i,j=1}^{3} (\mathbf{e}'_k \cdot \mathbf{e}_i)\, \Theta_{ij}\, (\mathbf{e}_j \cdot \mathbf{e}'_l) \\
&= \sum_{i,j=1}^{3} D_{ki}\, D_{lj}\, \Theta_{ij} \\
&= \int dm \sum_{i,j=1}^{3} D_{ki}\, D_{lj}\, \left(r^2\, \delta_{ij} - x_i\, x_j \right) \\
&= \int dm \left(r'^2\, \delta_{kl} - x'_k\, x'_l \right),
\end{aligned}
\tag{5.80}
$$

where the orthogonality condition (5.10) for the rotation matrix

$$
D_{ki} = \mathbf{e}'_k \cdot \mathbf{e}_i
\tag{5.81}
$$

was used. The relation

$$
x'_k = \sum_{i=1}^{3} D_{ki}\, x_i
\tag{5.82}
$$

provides the components x'_k of the given position vector in the new, rotated, frame, i.e., this corresponds to a (passive) rotation that does not change the length of the vector, $r^2 = r'^2$ [cf. Eq. (5.11)].

The last expression in (5.80) is the inertia tensor Θ'_{kl} in the new frame. In other words,

$$
\Theta'_{kl} = \mathbf{e}'_k \cdot \mathbf{\Theta} \cdot \mathbf{e}'_l,
\tag{5.83}
$$

and the dyad operator $\mathbf{\Theta}$ is indeed frame independent.

The relation between the tensors in the two frames is

$$
\boxed{\Theta'_{kl} = \sum_{i,j=1}^{3} D_{ki}\, D_{lj}\, \Theta_{ij}}
\tag{5.84}
$$

which shows that as far as rotations are concerned, Θ_{ij} behaves exactly like the product $x_i x_j$, i.e., each (Cartesian) index here transforms like the corresponding Cartesian component of a vector under the rotation, just like in Eq. (5.82). This is the defining property of a (Cartesian) tensor. In matrix form, this reads

$$
\boxed{\Theta' = D\, \Theta\, D^{\mathrm{T}}}.
\tag{5.85}
$$

This *similarity transformation* describes how the new inertia tensor is obtained from the old one under an orthogonal transformation. An important property of similarity transformations needed in Sec. 5.2.5 below is the fact that they leave the determinant of the transformed matrix invariant, that is, for any matrix A,

$$\det\left(DAD^{\mathrm{T}}\right) = \det(D)\,\det(A)\,\det(D^{\mathrm{T}}) = \det(A), \tag{5.86}$$

where $\det(D)\,\det(D^{\mathrm{T}}) = \det(DD^{\mathrm{T}}) = \det \mathbb{1} = 1$ was used.

5.2.3 Parallel-axis theorem

Let us consider a body system B in relation to the particular center-of-mass system B_{CM} obtained by parallel displacement of B such that the axes of B and B_{CM} are parallel. In other words, the respective B and B_{CM} coordinates x_i and x'_i are related by the constant displacement $x_i = x'_i + R_{Bi}$, where $\mathbf{R}_B = \sum_i R_{Bi}\mathbf{e}_i$ is the location of the center of mass in B. How is the inertia tensor Θ in B related to the B_{CM} tensor Θ^{CM}?

One has

$$\Theta_{ij} = \int dm\,\left(r^2\,\delta_{ij} - x_i x_j\right)$$

$$= \int dm\,\left[\sum_{k=1}^{3}(x'_k + R_{Bk})^2\,\delta_{ij} - (x'_i + R_{Bi})(x'_j + R_{Bj})\right]$$

$$= \overbrace{\int dm\,\left(r'^2\,\delta_{ij} - x'_i\,x'_j\right)}^{=\,\Theta^{\mathrm{CM}}_{ij}} + \overbrace{\int dm\,\left(R_B^2\,\delta_{ij} - R_{Bi}\,R_{Bj}\right)}^{=\,M}$$

$$+ \underbrace{\int dm\,\left[\sum_{k=1}^{3} 2x'_k R_{Bk}\,\delta_{ij} - x'_i\,R_{Bj} - x'_j\,R_{Bi}\right]}_{=\,0}. \tag{5.87}$$

The first term here is the CM tensor $\Theta^{\mathrm{CM}}_{ij}$, the second integral produces the total mass factor $M = \int dm$ for the otherwise constant integrand, and the third integral vanishes since $\int dm\,x'_i = 0$ for CM coordinates; hence,

$$\boxed{\Theta_{ij} = \Theta^{\mathrm{CM}}_{ij} + M\left(R_B^2\,\delta_{ij} - R_{Bi}\,R_{Bj}\right)}. \tag{5.88}$$

This is the *parallel-axis theorem*; it relates the inertia tensor in any body frame B to the corresponding CM tensor obtained by parallel displacement of the axes into the CM system. The second term here corresponds to the inertia tensor of a single point mass M at a distance $\mathbf{R}_B$ away from the origin of B, i.e., at the position of the center of mass.

The parallel-axis theorem can also be formulated for the dyad operator. Noting that for parallel axes, both B and B_{CM} have the same unit vectors, one immediately finds

$$\boxed{\boldsymbol{\Theta} = \boldsymbol{\Theta}^{\mathrm{CM}} + M\left(R_B^2\,\mathbf{1} - \mathbf{R}_B\mathbf{R}_B\right)}, \qquad (5.89)$$

where $\mathbf{1}$ is the unit dyad, Eq. (1.9), and $\mathbf{R}_B\mathbf{R}_B = \sum_{i,j}\mathbf{e}_i R_{Bi} R_{Bj}\mathbf{e}_j$ is a dyad that factorizes.

5.2.4 *Angular momentum*

The angular momentum depends on the reference point. In the following, we choose to describe the angular momentum with respect to the origin of the body system B, i.e.,

$$\mathbf{L} = \int dm\,\mathbf{r} \times \dot{\mathbf{r}}, \qquad (5.90)$$

where, as usual, the time derivative pertains to the inertial space system S. With (5.47), this becomes

$$\mathbf{L} = \int dm\,\mathbf{r} \times (\boldsymbol{\omega} \times \mathbf{r}). \qquad (5.91)$$

Evaluating the double cross product, we find

$$\mathbf{r} \times (\boldsymbol{\omega} \times \mathbf{r}) = \boldsymbol{\omega}\,r^2 - \mathbf{r}(\boldsymbol{\omega}\cdot\mathbf{r}) = \sum_{i,j=1}^{3}\left(r^2\,\delta_{ij} - x_i\,x_j\right)\omega_j\,\mathbf{e}_i, \qquad (5.92)$$

and therefore

$$\mathbf{L} = \sum_{i=1}^{3} L_i\,\mathbf{e}_i = \sum_{i,j=1}^{3}\mathbf{e}_i\,\Theta_{ij}\,\omega_j = \sum_{i,j=1}^{3}\mathbf{e}_i\,\Theta_{ij}\left(\mathbf{e}_j\cdot\boldsymbol{\omega}\right). \qquad (5.93)$$

With a column-vector representation

$$\vec{L} = \begin{pmatrix}\mathbf{L}\cdot\mathbf{e}_1 \\ \mathbf{L}\cdot\mathbf{e}_2 \\ \mathbf{L}\cdot\mathbf{e}_3\end{pmatrix} = \begin{pmatrix}L_1 \\ L_2 \\ L_3\end{pmatrix} \qquad (5.94)$$

for the vector $\mathbf{L}$, the angular momentum may be written in either one of the following forms,

$$\boxed{L_i = \sum_{j=1}^{3}\Theta_{ij}\,\omega_j, \qquad \vec{L} = \Theta\,\vec{\omega}, \qquad \mathbf{L} = \boldsymbol{\Theta}\cdot\boldsymbol{\omega}}. \qquad (5.95)$$

Only the last equation here, $\mathbf{L} = \boldsymbol{\Theta}\cdot\boldsymbol{\omega}$, is independent of any particular representation.

Moments of inertia

The *moment of inertia* with respect to a particular axis given by a unit vector $\mathbf{n}$ is defined by

$$I_{\mathbf{n}} = \mathbf{n} \cdot \boldsymbol{\Theta} \cdot \mathbf{n} = \sum_{i,j=1}^{3} \Theta_{ij}\, n_i\, n_j, \qquad (5.96)$$

where $n_i = \mathbf{n} \cdot \mathbf{e}_i$ are the components of $\mathbf{n}$ with respect to B. If $\mathbf{n}$ is one of the $\mathbf{e}_i$, one finds

$$I_{\mathbf{e}_i} = \mathbf{e}_i \cdot \boldsymbol{\Theta} \cdot \mathbf{e}_i = \Theta_{ii}, \qquad (5.97)$$

i.e., the diagonal elements Θ_{ii} of the inertia tensor are the moments of inertia with respect to the coordinate axes in B. For a rotation about a fixed axis $\mathbf{n}$, with angular velocity

$$\boldsymbol{\omega} = \omega_{\mathbf{n}}\, \mathbf{n}, \qquad (5.98)$$

the projection of the angular momentum onto that axis is

$$L_{\mathbf{n}} = \mathbf{n} \cdot \mathbf{L} = \mathbf{n} \cdot \boldsymbol{\Theta} \cdot \boldsymbol{\omega} = \mathbf{n} \cdot \boldsymbol{\Theta} \cdot \mathbf{n}\, \omega_{\mathbf{n}} = I_{\mathbf{n}}\, \omega_{\mathbf{n}}. \qquad (5.99)$$

In general, unless $\mathbf{n}$ is one of the principal axes of the rigid body (see the subsequent section), the angular momentum will have other components as well.

5.2.5   *Principal-axis transformation*

There exists a special orientation of the axes such that the inertia tensor becomes diagonal, i.e.,

$$\Theta_{ij} \;\longrightarrow\; \widehat{\Theta}_{ij} = I_i\, \delta_{ij}, \qquad \text{or} \qquad \widehat{\boldsymbol{\Theta}} = \begin{pmatrix} I_1 & 0 & 0 \\ 0 & I_2 & 0 \\ 0 & 0 & I_3 \end{pmatrix}. \qquad (5.100)$$

where the caret signifies this system. The axes of this special system are called *principal axes*; their unit vectors will be denoted by $\hat{\mathbf{e}}_i$. The inertia dyad,

$$\boldsymbol{\Theta} = \sum_{i=1}^{3} \hat{\mathbf{e}}_i\, I_i\, \hat{\mathbf{e}}_i, \qquad (5.101)$$

although written in terms of these unit vectors, does *not* need a caret since — as was shown in Sec. 5.2.2 — it is independent of any special orientation of the axes. The diagonal elements I_i of $\widehat{\Theta}$ are the *principal moments of inertia*. Because of

$$I_i = \int dm \left(r^2 - x_i^2 \right) = \int dm \left(x_1^2 + x_2^2 + x_3^2 - x_i^2 \right) > 0, \qquad (5.102)$$

they are always positive.[100] For symmetric bodies, the principal axes are aligned with the symmetry axes of the body.

The physical significance of the principal-axis system is that, if the rotation axis is aligned with one of the axes, i.e.,

$$\boldsymbol{\omega}^{(k)} = \omega\,\hat{\mathbf{e}}_k \qquad (5.103)$$

then the angular momentum becomes

$$\mathbf{L} = \boldsymbol{\Theta} \cdot \boldsymbol{\omega}^{(k)} = \sum_{i=1}^{3} \hat{\mathbf{e}}_i\, I_i \left(\hat{\mathbf{e}}_i \cdot \boldsymbol{\omega}^{(k)} \right) = \sum_{i=1}^{3} \hat{\mathbf{e}}_i\, I_i \underbrace{\left(\hat{\mathbf{e}}_i \cdot \hat{\mathbf{e}}_k \right)}_{=\delta_{ik}} \omega = I_k\, \boldsymbol{\omega}^{(k)}. \quad (5.104)$$

In other words, the angular momentum $\mathbf{L}$ is then parallel to $\boldsymbol{\omega}^{(k)}$, and therefore aligned with the rotation axis.

We show now by construction how to transform Θ_{ij} so that it becomes diagonal. Clearly, if it exists, the rotation that provides the desired result must have the form

$$P_{ij} = \hat{\mathbf{e}}_i \cdot \mathbf{e}_j, \qquad (5.105)$$

i.e., the *principal-axis transformation* becomes

$$\Theta_{kl} \quad \longrightarrow \quad \widehat{\Theta}_{kl} = \sum_{i,j=1}^{3} P_{ki} P_{lj} \Theta_{ij} = I_k\, \delta_{kl}. \qquad (5.106)$$

Multiplying this by P_{km}, summing over k, and using the orthogonality relation yields

$$\sum_{j=1}^{3} \left(\Theta_{mj} - I_l\, \delta_{mj} \right) P_{lj} = 0. \qquad (5.107)$$

With the column vector

$$\vec{p}_l = \begin{pmatrix} P_{l1} \\ P_{l2} \\ P_{l3} \end{pmatrix}, \qquad (5.108)$$

---

[100]Actually, for the very special case of a linear mass distribution along one of the axes, the corresponding principal moment of inertia would be zero. This corresponds, however, to a case with only two rotational degrees of freedom, and will be ignored here.

this can be written as a matrix equation

$$\boxed{(\Theta - I_l \mathbb{1})\,\vec{p}_l = 0, \qquad l = 1,\, 2,\, 3}$$ (5.109)

This is the *eigenvalue equation*[101] of the inertia matrix Θ; its three *eigenvalues* I_l are the principal moments of inertia and the *eigenvectors* $\vec{p}_l$ determine the rotation that diagonalizes Θ. To obtain nontrivial (i.e., nonzero) solutions, the condition

$$\boxed{\det\left(\Theta - I\mathbb{1}\right) = 0}$$ (5.110)

must be satisfied. The determinant is a third-order polynomial in I; its three roots I_1, I_2, and I_3 are the eigenvalues. The three orthogonal[102] eigenvectors $\vec{p}_i$ of (5.109), after proper normalization, provide the rotation matrix elements,

$$P_{ij} = (\vec{p}_i)_j,$$ (5.111)

where the right-hand side denotes the j-th element of the i-th eigenvector. In other words, the eigenvector

$$\vec{p}_i = \sum_{j=1}^{3} (\vec{p}_i)_j\, \mathbf{e}_j = \sum_{j=1}^{3} P_{ij}\, \mathbf{e}_j = \sum_{j=1}^{3} (\hat{\mathbf{e}}_i \cdot \mathbf{e}_j)\, \mathbf{e}_j = \hat{\mathbf{e}}_i \cdot \mathbf{1} = \hat{\mathbf{e}}_i$$ (5.112)

directly determines the principal axes in the representation of the original body system B.

Equations (5.109) and (5.110) provide the desired result that allow us to diagonalize Θ. Equation (5.110) is evaluated in a particular body system. Its eigenvalues I_i, however, are independent of rotations of that particular frame. This follows directly from the fact that the determinant is invariant under the similarity transformations (see Sec. 5.2.2) that relate the Θ's of rotated frames. Hence, for $\Theta' = D\Theta D^{\mathrm{T}}$, one has

$$0 = \det\left(\Theta' - I_l'\mathbb{1}\right) = \det\left(D\Theta D^{\mathrm{T}} - I_l'\mathbb{1}\right) = \det\left(D\left[\Theta - I_l'\mathbb{1}\right]D^{\mathrm{T}}\right)$$
$$= \det\left(\Theta - I_l'\mathbb{1}\right),$$ (5.113)

and thus $I_l' = I_l$ since this determinant provides the same characteristic polynomial as (5.110). We conclude:

$$\boxed{\text{The principal moments of inertia are uniquely determined by the choice of origin of the body system.}}$$ (5.114)

---

[101]The hybrid word eigenvalue is derived from the German *Eigenwert*, literally 'characteristic value'. The prefix *eigen*, meaning 'characteristic', 'proper', 'pertaining to', or 'property of', is attached to all related technical terms — eigenvector, eigensolution, eigenmode, etc.

[102]If all eigenvalues I_i are different, the eigenvectors will be orthogonal automatically. If some eigenvalues coincide, the corresponding eigenvectors can be made orthogonal by standard procedures.

According to the parallel-axis theorem (5.88), they can all be related to the principal moments of inertia in the center-of-mass system,

$$\boxed{I_i = I_i^{\text{CM}} + M d_i^2}\,,\tag{5.115}$$

where $d_i^2 = R_B^2 - R_{Bi}^2$ is the squared orthogonal distance between the x_i-axis and the center of mass. This is the simplified form of the parallel-axis theorem found in elementary textbooks. The principal moments of inertia (in particular, the CM values I_i^{CM}) are *properties* of the rigid body that determine its behavior under rotations, similar to the mass for linear motion.

5.3   Euler's Equations

For the description of the dynamics of a rigid body, we will choose an origin O_B of the body system B such that (5.70) holds true. Therefore, O_B is taken either as a fixed point at rest with respect to the inertial space frame S or as the center of mass.

In the first case, the motion is a pure rotation about an axis through O_B. This follows from *Euler's Theorem*:

$$\boxed{\begin{array}{l}\text{The general displacement of a rigid body with one point}\\ \text{fixed is a rotation about some axis through this point.}\end{array}}\tag{5.116}$$

This statement is intuitively obvious and we will not prove it here (for a proof, see [1]). A rigid body with one fixed point is called a *top*. A top thus possesses only the three rotational degrees of freedom. They may be described by, for example, the three Euler angles ϕ, θ, and ψ (see Fig. 5.3). It should be emphasized that the rotation axis of the top in general is not fixed, i.e., it may change dynamically as a function of time. (By contrast, a rigid body rotating about a *fixed* axis has only one degree of freedom; such a body is called a *rotor*.)

For the second case, according to Section 1.5.2.1, the center-of-mass motion, described by

$$M\ddot{\mathbf{R}} = \mathbf{F},\tag{5.117}$$

completely decouples from the equation of motion for the angular momentum. $\mathbf{F}$ is the total *external* force acting on the body. It can be written as

$$\mathbf{F} = \int d\mathbf{F}(\mathbf{r}),\tag{5.118}$$

where $d\mathbf{F}(\mathbf{r})$ is the external force that acts on the mass element dm at position $\mathbf{r}$; the integral extends over all $\mathbf{r}$ for which the mass density $\rho(\mathbf{r})$ is nonzero. This is similar to the notation (5.65) for a continuous mass density. We can go back to the discrete case treated in Sec. 1.5.2.1 by a straightforward procedure similar to (5.66).

In both cases, we may therefore concentrate solely on the rotation and employ

$$\boxed{\frac{d\mathbf{L}}{dt} = \mathbf{N} = \int \mathbf{r} \times d\mathbf{F}(\mathbf{r})} \tag{5.119}$$

in the body system B for the dynamical description of the rotations. In other words, all variables here and in

$$\mathbf{L} = \int dm\ (\mathbf{r} \times \dot{\mathbf{r}}) = \boldsymbol{\Theta} \cdot \boldsymbol{\omega} \tag{5.120}$$

pertain to the body system B.

In the equation of motion,

$$\frac{d}{dt}(\boldsymbol{\Theta} \cdot \boldsymbol{\omega}) = \mathbf{N}, \tag{5.121}$$

the time derivative is to be taken with respect to the external inertial frame, i.e., [cf. (1.253)]

$$\frac{d(\boldsymbol{\Theta} \cdot \boldsymbol{\omega})}{dt} = \frac{d_B(\boldsymbol{\Theta} \cdot \boldsymbol{\omega})}{dt} + \boldsymbol{\omega} \times (\boldsymbol{\Theta} \cdot \boldsymbol{\omega}), \tag{5.122}$$

where d_B/dt is the time derivative in B. In B, the inertia tensor is time-independent and hence we have

$$\boxed{\boldsymbol{\Theta} \cdot \dot{\boldsymbol{\omega}} + \boldsymbol{\omega} \times (\boldsymbol{\Theta} \cdot \boldsymbol{\omega}) = \mathbf{N}} \tag{5.123}$$

as the equation of motion governing the rotation of the rigid body in the body system B. Choosing now the special principal-axis system, where the inertia tensor is diagonal (cf. Sec. 5.2.5), this reduces to the *Euler equations of motion*,

$$\boxed{\begin{aligned}
I_1\dot{\omega}_1 + (I_3 - I_2)\,\omega_2\,\omega_3 &= N_1 \\
I_2\dot{\omega}_2 + (I_1 - I_3)\,\omega_3\,\omega_1 &= N_2 \\
I_3\dot{\omega}_3 + (I_2 - I_1)\,\omega_1\,\omega_2 &= N_3
\end{aligned}} \tag{5.124}$$

where the ω_i are given by (5.63), i.e.,

$$\boxed{\begin{aligned}
\omega_1 &= \dot{\phi}\sin\theta\sin\psi + \dot{\theta}\cos\psi, \\
\omega_2 &= \dot{\phi}\sin\theta\cos\psi - \dot{\theta}\sin\psi \\
\omega_3 &= \dot{\phi}\cos\theta + \dot{\psi}
\end{aligned}} \tag{5.125}$$

When put into the Euler equations, this results in three second-order differential equations for the three Euler angles $\phi(t)$, $\theta(t)$, and $\psi(t)$. These are the desired equations of motion for the rotation of a rigid body in the principal-axes frame.

These equations have one serious drawback. The components $N_i(t)$ of the external torque are required in the principal-axes body system. Their time-dependence, therefore, in general will depend on the motion of the body, and this may considerably complicate the solution of Euler's equations.

5.3.1 *Torque-free rotation*

The simplest case obtains when there is no external torque, $\mathbf{N} = 0$. This applies, for example, to a freely falling body, whose center of mass has been chosen as the origin of B.

Without forces, a translational motion would be uniform, with a constant velocity. To see whether a torque-free rotation is also uniform, i.e., whether $\omega_i = const$ $(i = 1, 2, 3)$ is a possible solution, let us consider the torque-free Euler equations,

$$I_1\dot{\omega}_1 + (I_3 - I_2)\,\omega_2\,\omega_3 = 0, \tag{5.126a}$$

$$I_2\dot{\omega}_2 + (I_1 - I_3)\,\omega_3\,\omega_1 = 0, \tag{5.126b}$$

$$I_3\dot{\omega}_3 + (I_2 - I_1)\,\omega_1\,\omega_2 = 0, \tag{5.126c}$$

and put all $\omega_i = const$; one then has

$$(I_3 - I_2)\,\omega_2\,\omega_3 = 0, \tag{5.127a}$$

$$(I_1 - I_3)\,\omega_3\,\omega_1 = 0, \tag{5.127b}$$

$$(I_2 - I_1)\,\omega_1\,\omega_2 = 0. \tag{5.127c}$$

Let us further assume that all principal moments of inertia are different,

$$I_i \neq I_j, \qquad \text{for} \quad i \neq j. \tag{5.128}$$

The only nontrivial solution then is that one of the components ω_i is a (nonzero) constant and the other two vanish,

$$\omega_i = const, \qquad \omega_j = \omega_k = 0, \tag{5.129}$$

where i, j, k are some permutation of $1, 2, 3$. In other words, one cannot find a solution for an arbitrary constant $\boldsymbol{\omega}$. We conclude, therefore, that, in general, *a torque-free rotation will only be uniform if the rotation axis is one of the principal axes.*

For a rotation about a principal axis, $\mathbf{L}$ is aligned with the rotation axis, according to (5.104), i.e., for $\boldsymbol{\omega} = \omega_i\,\hat{\mathbf{e}}_i$,

$$\mathbf{L} = I_i\,\omega_i\,\hat{\mathbf{e}}_i = const \quad \text{with respect to } S, \tag{5.130}$$

where in the last equality it was used that, for the torque-free rotation, the angular momentum is conserved. This means that the body rotates uniformly about the x_i-axis, and the corresponding principal-axis unit vector $\hat{\mathbf{e}}_i$ has a constant direction in the inertial frame S.

5.3.1.1   *Stability of uniform rotation*

We know from experience when tossing objects of various shapes into the air that some objects rotate rather nicely whereas others tend to get wobbly rather quickly. To investigate the stability of the uniform motion about a principal axis, say the x_1-axis, let us consider a rotation with nonzero, but very small, values of ω_2 and ω_3, i.e.,

$$\omega_1 = \omega_0 = const, \qquad \omega_2 \ll \omega_0, \qquad \omega_3 \ll \omega_0. \tag{5.131}$$

In the torque-free Euler equations (5.126), we can now neglect the small product term $\omega_2\omega_3$. The first equation then produces the solution

$$\dot{\omega}_1 = 0 \qquad \Rightarrow \qquad \omega_1 = \omega_0 = const. \tag{5.132}$$

With this solution, the second and third equations can then be decoupled by differentiating and substituting. This results in

$$\ddot{\omega}_2 + a\,\omega_0^2\,\omega_2 = 0, \qquad \ddot{\omega}_3 + a\,\omega_0^2\,\omega_3 = 0, \tag{5.133}$$

with

$$a = \frac{(I_1 - I_2)(I_1 - I_3)}{I_2 I_3}. \tag{5.134}$$

The behavior of the solutions is determined by the sign of a:

Case 1: I_1 is either the smallest or the largest of the three principal moments of inertia. Then $a > 0$ and the solutions are

$$\omega_2(t) = c_2 \cos\left(\sqrt{a}\omega_0 t + \varphi_1\right), \quad \omega_3(t) = c_3 \cos\left(\sqrt{a}\omega_0 t + \varphi_2\right). \tag{5.135}$$

Small amplitudes ($c_2, c_3 \ll \omega_0$) then lead to small oscillations around the stable solution $\boldsymbol{\omega} = (\omega_0, 0, 0)$.

Case 2: I_1 is the intermediate-size moment of inertia. Then $a < 0$ and the solutions are

$$\omega_i(t) = b_i\,e^{-\sqrt{|a|}t} + c_i\,e^{\sqrt{|a|}t}, \qquad i = 2, 3. \tag{5.136}$$

For the second term here, initially small amplitudes ($c_i \ll \omega_0$) increase exponentially, i.e, the solution $\boldsymbol{\omega} = (\omega_0, 0, 0)$ is not stable against small perturbations.

We conclude, therefore, that *only the rotations about the principal axes with the largest or smallest moments of inertia are stable.* For the constant angular momentum,

$$\mathbf{L} = I_1\, \omega_1\, \hat{\mathbf{e}}_1 + \mathcal{O}(\omega_2, \omega_3) = const, \tag{5.137}$$

the initially small term $\mathcal{O}(\omega_2, \omega_3)$ stays small, i.e., $\hat{\mathbf{e}}_1 \approx const$ in the external inertial frame. By contrast, for rotations about the axes with an intermediate moment of inertia, small perturbations of the initially ideal solution $\boldsymbol{\omega} = (\omega_0, 0, 0)$ will build up rapidly and destroy the uniform rotation.

5.3.2 *The torque-free symmetric top*

We will now investigate the torque-free motion of a body that is symmetric about one axis. The symmetry axis, taken to be the x_3-axis, is called the *figure axis*. The symmetry provides

$$I_3 \neq I_1 = I_2 \tag{5.138}$$

for the principal moments of inertia. A body with these properties is referred to as a *symmetric top*. As before, we choose as the origin either the center of mass or, if it exists, the fixed support point of the top. In either case, we treat only the rotation; the translational motion either decouples from the rotation or it is not present at all.

The Euler equations now are

$$I_1\, \dot{\omega}_1 - (I_1 - I_3)\, \omega_2\, \omega_3 = 0, \tag{5.139a}$$

$$I_1\, \dot{\omega}_2 + (I_1 - I_3)\, \omega_3\, \omega_1 = 0, \tag{5.139b}$$

$$I_3\, \dot{\omega}_3 = 0. \tag{5.139c}$$

The last equation immediately provides

$$\omega_3 = const = \omega_0. \tag{5.140}$$

Inserting this in the first two equations, decoupling by differentiation and substitution then yields

$$\ddot{\omega}_1 + \Omega^2 \omega_1 = 0, \qquad \text{and} \qquad \omega_2 = \frac{\dot{\omega}_1}{\Omega}, \tag{5.141}$$

with

$$\Omega = \frac{I_1 - I_3}{I_1}\omega_0. \tag{5.142}$$

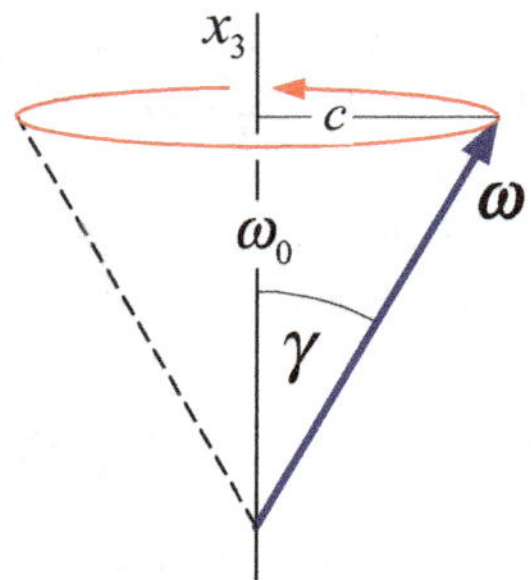

Fig. 5.5   The body cone of a torque-free symmetric top shows how the angular velocity vector $\boldsymbol{\omega}$ moves about the x_3 figure axis.

The solutions are

$$\omega_1(t) = c\sin(\Omega t + \psi_0), \qquad \omega_2(t) = c\cos(\Omega t + \psi_0). \tag{5.143}$$

The overall length of the vector $\boldsymbol{\omega}$ is a constant,

$$\boldsymbol{\omega}^2 = \omega_1^2 + \omega_2^2 + \omega_3^2 = c^2 + \omega_0^2 = const, \tag{5.144}$$

the projections of $\boldsymbol{\omega}$ on the x_3-axis and on the $x_1 x_2$-plane are also constants (ω_0 and c, respectively), and the vector $\boldsymbol{\omega}$ rotates with the (constant) angular velocity Ω. This means that $\boldsymbol{\omega}$ moves on a cone, the *body cone*, around the x_3 figure axis. The opening angle of the cone is

$$\gamma = \arctan\frac{c}{\omega_0} = const. \tag{5.145}$$

This geometry is shown in Fig. 5.5.

To calculate the Euler angles as functions of time, let us choose the external inertial coordinate system S so that its z-axis is aligned with the conserved angular momentum,

$$\mathbf{L} = L\mathbf{u}_3. \tag{5.146}$$

Using

$$\mathbf{u}_3 = \sin\theta\sin\psi\,\mathbf{e}_1 + \sin\theta\cos\psi\,\mathbf{e}_2 + \cos\theta\,\mathbf{e}_3, \tag{5.147}$$

which follows from (5.4) with (5.21), the angular-momentum vector is expressed in the body system as

$$\mathbf{L} \longrightarrow \vec{L} = L\begin{pmatrix} \mathbf{u}_3 \cdot \mathbf{e}_1 \\ \mathbf{u}_3 \cdot \mathbf{e}_2 \\ \mathbf{u}_3 \cdot \mathbf{e}_3 \end{pmatrix} = L\begin{pmatrix} \sin\theta\sin\psi \\ \sin\theta\cos\psi \\ \cos\theta \end{pmatrix} = \begin{pmatrix} I_1\,\omega_1 \\ I_1\,\omega_2 \\ I_3\,\omega_3 \end{pmatrix}. \tag{5.148}$$

With the ω_i as obtained above, the last equality yields three equations,

$$L\sin\theta\sin\psi = c\,I_1\sin(\Omega t + \psi_0) \tag{5.149a}$$

$$L\sin\theta\cos\psi = c\,I_1\cos(\Omega t + \psi_0) \tag{5.149b}$$

$$L\cos\theta = \omega_0\,I_3. \tag{5.149c}$$

From the last equation, we conclude

$$\theta(t) = \theta_0 = const,\qquad(5.150)$$

which in turn implies

$$\psi(t) = \Omega t + \psi_0\qquad(5.151)$$

for the other two equations. Moreover,

$$L \sin\theta_0 = c\,I_1,\qquad(5.152)$$

and hence

$$\tan\theta_0 = \frac{c\,I_1}{\omega_0\,I_3}.\qquad(5.153)$$

To obtain $\phi(t)$, we need to go back to either the first or second equation in (5.125) and insert the solutions just obtained for θ and ψ. The resulting differential equation,

$$\dot{\phi} = \frac{c}{\sin\theta_0},\qquad(5.154)$$

is solved by

$$\phi(t) = \frac{c}{\sin\theta_0}t + \phi_0.\qquad(5.155)$$

This result, together with (5.150) and (5.151), thus provides the complete solution for the torque-free symmetric top.

The resulting motion is shown in Fig. 5.6. To discuss this further, we recall:

θ: angle between the x_3 figure axis and the z-axis;

$\dot{\phi}$: angular velocity of the x_3 figure axis around the z-axis;

$\dot{\psi}$: angular velocity of the entire body around the x_3 figure axis

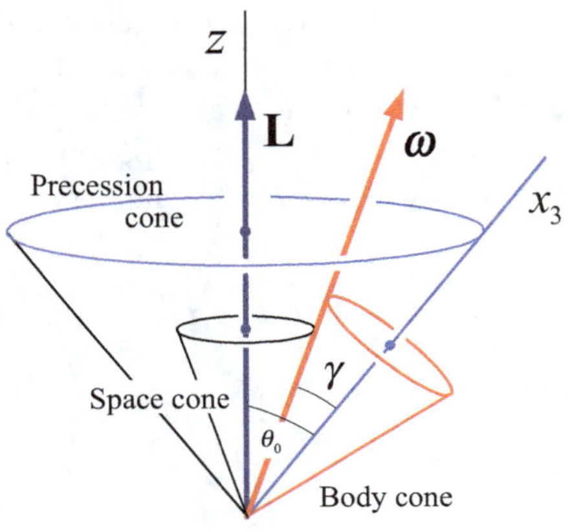

Fig. 5.6　Regular precession of the torque-free symmetric top. The angular momentum **L** (which is fixed in space), the rotation axis $\boldsymbol{\omega}$ and the x_3 figure axis are always in one plane. The rotation proceeds as if the body cone rolls around the space cone, making the figure axis precess around **L** on the precession cone. The body cone itself may be thought of as an actual model of the symmetric top.

(cf. Fig. 5.3). In view of $\theta = \theta_0 = const$, the figure axis traces out a cone, the *precession cone*, as it uniformly moves around the fixed z-axis of the space system with constant angular velocity $\dot{\phi} = c/\sin\theta_0$. From Eq. (5.61), we have

$$\boldsymbol{\omega} = \dot{\phi}\mathbf{u}_3 + \dot{\psi}\mathbf{e}_3, \tag{5.156}$$

i.e., $\boldsymbol{\omega}$ always lies in the plane of the fixed z-axis and the precessing figure axis, as shown in Fig. 5.6. Since both the angle θ_0 between the z-axis and the figure axis is constant, and the angle γ between $\boldsymbol{\omega}$ and the figure axis are constant, the angle between $\boldsymbol{\omega}$ and the z-axis must then also be a constant. Hence, $\boldsymbol{\omega}$ also traces out a cone, the *space cone*, as the body precesses about z. The motion can be thought of as if the body cone rolls around the space cone as the figure axis precesses. For this case, the figure axis does not exhibit any oscillation in θ; this motion is therefore called *regular*, or *simple*, *precession*. Oscillatory motion in θ, called *nutation*, will occur for the heavy top treated in the subsequent section.

5.4   The Heavy Top

As discussed, evaluating the equation of motion

$$\frac{d}{dt}(\boldsymbol{\Theta} \cdot \boldsymbol{\omega}) = \mathbf{N} \tag{5.157}$$

in the body system requires the knowledge of the torque $\mathbf{N}$ in the time-dependent body-system basis $\mathbf{e}_i(t)$. For nonvanishing torque, it may therefore be simpler to employ the Lagrange formalism instead.

As before, we choose either the center of mass or a fixed contact point as the origin of the body system. In the first case, the (trivial) center-of-mass motion decouples from the rotation and in the second case, only rotational motion remains. In either case, therefore, we restrict the discussion to rotations only. Choosing the Euler angles ϕ, θ, ψ as the corresponding independent generalized coordinates, the generic expression for the Lagrange function reads

$$L(\phi,\theta,\psi,\dot{\phi},\dot{\theta},\dot{\psi},t) = T_{\mathrm{rot}} - V = \frac{\boldsymbol{\omega} \cdot \boldsymbol{\Theta} \cdot \boldsymbol{\omega}}{2} - V(\phi,\theta,\psi,t). \tag{5.158}$$

Taking the rotation energy in the principal-axis system,

$$T_{\mathrm{rot}} = \frac{1}{2}\sum_{i=1}^{3} I_i\,\omega_i^2, \tag{5.159}$$

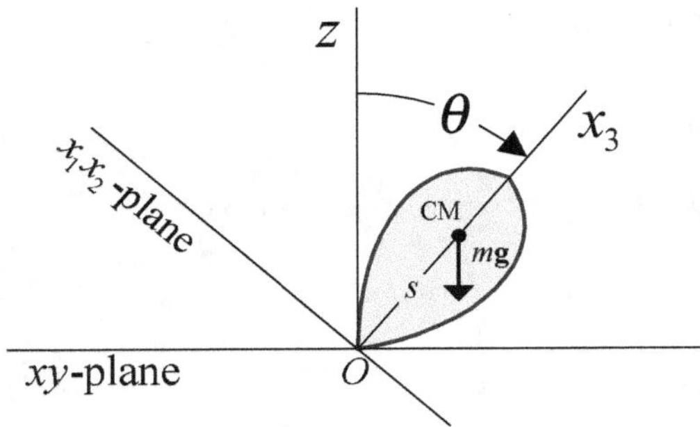

Fig. 5.7   Space and body coordinates for the heavy top with common, fixed origin O.

and employing Eq. (5.125), the Lagrange function becomes

$$L(\phi, \theta, \psi, \dot{\phi}, \dot{\theta}, \dot{\psi}, t) = \frac{I_1}{2} \left(\dot{\phi} \sin\theta \sin\psi + \dot{\theta} \cos\psi \right)^2$$
$$+ \frac{I_2}{2} \left(\dot{\phi} \sin\theta \cos\psi - \dot{\theta} \sin\psi \right)^2$$
$$+ \frac{I_3}{2} \left(\dot{\phi} \cos\theta + \dot{\psi} \right)^2 - V(\phi, \theta, \psi, t). \qquad (5.160)$$

For a given potential, the ensuing Lagrange equations then provide the three coupled second-order differential equations for the three Euler angles.

5.4.1   *The heavy symmetric top*

Let us consider the motion of a symmetric top of mass m under the influence of a (constant) gravitational acceleration g such that its contact point with the horizontal xy-plane of the space system is fixed. This contact point is taken as the common origin O of the space and the body systems; see Fig. 5.7. The figure axis of the body system is taken as the x_3-axis, i.e., $I_1 = I_2 \neq I_3$. This setup may be considered as a simplified model of a child's toy top, where the contact point does not change and friction is absent.

Taking the distance between the origin O and the center of mass CM along the figure axis as s, the potential energy of the top is $V = mgz = mgs \cos\theta$, where θ is the Euler angle between the z-axis of S and the x_3-axis of B. Hence, the Lagrange function is

$$L(\theta, \dot{\phi}, \dot{\theta}, \dot{\psi}) = \frac{I_1}{2} \left(\dot{\theta}^2 + \dot{\phi}^2 \sin^2\theta \right) + \frac{I_3}{2} \left(\dot{\phi} \cos\theta + \dot{\psi} \right)^2 - mgs \cos\theta.$$
$$(5.161)$$

In view of

$$\frac{\partial L}{\partial t} = 0, \qquad \frac{\partial L}{\partial \psi} = 0, \qquad \frac{\partial L}{\partial \phi} = 0, \qquad (5.162)$$

the following constants of motion hold true:

$$E = \frac{\partial L}{\partial \dot{\phi}}\dot{\phi} + \frac{\partial L}{\partial \dot{\psi}}\dot{\psi} + \frac{\partial L}{\partial \dot{\theta}}\dot{\theta} - L$$

$$= \frac{I_1}{2}\left(\dot{\theta}^2 + \dot{\phi}^2\,\sin^2\theta\right) + \frac{I_3}{2}\left(\dot{\phi}\,\cos\theta + \dot{\psi}\right)^2 + mgs\cos\theta$$

$$= const, \tag{5.163}$$

$$L_3 = \frac{\partial L}{\partial \dot{\psi}} = I_3\left(\dot{\psi} + \dot{\phi}\,\cos\theta\right) = const, \tag{5.164}$$

$$L_z = \frac{\partial L}{\partial \dot{\phi}} = I_1\dot{\phi}\,\sin^2\theta + L_3\,\cos\theta = const. \tag{5.165}$$

The first constant E is the total energy following from the time independence of L. The second and third constants L_3 and L_z are the angular momentum components in the body and space systems following from the independence of the rotation angles ψ and ϕ about the respective x_3- and the z-axes. All three constants of motion are determined by the initial conditions of the problem.

The motion exhibits the precession of the figure axis about the external, fixed, z-axis encountered already for the torque-free top. The angular velocity $d\phi/dt$ (cf. Fig. 5.3) of this precession is given via (5.165) as

$$\dot{\phi} = \frac{L_z - L_3\cos\theta}{I_1\sin^2\theta}. \tag{5.166}$$

Now, however, the precession is not simple since θ is not constant.

To discuss the motion in θ, note that the energy E can be written as

$$E = \frac{I_1}{2}\dot{\theta}^2 + U(\theta) = const, \tag{5.167}$$

where U is the effective potential

$$U(\theta) = mgs\cos\theta + \frac{(L_z - L_3\cos\theta)^2}{2I_1\sin^2\theta} + \frac{L_3^2}{2I_3}. \tag{5.168}$$

This expression for the total energy is a first-order differential equation completely analogous to the one-dimensional case discussed in Sec. 1.3.1. In the present case, the integral corresponding to (1.77),

$$t = t_0 + \int_{\theta_0}^{\theta} \frac{d\theta'}{\sqrt{\frac{2}{I_1}\left[E - U(\theta')\right]}}, \tag{5.169}$$

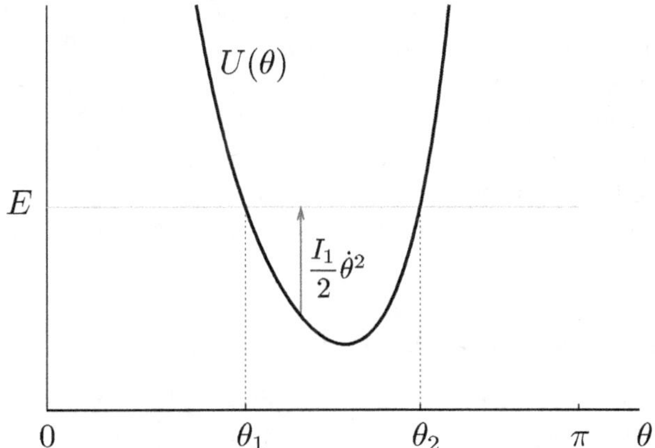

Fig. 5.8   Generic plot of the effective potential U governing the motion in θ. The kinetic energy $I_1\dot{\theta}^2/2$ of the θ motion will only be positive between the two turning points θ_1 and θ_2, i.e., motion outside of this range is not possible.

is an elliptic integral that cannot be solved in closed from. A numerical solution, however, will provide $t = t(\theta)$ and hence $\theta = \theta(t)$.

A qualitative understanding of the θ motion can be gained by employing the graphical method of Fig. 1.5. Figure 5.8 shows a generic plot of the effective potential $U(\theta)$. The physical condition $E - U(\theta) = I_1\dot{\theta}^2/2 > 0$ is only satisfied for a range of θ values between the two values θ_1 and θ_2. In other words, as the figure axis precesses about the fixed external z-axis, the angle θ between the moving figure axis and the fixed z-axis will oscillate between these two turning points. This oscillatory motion is called *nutation*. Note that, depending on the values of L_z and L_3 following from the initial conditions, the angular velocity of the precession may change its sign, according to (5.166), during the nutation motion.

In addition to precession and nutation, the heavy top undergoes rotation about the figure axis described by the angular velocity $d\psi/dt$. It is related to the angular velocity of the precession by

$$\dot{\psi} + \dot{\phi}\cos\theta = \frac{L_3}{I_3} = const, \tag{5.170}$$

according to (5.164).

A special case obtains if the energy E coincides with the minimum of the effective potential U (cf. Fig. 5.8). The two turning points then coincide, $\theta_1 = \theta_2$, and the precession becomes regular, without any nutation. The same is true if $g = 0$, i.e., if gravity is switched off; one then recovers the torque-free case discussed previously. An approximation of this limiting case is given by situations in which the rotation energy T_{rot} is very much larger than the potential energy V, i.e., $T_{\mathrm{rot}} \gg V$. For example, a toy top that is given a large initial spin will at first exhibit (almost) regular precession of the figure axis. Only after a while, when losses of rotational energy due to friction (not considered in the present simple treatment) will have slowed down the spin so that $T_{\mathrm{rot}} \approx V$, the nutation will increase and become more visible.

5.4.2   Earth as a rotating top

To discuss the rotation of Earth as a spinning top, we neglect possible mass fluctuations due to, for example, tidal changes, etc. In other words, Earth is taken as a rigid geoid with constant principal moments of inertia $I_3 \neq I_1 = I_2$, where the actual relative difference between the moments,

$$\frac{\Delta I}{I} = \frac{I_3 - I_1}{I_1} \approx \frac{1}{300}, \tag{5.171}$$

shows that the flattening of the pole regions is quite small, i.e., Earth is almost spherical.

5.4.2.1   Free rotation

The two major forces acting on Earth are due to the Sun and the Moon, i.e., Earth's center of mass is falling freely in their combined gravitational fields. If, at first, we neglect the corresponding (small) resulting torque, Earth behaves like the symmetric torque-free top described in Sec. 5.3.2. (In this respect, we emphasize once again that it is irrelevant that Earth would still be falling freely in the combined gravitational field since the rotational motion is unaffected by this.) If we take Earth's center of mass as the origin of the body system, the corresponding results can be taken over directly. The precession frequency of Earth's rotation axis $\boldsymbol{\omega}$ about Earth's symmetry axis thus follows from (5.142), i.e.,

$$\Omega = \frac{\Delta I}{I} \omega = \frac{1}{300} \frac{2\pi}{\mathrm{day}}. \tag{5.172}$$

In other words, it takes about ten months for the rotation axis to precess about the figure axis. For Earth, the directions of the two axes almost coincide; at the poles, the distance between them is only about 10 m.[103]

---

[103]It should be mentioned that, despite this small value, the fact that Earth in reality is not rigid leads already to noticeable deviations from the regular motion assumed in the present model treatment.

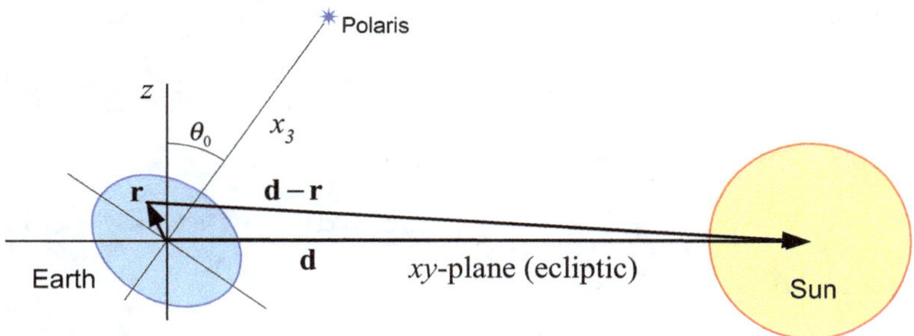

Fig. 5.9   Coordinates of the Earth-Sun system. Earth's figure axis is tilted by a constant angle $\theta_0 = 23.45°$ from the normal to the plane of the ecliptic. Currently, Earth's north pole is oriented towards the 'North Star', Polaris. This will remain true in the future for times small compared to the precession period of 25,800 years of Earth's figure axis about the space-fixed z-axis (see Sec. 5.4.2.3). Note that the orientation of Earth's figure axis with respect to the Sun's position shown here is only true for one point in time during the year (namely the summer solstice on the northern hemisphere, June 21st), i.e., while Earth's figure axis remains oriented towards Polaris, in the geocentric body system the Sun rotates about Earth in the plane of the ecliptic.

In the following, let us neglect the small difference between the directions of $\boldsymbol{\omega}$ and the x_3 figure axis, i.e., we consider Earth's rotation to occur about the figure axis. Since this is one of the principal axes, the angular momentum $\mathbf{L}$ then is aligned with this axis,

$$\mathbf{L} = L\,\mathbf{e}_3, \tag{5.173}$$

according to (5.104). Currently, this axis points towards the 'North Star', Polaris (see Fig. 5.9).

5.4.2.2   *Torque acting on Earth*

Next, we consider the torques due to the Sun and the Moon. We first will treat the torque due to the Sun in detail and then later estimate the effects of the Moon's influence based on the Sun's results.

Taking the distance of the Sun from Earth's CM as $\mathbf{d}$, the gravitational force $d\mathbf{F}$ on a mass element $dm = \rho(\mathbf{r})\,d^3r$ within Earth located a distance $\mathbf{r}$ from Earth's CM is (cf. Fig. 5.9)

$$d\mathbf{F} = d^3r\,\rho(\mathbf{r})\,Gm_s\frac{\mathbf{d-r}}{|\mathbf{d-r}|^3}, \tag{5.174}$$

where G is the gravitational constant, m_s the Sun's mass, and $\rho(\mathbf{r})$ the mass distribution of Earth.   The torque exerted by the Sun on

Earth then is

$$\mathbf{N}_S = \int \mathbf{r} \times d\mathbf{F} = Gm_s \int d^3 r \, \rho(\mathbf{r}) \, \mathbf{r} \times \frac{\mathbf{d} - \mathbf{r}}{|\mathbf{d} - \mathbf{r}|^3}. \tag{5.175}$$

Where $\rho(\mathbf{r}) \neq 0$, we have $|\mathbf{d}| \gg |\mathbf{r}|$ and hence

$$|\mathbf{d} - \mathbf{r}|^3 \approx d^3 - 3d(\mathbf{r} \cdot \mathbf{d}), \tag{5.176}$$

which leads to

$$\begin{aligned}
\mathbf{N}_S &= Gm_s \int d^3 r \, \rho(\mathbf{r}) \, \frac{\mathbf{r} \times \mathbf{d}}{d^3} \left(1 + 3\frac{\mathbf{r} \cdot \mathbf{d}}{d^2} \right) \\
&= Gm_s \int d^3 r \, \rho(\mathbf{r}) \sum_{i,j,k=1}^{3} \frac{\varepsilon_{ijk} x_j d_k}{d^3} \left(1 + 3\sum_{l=1}^{3} \frac{x_l d_l}{d^2} \right) \mathbf{e}_i \\
&= \frac{3Gm_s}{d^5} \int d^3 r \, \rho(\mathbf{r}) \sum_{i,j,k=1}^{3} \varepsilon_{ijk} x_j^2 d_j d_k \, \mathbf{e}_i. \tag{5.177}
\end{aligned}$$

The last equality follows from the fact that this integral is written in the body system with the origin at the center of mass of the geoid which is also the symmetry center. Since $\rho(\mathbf{r})$ is invariant under individual mirror images $x_i \to -x_i$ $(i = 1, 2, 3)$, all terms in the sum that are linear in x_i change sign under this symmetry operation. In view of the invariance of the integral under the mirror transformation, all linear contributions, therefore, vanish. Nonvanishing contributions arise only where x_j meets an x_l with $j = l$, and this provides the last expression.

To further evaluate this result in the geocentric body system, we approximate the Sun's orbit around the Earth by a circle with angular frequency $\Omega_s = 2\pi/1\,\mathrm{yr}$. Relative to the body system, the vector $\mathbf{d}(t)$ then is represented by

$$\mathbf{d} \quad \longrightarrow \quad \vec{d} = \begin{pmatrix} d_1 \\ d_2 \\ d_3 \end{pmatrix} = \begin{pmatrix} \cos\theta_0 & 0 & -\sin\theta_0 \\ 0 & 1 & 0 \\ \sin\theta_0 & 0 & \cos\theta_0 \end{pmatrix} \begin{pmatrix} d\cos\Omega_s t \\ d\sin\Omega_s t \\ 0 \end{pmatrix}$$

$$= d \begin{pmatrix} \cos\theta_0 \, \cos\Omega_s t \\ \sin\Omega_s t \\ \sin\theta_0 \, \cos\Omega_s t \end{pmatrix}. \tag{5.178}$$

This follows from expressing $\mathbf{d}$ first in coordinates where the xy-plane coincides with the ecliptic and then tilting the z-axis by an angle θ_0 around the y-axis into the x_3 figure axis. (The y-axis is chosen arbitrarily as the tilt axis. This merely defines the line of equinoxes[104] as the y-axis, i.e., the space system's positive x-axis is pointing toward the summer solstice position of the Sun; cf. Fig. 5.9.) Employing this representation in (5.177) and averaging over one period[105] yields

$$\langle \mathbf{N}_S \rangle = \frac{3Gm_S}{d^5} \overbrace{\frac{\Omega_S}{2\pi} \int_0^{2\pi/\Omega_S} dt}^{\text{averaging}} \int d^3r\, \rho(\mathbf{r})\, [(x_2^2 - x_3^2)\, d_2 d_3\, \mathbf{e}_1$$

$$+ (x_3^2 - x_1^2)\, d_3 d_1\, \mathbf{e}_2 + (x_1^2 - x_2^2)\, d_1 d_2\, \mathbf{e}_3]. \qquad (5.179)$$

Only the $\mathbf{e}_2$ component provides a nonvanishing time average, yielding a factor $\langle \cos^2 \Omega_S t \rangle = 1/2$. Hence,

$$\langle \mathbf{N}_S \rangle = \frac{3Gm_S}{2d^3} \cos\theta_0\, \sin\theta_0 \int d^3r\, \rho(\mathbf{r})\, (x_3^2 - x_1^2)\, \mathbf{e}_2$$

$$= -\frac{3Gm_S}{2d^3} \cos\theta_0\, \sin\theta_0\, (I_3 - I_1)\, \mathbf{e}_2$$

$$= \frac{3Gm_S}{2d^3} \cos\theta_0\, (I_3 - I_1)\, \mathbf{e}_3 \times \mathbf{e}_z, \qquad (5.180)$$

where $-\sin\theta_0\, \mathbf{e}_2$ was replaced by $\mathbf{e}_3 \times \mathbf{e}_z$ (cf. Fig. 5.9). This average torque thus is seen to be nonzero only because the spinning Earth possesses an equatorial bulge leading to $I_3 - I_1 > 0$.

5.4.2.3 *Precession of Earth's figure axis*

According to the basic equation (5.119) of rotational motion, this average torque results in an angular-momentum change according to

$$\frac{d\mathbf{L}}{dt} = \langle \mathbf{N}_S \rangle = \frac{3Gm_S}{2d^3} \cos\theta_0\, (I_3 - I_1)\, \mathbf{e}_3 \times \mathbf{e}_z. \qquad (5.181)$$

Strictly speaking, the angular momentum $\mathbf{L}$ appearing here should be the angular momentum $\langle \mathbf{L} \rangle$ averaged over one solar orbit. As mentioned already (see Sec. 5.4.2.1), we will neglect the slight difference between the rotation and the figure axes and employ $\mathbf{L} = L\mathbf{e}_3$ here.[106] Upon scalar multiplication

---

[104]The line of equinoxes is the intersection of Earth's equatorial plane with the ecliptic. At the beginning of spring and fall, the Sun is crossing the line of equinoxes.

[105]The time average $\langle f \rangle$ of a function $f(t)$ over a period τ is $\langle f \rangle = \frac{1}{\tau} \int_0^\tau dt\, f(t)$.

[106]In view of the fact that the precession of the rotation axis about the figure axis is about ten months [cf. Eq. (5.172)], which is not too far off from the averaging period of one year, the approximation $\langle \mathbf{L} \rangle \approx L\mathbf{e}_3$ is actually even better justified for the averaged angular momentum than for the instantaneous one.

by $\mathbf{L}$, we then find immediately from the right-hand side of (5.181),

$$0 = \mathbf{L} \cdot \frac{d\mathbf{L}}{dt} = \frac{1}{2} \frac{d\,(\mathbf{L} \cdot \mathbf{L})}{dt} = \frac{1}{2} \frac{dL^2}{dt}, \tag{5.182}$$

or $L = |\mathbf{L}| = const$. The time dependence of $\mathbf{L}(t)$, therefore, is solely due to the direction of $\mathbf{L}$,

$$\mathbf{L}(t) = L\mathbf{e}_3(t) = I_3 \omega \, \mathbf{e}_3(t). \tag{5.183}$$

L is expressed here by the principal moment of inertia I_3 of the figure axis and the angular velocity ω of Earth's rotation.

Next, we will show that the angle θ_0 between the figure axis and the space-fixed z axis is a constant. To this end, we take the dot product of Eq. (5.181) with $\mathbf{e}_z$ to arrive at

$$0 = \mathbf{e}_z \cdot \frac{d\mathbf{L}}{dt} = \frac{d\,(\mathbf{e}_z \cdot \mathbf{L})}{dt} \quad \Rightarrow \quad \mathbf{e}_z \cdot \mathbf{L} = L \cos \theta_0 = const. \tag{5.184}$$

Since $L = const$, this indeed implies $\theta_0 = const$. For Earth, this angle is

$$\theta_0 = 23.45^\circ \; ; \tag{5.185}$$

it determines the tilt of the figure axis with respect to the ecliptic (see Fig. 5.9). The latitudes $23.45^\circ\,N$ (*tropic of cancer*) and $23.45^\circ\,S$ (*tropic of capricorn*) are the northernmost and the southernmost latitudes, respectively, over which the Sun will be directly overhead at noon, marking the corresponding days of the summer and winter solstices.

The last result immediately implies that the only allowed locus of motion for $\mathbf{e}_3(t)$ is a cone of (fixed) opening angle θ_0 around the z-axis. Indeed, from

$$\frac{d\mathbf{L}}{dt} = I_3 \omega \frac{d\mathbf{e}_3}{dt} = \frac{3 G m_s}{2 d^3} \cos \theta_0 \, (I_3 - I_1) \, \mathbf{e}_3 \times \mathbf{e}_z, \tag{5.186}$$

it follows that

$$\frac{d\mathbf{e}_3}{dt} = \boldsymbol{\Omega}^{(s)}_{\mathrm{prec}} \times \mathbf{e}_3, \tag{5.187}$$

where

$$\boldsymbol{\Omega}^{(s)}_{\mathrm{prec}} = -\frac{I_3 - I_1}{I_3} \frac{3 G m_s}{2 \omega d^3} \cos \theta_0 \, \mathbf{e}_z \tag{5.188}$$

is the constant angular velocity of the figure axis $\mathbf{e}_3$ precessing about the z-axis. The minus sign here means that this precession is opposite the rotational motion of Earth about its axis.

Note that for the circular solar orbit assumed here, the gravitational attraction between Earth and Sun is balanced by the centrifugal force according to

$$G\frac{m_E m_S}{d^2} = \mu \Omega_S^2 d, \tag{5.189}$$

where

$$\mu = \frac{m_E m_S}{m_E + m_S} \tag{5.190}$$

is the reduced mass of the Earth-Sun system. The magnitude of the precession velocity then becomes

$$\Omega_{\text{prec}}^{(S)} = \frac{3}{2}\frac{I_3 - I_1}{I_3}\cos\theta_0 \frac{m_S}{m_S + m_E}\frac{\Omega_S^2}{\omega}. \tag{5.191}$$

The preceding result describes the Sun's influence. A similar result, along the same lines, can be obtained for the Moon's influence if we approximatively assume that the Moon's orbit lies in the plane of the ecliptic.[107] Therefore, in addition to the precessional velocity $\Omega_{\text{prec}}^{(S)}$ due to the Sun, we also have

$$\Omega_{\text{prec}}^{(M)} = \frac{3}{2}\frac{I_3 - I_1}{I_3}\cos\theta_0 \frac{m_M}{m_M + m_E}\frac{\Omega_M^2}{\omega} \tag{5.192}$$

for the precession due to the Moon, where m_M and Ω_M, respectively, are the mass and angular velocity of the Moon.

Under the assumptions made here, the magnitudes of the two angular velocities can simply be added to provide

$$\Omega_{\text{prec}} = \Omega_{\text{prec}}^{(M)} + \Omega_{\text{prec}}^{(S)} \approx 10^{-7}\omega. \tag{5.193}$$

We have used here $(I_3 - I_1)/I_3 \approx 1/300$ and $\cos\theta_0 \approx 0.92$ and

$$\frac{m}{m + m_E} \approx \begin{cases} 1 & m = m_S \ \text{(Sun)}, \\ 1/81 & m = m_M \ \text{(Moon)}, \end{cases} \tag{5.194}$$

and

$$\frac{\omega}{\Omega} = \begin{cases} 365 & \Omega = \Omega_S \ \text{(Sun)}, \\ 28 & \Omega = \Omega_M \ \text{(Moon)}. \end{cases} \tag{5.195}$$

In view of

$$\frac{m}{m + m_E}\frac{\Omega^2}{\omega^2} = \begin{cases} 7.5 \times 10^{-6} & \text{for the Sun}, \\ 15.7 \times 10^{-6} & \text{for the Moon}, \end{cases} \tag{5.196}$$

---

[107]The Moon's true orbital plane is tilted about 5° with respect to the ecliptic, so this is a reasonable approximation for the present semi-quantitative discussion.

the Moon's influence is actually twice as large as the Sun's since its close proximity to Earth more than compensates for its smaller mass.

The present semi-quantitative considerations thus provide a value of 10^7 days $\approx 27{,}400$ years for the period of this precession. This is in reasonable agreement with the actual observed value of about 25,800 years. This present result means that over this period of precession, also called a *Platonic year*, a succession of various stars will assume the role of the 'North Star' presently held by Polaris (for example, at around the year 13 500, it will be Vega's turn). Associated with this is a *precession of the equinoxes*, i.e., over the course of the precession the seasons will start earlier and earlier — and in 13,000 years summer on the northern hemisphere will start in December. The existence of this precession was first noted by the Greek mathematician and astronomer Hipparchos (about 190–125 BCE) already more than 2,000 years ago. To fully appreciate this achievement, one must realize that the equinoctial precession has a value of only $0.014°$/year, thus amounting to less than $1°$ over the entire lifetime of Hipparchos! He could not have made his discovery, therefore, without the use of existing observational data from 279 BCE and 431 BCE that provided a sufficiently long time base for firm conclusions.

It should be added that the actual motion of Earth's figure axis is subject to many additional influences. The experimental data exhibit a large number of nutation cycles of various frequencies that overlay each other and that modify the present results in detail. The cause of some of these cycles is still not entirely known.

Chapter 6

Small Oscillations

Many problems involve small displacements from a stable equilibrium position, with a system-dependent restoring force that seeks to drive the system back to equilibrium. The problems may also include additional external forces or internal friction, etc. In many instances, the system's response produces small oscillations or vibrations about the equilibrium point. Simple examples are a small ball rolling around the bottom of a bowl or an atom or molecule distorted by a weak external electric or magnetic field. In this part, we consider situations for systems where the unperturbed system can be described in terms of a potential. To illustrate the basic approach, we first look at the one-dimensional case.

6.1   Systems with One Degree of Freedom

Consider a generic system with one degree of freedom, described by the generalized coordinate q, experiencing a weak external perturbation. The Lagrange function,

$$L(q,\dot{q},t) = \underbrace{\frac{c(q)}{2}\dot{q}^2 - V(q)}_{=L_0(q,\dot{q})} - V_{\text{ext}}(q,t), \tag{6.1}$$

contains the Lagrangian $L_0(q,\dot{q})$ of the time-independent unperturbed system and the distorting potential $V_{\text{ext}}(q,t)$. The form of the kinetic energy follows from (2.97). The unperturbed system is said to be in equilibrium at $q = q_0$ if the generalized force acting on the system vanishes,

$$Q = -\left[\frac{dV}{dq}\right]_0, \tag{6.2}$$

where the index 0 signifies the derivative at the equilibrium point. The equilibrium is *stable* if a small perturbation of the system from equilibrium results in small bounded motion about the equilibrium.

Denoting the deviations from the equilibrium position by η,

$$q = q_0 + \eta, \tag{6.3}$$

and expanding the unperturbed potential around q_0,

$$V(q) = V(q_0) + \left[\frac{dV}{dq}\right]_0 \eta + \frac{1}{2}\left[\frac{d^2V}{dq^2}\right]_0 \eta^2 + \cdots$$

$$\approx V_0 + \frac{1}{2}k\eta^2, \tag{6.4}$$

the stability requirement of the equilibrium translates into the local-minimum condition for the potential, $k \equiv [d^2V/dq^2]_0 > 0$, in addition to $[dV/dq]_0 = 0$. It is assumed here that the displacement η is small enough so that higher-order terms are negligible. The constant term $V_0 = V(q_0)$ is irrelevant for the description of the problem and can be omitted from the Lagrange function. The remaining expression, $k\eta^2/2$, is just the harmonic-oscillator potential.

Similarly, expanding the kinetic energy provides

$$T(q, \dot{q}) = \frac{1}{2}\left(c(q_0) + \left[\frac{dc}{dq}\right]_0 \eta + \cdots\right)\dot{\eta}^2 \approx \frac{\mu}{2}\dot{\eta}^2, \tag{6.5}$$

where $\mu \equiv c(q_0)$ is a mass parameter (that, depending on circumstances, may or may not be equal to the usual mass m; for example, for a planar pendulum of length ℓ, where $q = \vartheta$, $\mu = m\ell^2$). Note that the higher-order terms of the expansion of $c(q)$ do not contribute since the overall $\dot{\eta}^2$ factor already provides the second order consistent with the expansion (6.4).

The expansion of the external potential yields

$$V_{\text{ext}}(q, t) = V_{\text{ext}}(q_0, t) + \left[\frac{\partial V_{\text{ext}}}{\partial q}\right]_0 \eta + \cdot$$

$$\approx V_{\text{ext}}(q_0, t) - f(t)\,\eta, \tag{6.6}$$

where the perturbation is considered so weak that only terms linear in η need to be taken into account. The first contribution, $V_{\text{ext}}(q_0, t)$, depends only on the time: it can be written as the time derivative of a function and consequently it can be omitted from the Lagrange function (see Sec. 3.2.1). The second term contains the generalized external force

$$Q_{\text{ext}}(q, t) = -\frac{\partial V_{\text{ext}}(q, t)}{\partial q} \xrightarrow{q=q_0} f(t), \tag{6.7}$$

where only the time dependence at the equilibrium position enters the description of small perturbations.

Collecting all pieces, the relevant Lagrange function for *small* displacements now reads

$$L(\eta, \dot{\eta}, t) = \frac{\mu}{2}\dot{\eta}^2 - \frac{k}{2}\eta^2 + f(t)\,\eta. \tag{6.8}$$

This is the well-known Lagrange function of the harmonic oscillator subject to an external driving force $f(t)$. The corresponding equation of motion,

$$\mu\ddot{\eta} + k\eta = f(t), \tag{6.9}$$

was considered already in Sec. 1.3, where we also discussed a number of closely related standard one-dimensional problems.

The present discussion shows that the treatment of harmonic motion is relevant beyond simple spring- or pendulum-type systems. Indeed, the behavior of many, many systems can be understood in terms of harmonic motion, damped harmonic motion, or driven harmonic motion, etc.

6.2 Systems with Many Degrees of Freedom

We consider small oscillations for a system with N degrees of freedom described by generalized coordinates $q_1, \ldots, q_N$, with an unperturbed time-independent Lagrange function

$$L_0 = L_0(q, \dot{q}) = \frac{1}{2}\sum_{i,j=1}^{N} c_{ij}(q)\,\dot{q}_i\dot{q}_j - V(q), \tag{6.10}$$

where, as in previous sections, q and $\dot{q}$ stand for the entire list of variables $q_1, \ldots, q_N$ and $\dot{q}_1, \ldots, \dot{q}_N$, respectively. Before introducing any external disturbance, let us first discuss in more detail the equilibrium situation of the unperturbed system.

Analogous to the one-dimensional case, all generalized forces Q_i vanish in equilibrium at $q_i = q_{0,i}$,

$$Q_i = -\left[\frac{\partial V}{\partial q_i}\right]_0 = 0. \tag{6.11}$$

Using

$$q_i = q_{0,i} + \eta_i \tag{6.12}$$

and expanding the potential and the kinetic energy around the equilibrium leads to

$$V(q) = V(q_0) + \sum_{i=1}^{N} \left[\frac{\partial V}{\partial q_i} \right]_0 \eta_i + \frac{1}{2} \sum_{i,j=1}^{N} \left[\frac{\partial^2 V}{\partial q_i \partial q_j} \right]_0 \eta_i \eta_j + \cdots$$

$$\approx V_0 + \frac{1}{2} \sum_{i,j=1}^{N} V_{ij} \, \eta_i \eta_j, \tag{6.13}$$

where

$$V_{ij} = V_{ji} = \left[\frac{\partial^2 V}{\partial q_i \partial q_j} \right]_0, \tag{6.14}$$

and

$$T(q, \dot{q}) \approx \frac{1}{2} \sum_{i,j=1}^{N} T_{ij} \, \dot{\eta}_i \dot{\eta}_j, \tag{6.15}$$

with

$$T_{ij} = T_{ji} = c_{ij}(q_0). \tag{6.16}$$

Dropping irrelevant constant terms, the Lagrange function then becomes

$$L_0 = L_0(\eta, \dot{\eta}) = \frac{1}{2} \sum_{i,j=1}^{N} \left(T_{ij} \, \dot{\eta}_i \dot{\eta}_j - V_{ij} \, \eta_i \eta_j \right), \tag{6.17}$$

leading to the Lagrange equations

$$\sum_{j=1}^{N} \left(T_{ij} \, \ddot{\eta}_j + V_{ij} \, \eta_j \right) = 0, \qquad i = 1, \ldots, N. \tag{6.18}$$

These equations of motion are a system of N linear homogeneous second-order differential equations with constant coefficients. They describe the behavior of the unperturbed system around the equilibrium positions.

It is convenient to express the preceding equations in a matrix formalism. Employing matrices

$$\mathsf{T} = (T_{ij}) \qquad \text{and} \qquad \mathsf{V} = (V_{ij}), \tag{6.19}$$

and introducing a column vector $\boldsymbol{\eta}$, and its transposed, by

$$\boldsymbol{\eta} = \begin{pmatrix} \eta_1 \\ \eta_2 \\ \vdots \\ \eta_N \end{pmatrix}, \qquad \boldsymbol{\eta}^{\mathrm{T}} = (\eta_1, \, \eta_2, \, \cdots, \, \eta_N), \tag{6.20}$$

the kinetic and potential energies are given by

$$T = \frac{1}{2}\dot{\boldsymbol{\eta}}^{\mathrm{T}}\mathsf{T}\dot{\boldsymbol{\eta}} \qquad \text{and} \qquad V = \frac{1}{2}\boldsymbol{\eta}^{\mathrm{T}}\mathsf{V}\boldsymbol{\eta}, \tag{6.21}$$

where the irrelevant constant part of the potential is omitted. The Lagrange function,

$$\boxed{L_0 = \frac{1}{2}\left(\dot{\boldsymbol{\eta}}^{\mathrm{T}}\mathsf{T}\dot{\boldsymbol{\eta}} - \boldsymbol{\eta}^{\mathrm{T}}\mathsf{V}\boldsymbol{\eta}\right)}, \tag{6.22}$$

and the equations of motion,

$$\boxed{\mathsf{T}\ddot{\boldsymbol{\eta}} + \mathsf{V}\boldsymbol{\eta} = 0}, \tag{6.23}$$

can then also be written in this compact notation.

The complex ansatz

$$\boldsymbol{\eta} = \mathbf{A}\,e^{-i\omega t}, \tag{6.24}$$

where $\mathbf{A}$ is a column vector formed of N constant amplitudes A_i, provides a solution for (6.23). Of course, since the original equation (6.23) is real, the physically relevant solutions are given only by the real part; however, it is more convenient to work with the complex ansatz. Employing this ansatz in (6.23) leads to

$$\left(\mathsf{V} - \omega^2\mathsf{T}\right)\mathbf{A} = 0. \tag{6.25}$$

This matrix equation constitutes a set of N linear homogeneous equations for the N amplitudes A_i. Nontrivial solutions, therefore, are only obtained if the determinant of the N-dimensional coefficient matrix vanishes,

$$\det\left(\mathsf{V} - \omega^2\mathsf{T}\right) = 0. \tag{6.26}$$

This is an N-th order polynomial in ω^2, also called the *characteristic polynomial*, with N roots $\omega_1^2, \omega_2^2, \ldots, \omega_N^2$. For these eigenvalues, the *generalized eigenvalue equation*

$$\left(\mathsf{V} - \omega_k^2\mathsf{T}\right)\mathbf{A}_k = 0, \qquad k = 1,\ldots,N, \tag{6.27}$$

has N nontrivial (i.e., non-zero) solution vectors $\mathbf{A}_k$.

It can be shown that

$$(\omega_k^2)^* = \omega_k^2, \quad \text{and} \quad \omega_k^2 \geq 0, \qquad k = 1,\ldots,N, \tag{6.28}$$

i.e., the frequencies $\pm\omega_k = \pm\sqrt{\omega_k^2}$ are real. The proof is rather technical and not very illuminating, and it will be omitted here.[108] However, its result

---

[108]Essentially, the proof hinges on the fact that both the kinetic energy $T = \dot{\boldsymbol{\eta}}^{\mathrm{T}}\mathsf{T}\dot{\boldsymbol{\eta}}/2$ and the potential $V = \boldsymbol{\eta}^{\mathrm{T}}\mathsf{V}\boldsymbol{\eta}/2$ are positive bilinear forms for stable equilibrium situations. For details, see, e.g., [1].

can be made plausible quite easily on physical grounds: To describe physically meaningful solutions around a stable equilibrium, the ansatz (6.24) must remain bounded for all times. If a solution ω_k^2 has a nonvanishing imaginary part or is negative real, then ω_k or $-\omega_k$ has a positive imaginary part. This would imply an exponential solution (6.24) rising beyond any bounds, thus contradicting the stability and boundedness assumption. Hence, (6.28) must hold true.

Let us assume at first that $\omega_k^2 > 0$. Then for every ω_k^2, there are two solutions, one with $+\omega_k$ and one with $-\omega_k$. The corresponding physical solutions,

$$\mathrm{Re}\left(\mathbf{A}_k\, e^{\pm i\omega_k t}\right) = (\mathrm{Re}\,\mathbf{A}_k)\cos(\omega_k t) \mp (\mathrm{Im}\,\mathbf{A}_k)\sin(\omega_k t)$$
$$= \hat{\mathbf{A}}_k \cos(\omega_k t + \phi_k), \tag{6.29}$$

can, without loss of generality, always be written in the last form, using a positive frequency only, with a phase ϕ_k and an amplitude vector

$$\hat{\mathbf{A}}_k = \begin{pmatrix} a_{1k} \\ a_{2k} \\ \vdots \\ a_{Nk} \end{pmatrix} \tag{6.30}$$

with *real* elements a_{ik}.

The general physical solution of (6.23) then is an arbitrary linear combination,

$$\boxed{\boldsymbol{\eta} = \sum_{k=1}^{N} \hat{\mathbf{A}}_k b_k \cos(\omega_k t + \phi_k)} \;, \tag{6.31}$$

where the b_k and ϕ_k are the $2N$ real integration constants of (6.23) determined by the initial conditions. Since (6.27) is homogeneous, at least one component of its solution vector can be chosen freely.[109] Using a *real* non-zero seed value for this undetermined component ensures that the desired solutions of (6.27) are real, i.e., we have

$$\boxed{(\mathsf{V} - \omega_k^2 \mathsf{T})\hat{\mathbf{A}}_k = 0, \qquad \hat{\mathbf{A}}_k \text{ real.}} \tag{6.32}$$

The actual normalization of the $\hat{\mathbf{A}}_k$ is irrelevant since only the product $\hat{\mathbf{A}}_k b_k$ has any physical meaning; any 'missing' factor in $\hat{\mathbf{A}}_k$, therefore, will be

---

[109]If r of the eigenvalues are degenerate, the eigenvalue system has rank $N - r$, and then r (real) components may be chosen freely.

compensated by the determination of b_k according to the initial conditions. In view of this, we may then *choose* the normalization[110] according to

$$\hat{\mathbf{A}}_k^{\mathrm{T}} \mathsf{T} \hat{\mathbf{A}}_k = \sum_{i,j=1}^{N} T_{ij} a_{ik} a_{jk} = 1, \qquad k = 1, \ldots N. \tag{6.33}$$

This particular choice is made here only to simplify the following formal presentation (so that one does not have to keep track of the normalization factors in the following equations). We emphasize once more that since only the product $\hat{\mathbf{A}}_k b_k$ is physically relevant, in practice one may skip the formal step of actually normalizing the $\hat{\mathbf{A}}_k$ and simply choose integration constants b_k appropriate for the given initial conditions. [This is illustrated in Eqs. (6.87) and (6.88) for the example of the triatomic molecule given in Sec. 6.3.3.]

If one of the roots of the characteristic polynomial vanishes, $\omega_0 = 0$, the harmonic ansatz (6.24) is not appropriate. (The index 0 is arbitrarily chosen here as corresponding to a vanishing frequency.) Indeed, since the corresponding equation of motion is simply

$$\mathsf{T} \ddot{\boldsymbol{\eta}}_0 = 0, \tag{6.34}$$

this is solved by the ansatz

$$\boldsymbol{\eta}_0 = \hat{\mathbf{A}}_0 \left(b_0 t + c_0 \right), \tag{6.35}$$

with two real integration constants b_0 and c_0. The real amplitude vector $\hat{\mathbf{A}}_0$ is a nontrivial solution of

$$\mathsf{V} \hat{\mathbf{A}}_0 = 0, \tag{6.36}$$

again normalized according to (6.33). This situation, corresponding to a uniform translation, arises in modes where the potential energy vanishes, $V = \boldsymbol{\eta}_0^{\mathrm{T}} \mathsf{V} \boldsymbol{\eta}_0 / 2 = 0$, and the system as a whole can move without altering the system's potential energy. This solution is of no interest for the vibrational modes of the system, and it can be avoided from the outset by restricting the discussion to the center-of-mass system. (See also the subsequent discussion of a triatomic molecule in Sec. 6.3.3.)

---

[110]*Choosing the normalization* means that we first calculate the (positive!) numbers $t_k \equiv \hat{\mathbf{A}}_k^{\mathrm{T}} \mathsf{T} \hat{\mathbf{A}}_k$ for each eigenvector $\hat{\mathbf{A}}_k$ as they follow from the initial choice of its seed component and then rescale by putting $\hat{\mathbf{A}}_k / \sqrt{t_k} \to \hat{\mathbf{A}}_k$. It is the rescaled vectors that enter (6.33).

6.2.1 *Normal coordinates*

The generalized coordinates η_i in the preceding considerations are not unique. There exists, however, a special, unique (up to arbitrary normalization factors) set of coordinates in which all degrees of freedom are completely decoupled. Introducing such *normal coordinates* ξ_k is closely related to the principal-axis transformations encountered in Sec. 5.2.5 for rigid bodies.

The existence of coordinates ξ_k that describe the normal modes of the vibrations — i.e., modes in which each ξ_k is associated with one and only one frequency ω_k, instead of mixing them, as in (6.31) — is immediately obvious if we work our way back from the general solution (6.31). Written explicitly it reads

$$\eta_i = \sum_{k=1}^{N} a_{ik}\Big[b_k\cos(\omega_k t + \phi_k)\Big].\tag{6.37}$$

This evidently corresponds to a linear coordinate transformation

$$\eta_i = \sum_{k=1}^{N} a_{ik}\,\xi_k, \qquad \text{or} \qquad \boldsymbol{\eta} = \mathsf{A}\boldsymbol{\xi},\tag{6.38}$$

where the

$$\xi_k = b_k\cos(\omega_k t + \phi_k)\tag{6.39}$$

are the general solutions[111] of single-mode equations of motion

$$\ddot{\xi}_k + \omega_k^2\xi_k = 0, \qquad k = 1,\ldots,N.\tag{6.40}$$

The required transformation matrix,

$$\mathsf{A} = \Big(\hat{\mathbf{A}}_1,\ \hat{\mathbf{A}}_2,\ \ldots,\ \hat{\mathbf{A}}_N\Big) = \begin{pmatrix} a_{11} & a_{12} & \ldots & a_{1N} \\ a_{21} & a_{22} & \ldots & a_{2N} \\ \vdots & \vdots & \ddots & \vdots \\ a_{N1} & a_{N2} & \ldots & a_{NN} \end{pmatrix},\tag{6.41}$$

is formed by concatenating the solution column vectors of (6.32) into a matrix.

Formally, this result is obtained by the following procedure. Note that with (6.32), the transposed relation reads

$$\hat{\mathbf{A}}_l^{\mathrm{T}}\big(\mathsf{V} - \omega_l^2\mathsf{T}\big) = 0,\tag{6.42}$$

---

[111]As discussed above, this presumes that $\omega_k \neq 0$. For $\omega_k = 0$, the general solution would be $\xi_k = b_k t + c_k$.

where the symmetry of T and V was used. Multiplying this equation with $\hat{\mathbf{A}}_k$ from the right and (6.32) with $\hat{\mathbf{A}}_l^{\mathsf{T}}$ from the left, and subtracting the two expressions produces

$$\left(\omega_k^2 - \omega_l^2\right) \hat{\mathbf{A}}_l^{\mathsf{T}} \mathsf{T} \hat{\mathbf{A}}_k = 0. \tag{6.43}$$

Assuming non-degeneracy of the eigenvalues, i.e., $\omega_k^2 \neq \omega_l^2$ for $k \neq l$, this implies that $\hat{\mathbf{A}}_l^{\mathsf{T}} \mathsf{T} \hat{\mathbf{A}}_k$ vanishes for $k \neq l$. Combining this with our normalization choice (6.33), we thus have

$$\hat{\mathbf{A}}_l^{\mathsf{T}} \mathsf{T} \hat{\mathbf{A}}_k = \sum_{i,j=1}^{N} T_{ij} a_{il} a_{jk} = \delta_{kl}, \tag{6.44}$$

If, for $k \neq l$, the eigenvalues are equal, $\omega_l^2 = \omega_k^2$, then this reasoning breaks down. However, in this case, any linear combination $\alpha \hat{\mathbf{A}}_k + \beta \hat{\mathbf{A}}_l$ will also be an eigenvector of (6.32) for the same eigenvalue. By standard orthogonalization procedures, it is always possible to choose two linear combinations that replace the original $\hat{\mathbf{A}}_k$ and $\hat{\mathbf{A}}_l$ such that (6.44) remains true. This procedure is also possible for any number of degenerate eigenvalues. (See, e.g., [1] for the details of the orthogonalization procedure.) In this sense, therefore, the result (6.44) can be taken to be generally valid.

Equation (6.44) is a *generalized orthogonality relation* for the eigenvectors $\hat{\mathbf{A}}_k$. For the matrix A it reads

$$\mathsf{A}^{\mathsf{T}} \mathsf{T} \mathsf{A} = \mathbb{1}, \tag{6.45}$$

where $\mathbb{1}$ is the $N \times N$ unit matrix. From the eigenvalue equation (6.32), it follows then immediately that

$$\mathsf{A}^{\mathsf{T}} \mathsf{V} \mathsf{A} = \mathsf{A}^{\mathsf{T}} \mathsf{T} \mathsf{A}\, \mathsf{w}^2 = \mathsf{w}^2 \qquad \text{(diagonal)}, \tag{6.46}$$

where the diagonal matrix

$$\mathsf{w} = (\omega_k \delta_{kl}) = \begin{pmatrix} \omega_1 & 0 & \cdots & 0 \\ 0 & \omega_2 & \cdots & 0 \\ \vdots & \vdots & \ddots & \vdots \\ 0 & 0 & \cdots & \omega_N \end{pmatrix} \tag{6.47}$$

was introduced for the eigenvalues. Of course, since w is diagonal, so is w^2. The transformation A thus *simultaneously diagonalizes both* the kinetic-energy matrix T and the potential matrix V. This decouples all frequency modes and provides the definition of the corresponding normal coordinates in (6.38). The reverse of the transformation (6.38) is then

$$\boldsymbol{\xi} = \mathsf{A}^{\mathsf{T}} \mathsf{T} \boldsymbol{\eta}. \tag{6.48}$$

To obtain the Lagrange function in the new coordinates, we use (6.38) in (6.22), i.e.,

$$
\begin{aligned}
L_0 &= \frac{1}{2}\left(\dot{\boldsymbol{\eta}}^{\mathrm{T}}\mathsf{T}\dot{\boldsymbol{\eta}} - \boldsymbol{\eta}^{\mathrm{T}}\mathsf{V}\boldsymbol{\eta}\right) = \frac{1}{2}\left[(\mathsf{A}\dot{\boldsymbol{\xi}})^{\mathrm{T}}\mathsf{T}(\mathsf{A}\dot{\boldsymbol{\xi}}) - (\mathsf{A}\boldsymbol{\xi})^{\mathrm{T}}\mathsf{V}(\mathsf{A}\boldsymbol{\xi})\right] \\
&= \frac{1}{2}\left[\dot{\boldsymbol{\xi}}^{\mathrm{T}}(\mathsf{A}^{\mathrm{T}}\mathsf{T}\mathsf{A})\dot{\boldsymbol{\xi}} - \boldsymbol{\xi}^{\mathrm{T}}(\mathsf{A}^{\mathrm{T}}\mathsf{V}\mathsf{A})\boldsymbol{\xi}\right] \\
&= \frac{1}{2}\left[\dot{\boldsymbol{\xi}}^{\mathrm{T}}\dot{\boldsymbol{\xi}} - \boldsymbol{\xi}^{\mathrm{T}}\mathsf{w}^2\boldsymbol{\xi}\right] \\
&= \frac{1}{2}\left[\dot{\boldsymbol{\xi}}^{\mathrm{T}}\dot{\boldsymbol{\xi}} - (\mathsf{w}\boldsymbol{\xi})^{\mathrm{T}}(\mathsf{w}\boldsymbol{\xi})\right] \\
&= \frac{1}{2}\sum_{k=1}^{N}\left(\dot{\xi}_k^2 - \omega_k^2\xi_k^2\right).
\end{aligned}
\tag{6.49}
$$

The final expression for the Lagrange function thus is a linear combination of single-mode Lagrange functions $L_k = (\dot{\xi}_k^2 - \omega_k^2\xi_k^2)/2$ for each ω_k. The corresponding Lagrange equations were given already in (6.40), with the general solutions (6.39).

6.2.2   *External forces*

To incorporate external forces, we proceed now as in the one-dimensional case by adding a perturbation potential to the original Lagrange function,

$$
L = L_0 - V_{\text{ext}}(q, t),
\tag{6.50}
$$

and expanding it around the equilibrium position of the unperturbed problem,

$$
V_{\text{ext}}(q, t) = V_{\text{ext}}(q_0, t) - \sum_{i=1}^{N} f_i(t)\,\eta_i + \cdots,
\tag{6.51}
$$

where

$$
f_i(t) = -\left[\frac{\partial V_{\text{ext}}(q, t)}{\partial q_i}\right]_0.
\tag{6.52}
$$

Dropping physically irrelevant terms, the Lagrange function then becomes

$$
\begin{aligned}
L &= L_0 + \sum_{i=1}^{N} f_i(t)\,\eta_i \\
&= \sum_{k=1}^{N}\left(\frac{1}{2}\dot{\xi}_k^2 - \frac{1}{2}\omega_k^2\xi_k^2 + F_k(t)\,\xi_k\right)
\end{aligned}
\tag{6.53}
$$

with the generalized force

$$F_k(t) = \sum_{i=1}^{N} f_i(t)a_{ik} \qquad (6.54)$$

appropriate for the normal coordinates. This leads to the equations of motion

$$\ddot{\xi}_k + \omega_k^2 \xi_k = F_k(t), \qquad k = 1,\ldots,N, \qquad (6.55)$$

describing N *independent* forced oscillations, each analogous to the one-dimensional case (6.9). At this stage, one may also add dissipative forces (cf. Sec. 1.3).

6.3   Applications

The formalism will now be illustrated by looking at three examples. First, we consider the generic two-dimensional motion of a point mass to clarify the geometric meaning of the normal coordinates in terms of the principal axes of equipotential surfaces. Then, we treat the motion of two coupled pendula and, finally, the vibrational modes of a triatomic molecule.

6.3.1   *Two-dimensional motion*

The Lagrange function for the two-dimensional motion of a point mass is

$$L = \frac{m}{2}\left(\dot{x}_1^2 + \dot{x}_2^2\right) - V(x_1, x_2). \qquad (6.56)$$

Let us assume that the potential has a minimum at (x_{10}, x_{20}). Defining the displacements from that position by

$$\eta_i = x_i - x_{i0}, \qquad i = 1, 2, \qquad (6.57)$$

the potential can be expanded as

$$V = V_0 + \frac{1}{2}\left(V_{11}\eta_1^2 + V_{22}\eta_2^2 + 2V_{12}\eta_1\eta_2\right) + \cdots, \qquad (6.58)$$

where $V_0 = V(x_{10}, x_{20})$ and

$$V_{ij} = V_{ji} = \left[\frac{\partial^2 V}{\partial x_i \partial x_j}\right]_0, \qquad i,j = 1,2. \qquad (6.59)$$

Hence, the Lagrange function for small displacements is

$$L = \frac{1}{2}\sum_{i,j=1}^{2}\left(T_{ij}\,\dot{\eta}_i\dot{\eta}_j - V_{ij}\,\eta_i\eta_j\right), \qquad (6.60)$$

with $T_{ij} = m\delta_{ij}$.

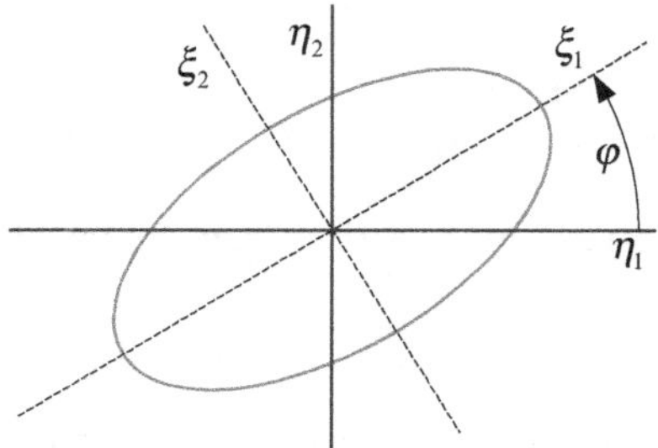

Fig. 6.1　Orientation of an elliptic equipotential line with respect to the original coordinates (η_1, η_2) and the normal coordinates (ξ_1, ξ_2) aligned with the principal axes of the ellipse. The origin is at the minimum position of the potential at (x_{10}, x_{20}).

This is a particularly simple example since the kinetic-energy matrix, $\mathsf{T} = m\mathbb{1}$, is not just diagonal already, but it is even a multiple of the unit matrix. Decoupling the problem by diagonalization, therefore, involves only the potential matrix V. In view of this simplification, we will deviate from the general procedure outlined above. Instead, we choose an approach that will show the geometric meaning of the transformation to normal coordinates.

The potential matrix V is real and symmetric, and every real and symmetric matrix can be diagonalized by an orthogonal matrix (see Sec. 5.1.1). Therefore, there exists an orthogonal matrix R with

$$\mathsf{V} = \begin{pmatrix} V_{11} & V_{12} \\ V_{21} & V_{22} \end{pmatrix} \quad \longrightarrow \quad \mathsf{R}^{\mathrm{T}}\mathsf{V}\mathsf{R} = \begin{pmatrix} k_1 & 0 \\ 0 & k_2 \end{pmatrix}. \tag{6.61}$$

Moreover, in view of the defining property of an orthogonal matrix, $\mathsf{R}^{\mathrm{T}}\mathsf{R} = \mathbb{1}$, one finds

$$\mathsf{T} = m\mathbb{1} \quad \longrightarrow \quad \mathsf{R}^{\mathrm{T}}\mathsf{T}\mathsf{R} = m\mathsf{R}^{\mathrm{T}}\mathbb{1}\mathsf{R} = m\mathbb{1} = \mathsf{T}, \tag{6.62}$$

i.e., this transformation does not affect the kinetic-energy matrix at all.

Geometrically, it is quite obvious how to choose this orthogonal transformation. It follows from (6.58) that the equipotential lines around the minimum are given by

$$\sum_{i,j=1}^{2} V_{ij}\, \eta_i \eta_j = const > 0. \tag{6.63}$$

The geometric locus defined by this equation is an ellipse centered around the equilibrium position $\eta_1 = \eta_2 = 0$. Expressing this ellipse in coordinates ξ_1, ξ_2 aligned with its principal axes corresponds to a transformation that diagonalizes V. This is achieved by a rotation R through an appropriately chosen angle φ (see Fig. 6.1), i.e.,

$$\begin{pmatrix} \eta_1 \\ \eta_2 \end{pmatrix} = \begin{pmatrix} \cos\varphi & -\sin\varphi \\ \sin\varphi & \cos\varphi \end{pmatrix} \begin{pmatrix} \xi_1 \\ \xi_2 \end{pmatrix}, \quad \text{or} \quad \boldsymbol{\eta} = \mathsf{R}\boldsymbol{\xi}, \tag{6.64}$$

where $\boldsymbol{\eta}$ and $\boldsymbol{\xi}$ denote the respective column vectors. This principal-axis transformation then results in

$$\sum_{i,j=1}^{2} V_{ij}\,\eta_i\eta_j = \boldsymbol{\eta}^{\mathrm{T}}\mathsf{V}\boldsymbol{\eta} = (\mathsf{R}\boldsymbol{\xi})^{\mathrm{T}}\mathsf{V}(\mathsf{R}\boldsymbol{\xi}) = \boldsymbol{\xi}^{\mathrm{T}}(\mathsf{R}^{\mathrm{T}}\mathsf{V}\mathsf{R})\boldsymbol{\xi}$$

$$= k_1\xi_1^2 + k_2\xi_2^2 = const > 0. \tag{6.65}$$

Since the minimum of the potential lies at $\xi_1 = \xi_2 = 0$, both k_1 and k_2 must be positive on physical grounds. The last expression, therefore, is the ellipse equation for the principal-axes coordinates. Geometrically, k_1 and k_2 are directly related to the lengths of the symmetry axes of the ellipse (cf. Fig. 6.1).

In the new coordinates, the Lagrange function now reads

$$L(\xi,\dot{\xi}) = \frac{m}{2}\sum_{i=1}^{2}\left(\dot{\xi}_i^2 - \omega_i^2\xi_i^2\right), \qquad \text{with} \qquad \omega_i^2 = \frac{k_i}{m}. \tag{6.66}$$

The eigenmodes of this system, therefore, are oscillations along the principal axes of the equipotential ellipse.

In higher-dimensional problems, the equipotential 'lines' become closed multidimensional (hyper)surfaces. For symmetric surfaces, the normal coordinates align with the symmetry axes of these surfaces.

6.3.2 Coupled pendula

Consider two identical planar mathematical pendula coupled by a massless spring as shown in Fig. 6.2. The Lagrange function for this setup is given by the sum of the Lagrange functions for the individual pendula [see Eq. (2.126)] and by the potential energy of the spring. For small amplitudes, where $\cos\vartheta \approx 1 - \vartheta^2/2$ and the spring displacement is proportional to $\ell|\vartheta_1 - \vartheta_2|$, this reads

$$L = \underbrace{\frac{m\ell^2}{2}\dot{\vartheta}_1^2 - \frac{mgl}{2}\vartheta_1^2}_{\text{pendulum 1}} + \underbrace{\frac{m\ell^2}{2}\dot{\vartheta}_2^2 - \frac{mgl}{2}\vartheta_2^2}_{\text{pendulum 2}} - \underbrace{\frac{k\ell^2}{2}\left(\vartheta_1 - \vartheta_2\right)^2}_{\text{spring potential}}. \tag{6.67}$$

The variable dependence of the potential suggests to employ new coordinates

$$\xi_1 = \frac{\vartheta_1 + \vartheta_2}{2}, \qquad \xi_2 = \frac{\vartheta_1 - \vartheta_2}{2}. \tag{6.68}$$

With the reverse transformation

$$\vartheta_1 = \xi_1 + \xi_2 \qquad \text{and} \qquad \vartheta_2 = \xi_1 - \xi_2, \tag{6.69}$$

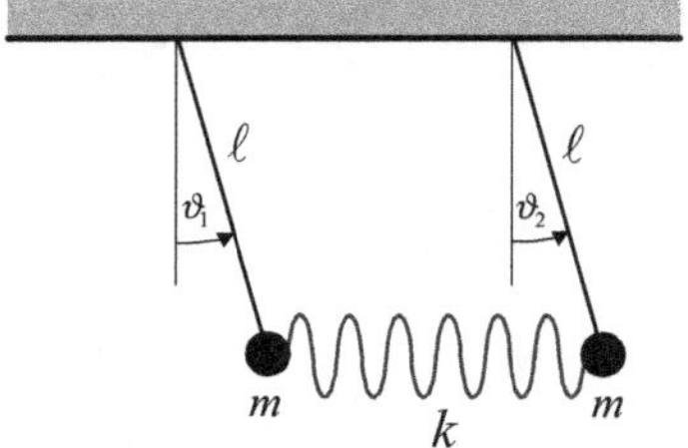

Fig. 6.2   Two identical planar pendula (mass m, length ℓ) suspended in the field of gravity and coupled by a massless spring (constant k). The spring is relaxed when both pendulum masses are at their lowest point. Hence, the equilibrium position is given by $\vartheta_1 = \vartheta_2 = 0$.

the Lagrange function then becomes

$$L = m\ell^2 \sum_{i=1}^{2} \left(\dot{\xi}_i^2 - \omega_i^2 \xi_i^2 \right),$$

(6.70)

with

$$\omega_1 = \sqrt{\frac{g}{\ell}} \quad \text{and} \quad \omega_2 = \sqrt{\frac{g}{\ell} + \frac{2k}{m}}.$$

(6.71)

The two eigenmodes completely decouple here and, hence, the variables ξ_i are the normal coordinates for this problem. Note that the spring's influence, via k, only appears for the second eigenmode.

The eigenmode solutions are the standard solutions for the simple harmonic oscillator,

$$\xi_1(t) = a\cos(\omega_1 t + \phi_1) \quad \text{and} \quad \xi_2(t) = b\cos(\omega_2 t + \phi_2),$$

(6.72)

which are then superposed according to (6.69) to produce the general solution

$$\vartheta_1(t) = a\cos(\omega_1 t + \phi_1) + b\cos(\omega_2 t + \phi_2),$$
$$\vartheta_2(t) = a\cos(\omega_1 t + \phi_1) - b\cos(\omega_2 t + \phi_2).$$

(6.73)

The two amplitudes a, b and the two phases ϕ_1, ϕ_2 are determined by four initial conditions ϑ_{10}, ϑ_{20}, $\dot{\vartheta}_{10}$, and $\dot{\vartheta}_{20}$.

To discuss this further, let us consider the situations when only one of these modes is activated. First, for the case where mode 2 is switched off ($b = 0$), the two pendula *swing in phase*, i.e.,

$$\vartheta_1(t) = \vartheta_2(t) = a\cos(\omega_1 t + \phi_1).$$

(6.74)

The spring remains relaxed at all times; it has no influence on the motion and therefore ω_1 does not depend on k. For the second case, where mode 1 is switched off ($a = 0$), the pendula *swing 180° out of phase*,

$$\vartheta_1(t) = -\vartheta_2(t) = b\cos(\omega_2 t + \phi_2).$$

(6.75)

The spring gets stretched or compressed as the pendula move away from each other or approach each other, respectively, and the corresponding common frequency is determined by a combination of the pendulum and spring constants.

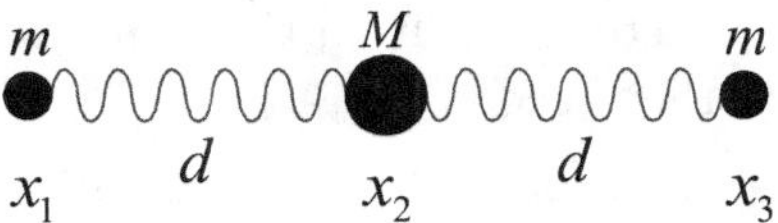

Fig. 6.3   Model of a linear triatomic molecule with one center atom of mass M flanked by two identical atoms of mass m. The force between adjacent atoms is taken to be a spring-type harmonic force, with an equilibrium distance d.

6.3.3   *Vibrational modes of a triatomic molecule*

The relatively simple situations of the previous two examples made it possible to obtain the solutions without actually going through the details of the solution procedure outlined in Sec. 6.2. The next example, a model for a triatomic molecule, is more complicated and will be treated in detail.

Consider the linear molecule depicted in Fig. 6.3, with equilibrium positions for the three atoms at x_{i0}. Introducing the displacements

$$\eta_i = x_i - x_{i0}, \tag{6.76}$$

the harmonic-oscillator potential for this configuration is

$$
\begin{aligned}
V(x_1, x_2, x_3) &= \frac{k}{2}\left[(x_2 - x_1 - d)^2 + (x_3 - x_2 - d)^2\right] \\
&= \frac{k}{2}\left[(\eta_2 - \eta_1)^2 + (\eta_3 - \eta_2)^2\right],
\end{aligned}
\tag{6.77}
$$

where the equilibrium distance,

$$d = x_{20} - x_{10} = x_{30} - x_{20}, \tag{6.78}$$

was used. The kinetic energy is

$$
\begin{aligned}
T(\dot{x}_1, \dot{x}_2, \dot{x}_3) &= \frac{m}{2}\left(\dot{x}_1^2 + \dot{x}_3^2\right) + \frac{M}{2}\dot{x}_2^2 \\
&= \frac{m}{2}\left(\dot{\eta}_1^2 + \dot{\eta}_3^2\right) + \frac{M}{2}\dot{\eta}_2^2.
\end{aligned}
\tag{6.79}
$$

The expressions for V and T already have the standard quadratic form for the kinetic and potential energies; expansions are not necessary. Reading off the corresponding matrices directly,

$$
\mathsf{T} = \begin{pmatrix} m & 0 & 0 \\ 0 & M & 0 \\ 0 & 0 & m \end{pmatrix}, \qquad
\mathsf{V} = \begin{pmatrix} k & -k & 0 \\ -k & 2k & -k \\ 0 & -k & k \end{pmatrix}, \tag{6.80}
$$

the standard form for the Lagrange function is obtained,

$$L(\eta, \dot{\eta}) = \frac{1}{2}\sum_{i,j=1}^{3}\left(T_{ij}\,\dot{\eta}_i\dot{\eta}_j - V_{ij}\,\eta_i\eta_j\right) = \frac{1}{2}\left(\dot{\boldsymbol{\eta}}^{\mathrm{T}}\mathsf{T}\dot{\boldsymbol{\eta}} - \boldsymbol{\eta}^{\mathrm{T}}\mathsf{V}\boldsymbol{\eta}\right), \tag{6.81}$$

where $\boldsymbol{\eta}$ is the column vector built from the η_i in the usual manner. The characteristic equation then reads

$$\det(\mathsf{V} - \omega^2\mathsf{T}) = \begin{vmatrix} k - \omega^2 m & -k & 0 \\ -k & 2k - \omega^2 M & -k \\ 0 & -k & k - \omega^2 m \end{vmatrix}$$

$$= (k - \omega^2 m)^2 (2k - \omega^2 M) - 2k^2 (k - \omega^2 m)$$

$$= \omega^2 \left[k - \omega^2 m \right] \left[\omega^2 m M - k(2m + M) \right] = 0, \qquad (6.82)$$

with solutions

$$\omega_1 = 0, \qquad \omega_2 = \sqrt{\frac{k}{m}}, \qquad \omega_3 = \sqrt{\frac{k}{m}\left(1 + \frac{2m}{M}\right)}. \qquad (6.83)$$

The three eigenvectors $\hat{\mathsf{A}}_k$ are determined by the equations

$$(\mathsf{V} - \omega_1^2\mathsf{T})\hat{\mathsf{A}}_1 = \mathsf{V}\hat{\mathsf{A}}_1 = \begin{pmatrix} k & -k & 0 \\ -k & 2k & -k \\ 0 & -k & k \end{pmatrix} \begin{pmatrix} a_{11} \\ a_{21} \\ a_{31} \end{pmatrix} = 0, \qquad (6.84)$$

$$(\mathsf{V} - \omega_2^2\mathsf{T})\hat{\mathsf{A}}_2 = \begin{pmatrix} 0 & -k & 0 \\ -k & 2k - k\frac{M}{m} & -k \\ 0 & -k & 0 \end{pmatrix} \begin{pmatrix} a_{12} \\ a_{22} \\ a_{32} \end{pmatrix} = 0, \qquad (6.85)$$

$$(\mathsf{V} - \omega_3^2\mathsf{T})\hat{\mathsf{A}}_3 = \begin{pmatrix} 2k\frac{m}{M} & -k & 0 \\ -k & -k\frac{M}{m} & -k \\ 0 & -k & 2k\frac{m}{M} \end{pmatrix} \begin{pmatrix} a_{13} \\ a_{23} \\ a_{33} \end{pmatrix} = 0. \qquad (6.86)$$

Choosing the respective first components of the solutions as $a_{11} = a_{12} = a_{13} = 1$, the (*unnormalized*) eigenvectors are found as

$$\hat{\mathbf{A}}_1 = \begin{pmatrix} 1 \\ 1 \\ 1 \end{pmatrix}, \qquad \hat{\mathbf{A}}_2 = \begin{pmatrix} 1 \\ 0 \\ -1 \end{pmatrix}, \qquad \hat{\mathbf{A}}_3 = \begin{pmatrix} 1 \\ -2\frac{m}{M} \\ 1 \end{pmatrix}. \qquad (6.87)$$

The general solution vector then is given by

$$\boldsymbol{\eta} = \hat{\mathbf{A}}_1 b_1 \left(t + t_1\right) + \hat{\mathbf{A}}_2 b_2 \cos(\omega_2 t + \phi_2) + \hat{\mathbf{A}}_3 b_3 \cos(\omega_3 t + \phi_3), \qquad (6.88)$$

with the six real constants b_1, b_2, b_3, t_1, ϕ_2, and ϕ_3 determined by two initial conditions for each of the three atoms.

The first eigenmode, $\omega_1 = 0$, corresponds to a uniform motion of the molecule as a whole. This can easily be seen by switching off the other two modes (put $b_2 = b_3 = 0$). The fact that all components of $\hat{\mathbf{A}}_1$ are identical then means that the displacements of all three atoms, $\eta_i = b_1 \left(t + t_1\right)$,

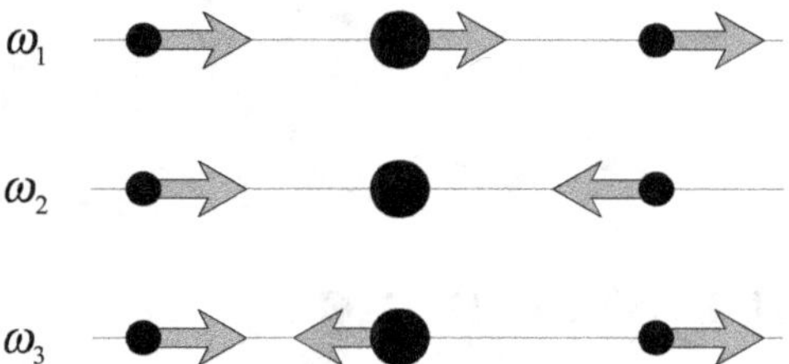

Fig. 6.4 Eigenmodes of the linear triatomic molecule shown in Fig. 6.3. For $\omega_1 = 0$, the molecule is translated as a whole. For ω_2, the outer atoms oscillate symmetrically about the stationary central atom. For ω_3, the outer atoms are in phase with each other, but out of phase by $180°$ with respect to the central atom.

describe the same uniform motion. The reason for this behavior is found in the fact that the center of mass can be expressed in terms of the three degrees of freedom considered here, i.e.,

$$\eta_{\text{CM}} = \frac{m(\eta_1 + \eta_3) + M\eta_2}{2m + M}. \tag{6.89}$$

Its motion, therefore, follows from the Lagrange function. However, there is no restoring force associated with this displacement that would result in an oscillatory mode and hence, the associated frequency is zero. If we express the problem in center-of-mass coordinates to start with, it has only two degrees of freedom and this translational mode does not occur.

Putting $b_1 = b_3 = 0$, we can isolate the second mode, associated with ω_2. The third component of $\hat{\mathbf{A}}_2$ is the negative of the first and the second component is zero. The central atom, therefore, is at rest here and the outer atoms oscillate symmetrically about this center. For the third mode, with ω_3, we put $b_1 = b_2 = 0$ and find from $\hat{\mathbf{A}}_3$ that the outer atoms vibrate in the same direction, while the central atom moves in the opposite direction. In the latter two modes, the center of mass stays at rest, of course. All three eigenmodes are depicted in Fig. 6.4.

Note here that, as alluded to in the discussion of Eq. (6.33), it was not necessary to explicitly carry out the step of normalizing the eigenvectors $\hat{\mathbf{A}}_i$ since only the products $\hat{\mathbf{A}}_i b_i$ enter the physical trajectories, i.e., any explicit choice of normalization for the $\hat{\mathbf{A}}_i$ will be compensated by the b_i such that the products $\hat{\mathbf{A}}_i b_i$ remain the same.

Chapter 7

Continuum Mechanics

The central objective of the description of mechanical phenomena given in the preceding chapters was to obtain equations of motions for point masses whose solutions — the trajectories — allow one to follow the time evolution of individual particles (or, as in the case of rigid bodies, a whole collection of particles whose relative positions are fixed). There is, however, a large class of problems where this is not a very useful answer. Consider, for example, an elastic string stretched taut between two fixed points. If the string is plucked, it vibrates. In order to describe this vibration, what we need is a description of the (continuous) mass distribution of the string as a function of position and time. This is an example for the type of problems treated in *continuum mechanics*. In general, this area of mechanics deals with the dynamics of elastic bodies, of liquids, and of gases. The description aims to provide equations for *fields*, rather than trajectories.

A field, in general, is a function that associates an appropriate mathematical quantity — a scalar, a vector, a matrix, etc. — with each point in space and time. Examples are temperature fields $T(\mathbf{r}, t)$ of the atmosphere, velocity fields $\mathbf{v}(\mathbf{r}, t)$ of currents in fluids, pressure fields $P(\mathbf{r}, t)$ in a gas, electric and magnetic fields $\mathbf{E}(\mathbf{r}, t)$, $\mathbf{B}(\mathbf{r}, t)$ of electromagnetic waves, etc.

In the following, we will explore the basic concepts of continuum mechanics by treating a vibrating string and a non-rigid beam in an external force field. Furthermore, we give an elementary introduction to fluid dynamics neglecting friction effects due to viscosity.

7.1   Vibrating String

The field associated with the vibrating string is a particularly simple example: Taking the string (of length ℓ) to be stretched along the x-axis

Fig. 7.1 Vibrating string with transverse displacement $u(x,t)$ at point x.

initially, with fixed endpoints at $x = 0$ and $x = \ell$, it describes the lateral (or transverse) displacement u at each point x for a particular time t (see Fig. 7.1). The function $u(x,t)$, therefore, provides a complete description of the vibrational motion in a given plane.

To derive the equations of motion that govern the vibration (neglecting gravity), we assume a uniform distribution of the string's mass along its length. Moreover, we shall consider only small displacements, and simulate the elastic properties of the string by spring-like forces.

First, we divide the string's length into N identical pieces of length $\Delta x = \ell/N$. For sufficiently large N, the string can be considered as a collection of (identical) point masses Δm located at

$$x_i = \left(i - \tfrac{1}{2}\right) \Delta x, \qquad i = 1, \ldots, N, \tag{7.1}$$

whose lateral displacement $u_i(t)$ is given by

$$u_i(t) = u(x_i, t). \tag{7.2}$$

The total kinetic energy of this system of N point masses is

$$T = \sum_{i=1}^{N} \frac{\Delta m}{2} \left(\frac{du_i}{dt}\right)^2 = \sum_{i=1}^{N} \frac{\rho \Delta x}{2} \dot{u}_i^2, \tag{7.3}$$

where $\rho = \Delta m/\Delta x$ is the homogeneous linear mass density of the string.

As readily read off Fig. (7.2), the lateral string displacement increases the distance between the i-th and the $(i+1)$-st point mass by

$$\Delta s - \Delta x = \sqrt{(\Delta x)^2 + (u_{i+1} - u_i)^2} - \Delta x$$

$$= \Delta x \sqrt{1 + \frac{(u_{i+1} - u_i)^2}{(\Delta x)^2}} - \Delta x$$

$$\approx \Delta x \, \frac{(u_{i+1} - u_i)^2}{2(\Delta x)^2}, \tag{7.4}$$

where it was assumed that $|u_{i+1} - u_i| \ll \Delta x$ to allow for a first-order expansion of the square root. Further assuming spring-type elastic forces,

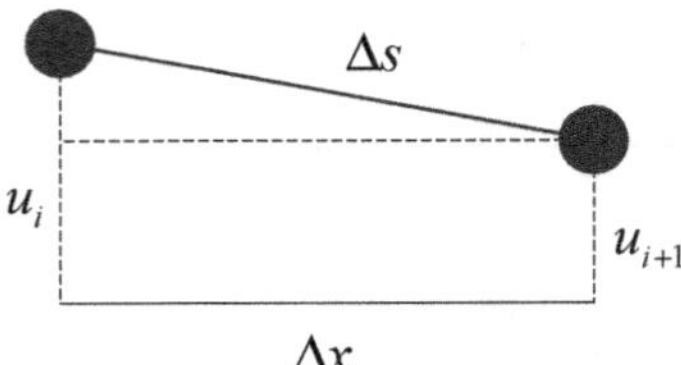

Fig. 7.2   Length change of the string between transverse displacements u_i and u_{i+1}.

with an equilibrium string tension K, this correspondingly increases the potential energy by the work $K(\Delta s - \Delta x)$ required to bring about this change, producing

$$V = \sum_{i=1}^{N-1} K\Delta x \frac{(u_{i+1} - u_i)^2}{2(\Delta x)^2} = \frac{K}{2} \sum_{i=1}^{N-1} \Delta x \left(\frac{u_{i+1} - u_i}{\Delta x} \right)^2, \qquad (7.5)$$

for the total potential energy. Note that this sums up only the potential energies between adjacent mass elements and ignores the potential energies of the masses in the two end intervals with respect to the adjacent end points. However, in the limit $N \to \infty$, these contributions vanish since the end points are kept fixed.

Taking now the limit $N \to \infty$, the kinetic and the potential energies respectively become

$$T = \lim_{N\to\infty} \sum_{i=1}^{N} \frac{\rho\Delta x}{2} \dot{u}_i^2 = \frac{\rho}{2} \lim_{N\to\infty} \sum_{i=1}^{N} \Delta x \left(\frac{du(x_i,t)}{dt} \right)^2$$

$$= \frac{\rho}{2} \int_0^{\ell} dx \left(\frac{\partial u(x,t)}{\partial t} \right)^2, \qquad (7.6)$$

and

$$V = \lim_{N\to\infty} \frac{K}{2} \sum_{i=1}^{N-1} \Delta x \left(\frac{u_{i+1} - u_i}{\Delta x} \right)^2 = \frac{K}{2} \int_0^{\ell} dx \left(\frac{\partial u(x,t)}{\partial x} \right)^2, \qquad (7.7)$$

and the Lagrange function can then be written as

$$L = T - V = \int_0^{\ell} dx \, \mathscr{L}(\dot{u}, u'), \qquad (7.8)$$

with the *Lagrange density*

$$\mathscr{L}(\dot{u}, u') = \frac{\rho}{2} \left(\frac{\partial u(x,t)}{\partial t} \right)^2 - \frac{K}{2} \left(\frac{\partial u(x,t)}{\partial x} \right)^2 = \frac{\rho}{2} \dot{u}^2 - \frac{K}{2} u'^2, \qquad (7.9)$$

with

$$\dot{u} = \frac{\partial u}{\partial t}, \qquad u' = \frac{\partial u}{\partial x} \qquad (7.10)$$

abbreviating now the partial derivatives.

Hamilton's principle, $\delta S = \delta \int L \, dt = 0$, applied to the present case leads to

$$\delta S = \delta \int_{t_i}^{t_f} dt \, L = \delta \int_{t_i}^{t_f} dt \int_0^\ell dx \, \mathscr{L}(\dot{u}, u') = 0. \qquad (7.11)$$

The generalized Euler-Lagrange equation of variational calculus for the Lagrange density then reads [cf. Eq. (3.31)]

$$\frac{\partial \mathscr{L}}{\partial u} - \frac{\partial}{\partial t} \frac{\partial \mathscr{L}}{\partial \dot{u}} - \frac{\partial}{\partial x} \frac{\partial \mathscr{L}}{\partial u'} = 0, \qquad (7.12)$$

which determines the function $u(x,t)$ that makes δS vanish. Explicitly this *wave equation for the force-free string* reads

$$\boxed{\frac{1}{c^2} \frac{\partial^2 u(x,t)}{\partial t^2} - \frac{\partial^2 u(x,t)}{\partial x^2} = 0}, \qquad (7.13)$$

where the *wave velocity*

$$c = \sqrt{\frac{K}{\rho}} \qquad (7.14)$$

was introduced.

The wave equation (7.13) is a *partial differential equation* that requires specification of *boundary conditions* to determine the solutions. Since the string is kept fixed at the end points, one has

$$u(0,t) = u(\ell,t) = 0. \qquad (7.15)$$

The derivation of the Euler–Lagrange equation assumes the solution is specified at the end points, i.e.,

$$u(x,t_1) = F(x), \qquad \text{and} \qquad u(x,t_2) = H(x), \qquad (7.16)$$

where $F(x)$ and $H(x)$ are some given (time-independent) functions that described a 'snap-shot' picture of the string at initial and final times, respectively. Equivalently, one may also specify u and its time derivative $\dot{u}$ at one of the end points, for example, the initial time,

$$u(x,t_1) = F(x), \qquad \text{and} \qquad \dot{u}(x,t_1) = G(x). \qquad (7.17)$$

7.1.1   *Solving the one-dimensional wave equation*

To solve the wave equation (7.13), we first note that if u_1 and u_2 are solutions of the equation, so is a linear combination $u = a_1 u_1 + a_2 u_2$ with arbitrary coefficients a_1 and a_2. This follows from the *linearity* of the partial differential equation. In the following, we will determine a complete set of solutions $u_n(x,t)$ of (7.13) that satisfy the boundary conditions (7.15). The complete solution will then be given by the linear combination $\sum_n a_n u_n(x,t)$ with coefficients a_n specified such that the initial conditions (7.17) hold true.

For partial differential equations that contain no mixed derivatives a standard solution strategy is the *separation of variables*. Employing in (7.13) the separation ansatz

$$u(x,t) = v(x)g(t) \tag{7.18}$$

leads to

$$\frac{v''(x)}{v(x)} = \frac{1}{c^2}\frac{\ddot{g}(t)}{g(t)}, \tag{7.19}$$

where the primes and dots now correspond to ordinary differentiations of the single-variable functions v and g with respect to x and t, respectively. Since x and t are independent of each other, the left- and the right-hand sides of this equation vary independently and therefore must be equal to a common constant, which we denote by $-k^2$. Hence

$$v''(x) + k^2 v(x) = 0, \tag{7.20a}$$

$$\ddot{g}(t) + \omega^2 g(t) = 0, \qquad \text{with} \qquad \omega = ck, \tag{7.20b}$$

provide two ordinary differential equations equivalent to the original partial differential equation.

The general solution for $v(x)$ is

$$v(x) = c_1 \sin(kx) + c_2 \cos(kx). \tag{7.21}$$

To satisfy the boundary condition $v(0) = v(\ell) = 0$, one requires $c_2 = 0$ and $k\ell$ must be an integer multiple of π, i.e., $k\ell = n\pi$, with $n = 0, \pm 1, \pm 2, \dots$. We can exclude $n = 0$ since this only provides the trivial solution $u = 0$. Moreover, absorbing the overall minus sign arising from negative n values into c_1, we can restrict the considerations to positive n's. We then have

$$k \to k_n = \frac{n\pi}{\ell}, \qquad v_n(x) = c_1 \sin(k_n x), \qquad n = 1, 2, \dots, \tag{7.22}$$

where the index n distinguishes the solutions for different n values.

The differential equation (7.20b) now also depends on n,

$$\ddot{g}_n(t) + \omega_n^2\, g_n(t) = 0, \qquad \text{with} \qquad \omega_n = ck_n, \tag{7.23}$$

with a general solution

$$g_n(t) = a_n \sin(\omega_n t) + b_n \cos(\omega_n t). \tag{7.24}$$

Combining this with $v_n(x)$ then provides

$$u_n(x,t) = c_1 \sin(k_n x)\left[a_n \sin(\omega_n t) + b_n \cos(\omega_n t)\right] \tag{7.25}$$

as a solution of the wave equation. Absorbing the overall constant c_1 into the, as yet undetermined, coefficients a_n and b_n, the general solution is then given by a linear combination

$$u(x,t) = \sum_{n=1}^{\infty} \sin(k_n x)\left[a_n \sin(\omega_n t) + b_n \cos(\omega_n t)\right]. \tag{7.26}$$

To determine the coefficients here, the initial conditions,

$$u(x,0) = \sum_{n=1}^{\infty} b_n \sin(k_n x) = F(x), \tag{7.27a}$$

$$\dot{u}(x,0) = \sum_{n=1}^{\infty} a_n \omega_n \sin(k_n x) = G(x), \tag{7.27b}$$

must be satisfied. The initial time was chosen as $t_1 = 0$. Since the initial conditions must reproduce the boundary conditions, only functions F and G with $F(0) = F(\ell) = 0$ and $G(0) = G(\ell) = 0$ are permitted as possible initial conditions. Equation (7.27) provides expansions of the functions $F(x)$ and $G(x)$ into *Fourier series*, with coefficients

$$a_n = \frac{2}{\ell \omega_n} \int_0^{\ell} dx\, G(x)\, \sin(k_n x), \tag{7.28a}$$

$$b_n = \frac{2}{\ell} \int_0^{\ell} dx\, F(x)\, \sin(k_n x). \tag{7.28b}$$

These expressions follow from the *theory of Fourier series*.[112]

Equivalently, the general solution can also be written in the form

$$u(x,t) = \sum_{n=1}^{\infty} A_n \sin(k_n x) \sin\left(\omega_n t + \phi_n\right), \tag{7.29}$$

---

[112]See, for example, Refs. [8,9] or any other standard textbook on mathematical physics.

as a superposition of so-called *eigenmode* solutions

$$u_n(x,t) = A_n \sin(k_n x) \sin(\omega_n t + \phi_n),\tag{7.30}$$

where ω_n is called the *eigenfrequency*, and ϕ_n is the phase of the eigenmode.

If the string is in an eigenmode, there are fixed nodes at regular intervals along the length of the string where $u(x,t) = 0$ at all times. Such a vibrational mode is called a *standing wave*, and the *wave length*

$$\lambda_n = \frac{2\pi}{k_n}\tag{7.31}$$

of this eigenmode is twice the distance between two adjacent nodes. The lowest mode for $n = 1$, called the *fundamental mode*, has two nodes, i.e., the two fixed end points. The n-th mode has $n + 1$ nodes. (Figure 7.1, for example, corresponds to the eigenmode with $n = 3$.) The *fundamental*, or *first harmonic, frequency*

$$\omega_1 = \frac{\pi c}{\ell} = \frac{\pi}{\ell}\sqrt{\frac{K}{\rho}},\tag{7.32}$$

can be changed by altering the string tension K. This happens when one tunes, for example, a violin. All higher harmonics then change accordingly as well. When the string vibrating in an eigenmode is coupled with a resonant body and the air, one hears a tone at the frequency of the corresponding harmonic. The string's vibrational state may also correspond to a superposition of eigenmodes that produce *wave packets* that travel back and forth between the two fixed end points.

7.1.2 Generalizations

External forces: The preceding case assumes that there are no forces acting on the string. If there is a force, it can be written in terms of a *force density* $f(x,t)$ describing the force per length, i.e., $f(x,t)\Delta x$ is the force acting on the mass element Δm contained within the interval Δx centered around x. If f acts perpendicular to the string, the corresponding displacement $u(x,t)$ of each mass element Δm reduces the potential energy by $\Delta U = -u\,f\,\Delta x$. In other words, the Lagrange density becomes now

$$\mathscr{L}(u,\dot{u},u',x,t) = \frac{\rho}{2}\dot{u}^2 - \frac{K}{2}u'^2 + u\,f.\tag{7.33}$$

In contrast to (7.9), $\mathscr{L}$ depends now explicitly on u and, via f, also on x and t. The corresponding *inhomogeneous wave equation* is then found from (7.12) as

$$\frac{\partial^2 u(x,t)}{\partial x^2} - \frac{1}{c^2}\frac{\partial^2 u(x,t)}{\partial t^2} = -\frac{f(x,t)}{K}.\tag{7.34}$$

An example of such an external force would be gravity; for Fig. 7.1, with gravity acting downward, it would be written as $f = -\rho g$.

Two- and three-dimensional generalizations: Generalized to two dimensions, the string becomes a membrane whose equilibrium state is some area in the xy-plane, with a fixed perimeter along some closed-loop curve. Analogous to (7.13), the transverse vibrations $u(x, y, t)$ are described by a two-dimensional wave equation of the form

$$\frac{\partial^2 u(x, y, t)}{\partial x^2} + \frac{\partial^2 u(x, y, t)}{\partial y^2} - \frac{1}{c^2} \frac{\partial^2 u(x, y, t)}{\partial t^2} = 0. \tag{7.35}$$

In three dimension, the corresponding field $u(x, y, z, t)$ satisfies

$$\frac{\partial^2 u(x, y, z, t)}{\partial x^2} + \frac{\partial^2 u(x, y, z, t)}{\partial y^2} + \frac{\partial^2 u(x, y, z, t)}{\partial z^2} - \frac{1}{c^2} \frac{\partial^2 u(x, y, z, t)}{\partial t^2} = 0, \tag{7.36}$$

or

$$\Delta u = \frac{1}{c^2} \frac{\partial^2 u}{\partial t^2}, \tag{7.37}$$

where $\Delta = \nabla \cdot \nabla$ is the Laplace operator. This describes, for example, sound waves within a rigid body. They can be transverse or longitudinal. Longitudinal sound waves also occur in gases and fluids, where $u = \rho - \rho_0$ describes the density fluctuations around the equilibrium density ρ_0. A similar wave equation also obtains for the electromagnetic field without any currents and charges; c then is the speed of light.

The wave equations discussed here are the equations of motions for the respective fields u; they are examples of *linear field equations*. Linear equations usually permit relatively straightforward solutions. Many fundamental field equations of physics are linear (examples are the Maxwell, the Schrödinger, and the Dirac equations). *Nonlinear field equations*, however, are also possible; they typically are much more complicated to solve. Section 7.3, for example, presents an introduction to the nonlinear equations of fluid dynamics.

7.2 Beam in an External Force Field

Take a horizontal flexible beam, or plank, of length ℓ initially covering the interval $[0, \ell]$ along the x-axis (similar to the string of the previous example). We consider the lateral displacement (bending) of the beam subject

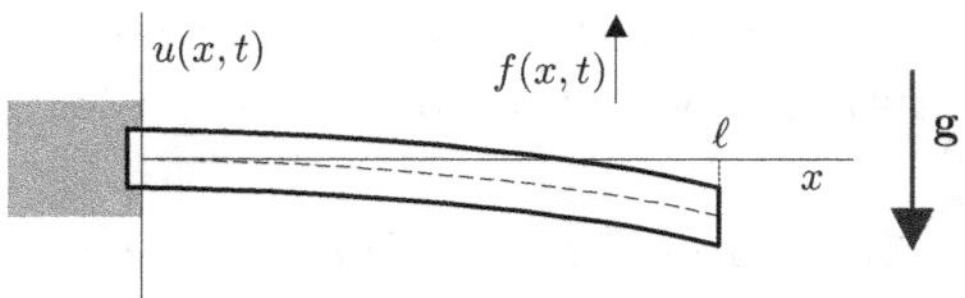

Fig. 7.3   Example of a beam bent in the gravitational field **g** when one end is fixed and the other is free. The displacement of the (dashed) centerline from the horizontal force-free position is described by $u(x,t)$. Assuming a uniform mass distribution, the linear force density $f(x,t)$ due to the gravitational force in this example is $f(x,t) = -mg/\ell$, where m is the total mass of the beam and ℓ its length. It is assumed that $|u'| \ll 1$ so that longitudinal mass displacements can be neglected.

to an external force density $f(x,t)$ describing the force per length perpendicular to the beam. As in the previous example, we denote the transversal displacement by $u(x,t)$, with u describing the centerline of the beam (see Fig. 7.3), and the partial derivatives with respect to t and x by $\dot{u}$ and u', respectively. Assuming that $|u'| \ll 1$, we can neglect the longitudinal displacements of mass elements, i.e., we consider only transverse contributions to the energy. The kinetic energy then has the same form as for the vibrating string, Eq. (7.6),

$$T = \frac{\rho}{2} \int\limits_0^\ell dx \left(\frac{\partial u(x,t)}{\partial t} \right)^2 , \qquad (7.38)$$

where $\rho = m/\ell$ is the beam's constant linear mass density along x.

The potential energy contains two contributions: One due to the (internal) elastic properties of the beam when it is distorted away from a straight line, V_b, and another contribution due to the external force, V_f.

The potential energy contribution arising from bending the beam out of a straight-line equilibrium $u' = const$ can be expressed in terms of the second derivative u''. This is easily seen by considering a beam segment, like in Fig. 7.4, and calculating the local deviation from a straight line. Using the notation of Fig. 7.4, we find

$$s_{12} + s_{23} - s_{13} = \sqrt{\Delta x^2 + (u_2 - u_1)^2} + \sqrt{\Delta x^2 + (u_3 - u_2)^2}$$
$$- \sqrt{4\Delta x^2 + (u_3 - u_1)^2}$$
$$\approx \frac{\Delta x}{4} \left(\frac{u_3 - u_2}{\Delta x} - \frac{u_2 - u_1}{\Delta x} \right)^2 , \qquad (7.39)$$

where $|u'| \ll 1$ was used to expand the square roots, similar to (7.4). For small Δx, the expression in the parentheses is

$$\frac{u_3 - u_2}{\Delta x} - \frac{u_2 - u_1}{\Delta x} \quad \longrightarrow \quad \Delta u' \approx \frac{\partial^2 u}{\partial x^2} \Delta x. \qquad (7.40)$$

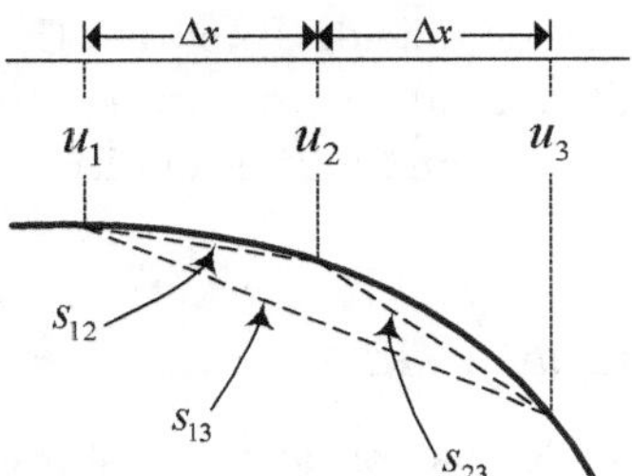

Fig. 7.4   Local curvature of the centerline of a beam segment.

A given change $\Delta u'$ will contribute more to the potential energy if it occurs over a smaller interval Δx, and less if it occurs over a larger interval, since the local radius of curvature (describing how much the beam is bent) will be smaller or larger, respectively. Therefore, only the relative change $\Delta u'/\Delta x \approx \partial^2 u/\partial x^2$ is relevant for the calculation of the potential energy. Hence, the potential energy contribution due to bending the beam away from the straight-line equilibrium described by $u' = const$ is given by

$$V_b = \frac{b}{2} \int_0^\ell dx \left(\frac{\partial^2 u(x,t)}{\partial x^2} \right)^2, \tag{7.41}$$

where b is a (positive) parameter depending on the beam's material and its geometric dimensions.

If the external force $f(x,t)\Delta x$ acts on the mass element Δm at x along the displacement $u(x,t)$, it will reduce the potential energy by $\Delta V_f = -u(x,t)f(x,t)\Delta x$. The integral

$$V_f = - \int_0^\ell dx\, u(x,t) f(x,t), \tag{7.42}$$

therefore, describes the entire contribution due to the external force.

The Lagrangian density $\mathscr{L}$ following from the Lagrange function $L = T - V_b - V_f = \int dx\,\mathscr{L}$ then is given by

$$\boxed{\mathscr{L}(u, \dot{u}, u'', x, t) = \frac{\rho}{2}\dot{u}^2 - \frac{b}{2}u''^2 + uf(x,t)}. \tag{7.43}$$

The Euler–Lagrange equation for this density [cf. Eq. (3.33)],

$$\frac{\partial}{\partial t}\frac{\partial\mathscr{L}}{\partial\dot{u}} - \frac{\partial^2}{\partial x^2}\frac{\partial\mathscr{L}}{\partial u''} = \frac{\partial\mathscr{L}}{\partial u}, \tag{7.44}$$

leads to the field equation

$$\boxed{\rho\ddot{u}(x,t) + bu''''(x,t) = f(x,t)} \tag{7.45}$$

describing the motion of the beam in an external force field. To obtain the complete solution for a given situation, this differential equation needs to be supplemented by appropriate boundary conditions and initial conditions $u(x,0)$ and $\dot{u}(x,0)$.

7.2.1   *Static sagging in the gravitational field*

To illustrate the solution of this field equation, let us consider the static situation of a beam sagging under its own weight in the gravitational field. The force density for this case is constant, i.e.,

$$f(x,t) = -\rho g \tag{7.46}$$

(cf. Fig. 7.3). The static equilibrium situation $u = u(x)$ is determined by minimizing the potential energy,

$$V = V_b + V_f = \int_0^\ell dx \left(\frac{b}{2} u''^2 + \rho g u \right) = \text{minimal.} \tag{7.47}$$

From Hamilton's principle in the form $\delta V = 0$, one then immediately finds the equilibrium condition

$$u''''(x) = -\frac{\rho g}{b} \equiv -24A. \tag{7.48}$$

This also follows from the field equation (7.45) with $\dot{u}(x,t) = 0$. The general solution is given by the fourth-order polynomial

$$u(x) = -Ax^4 + C_3 x^3 + C_2 x^2 + C_1 x + C_0. \tag{7.49}$$

The four constants C_i are determined by the boundary conditions for the beam; for the present static case, there are no additional initial conditions.

The appropriate boundary conditions follow from the specific way the beam is supported. For supports at the ends of the beam, Fig. 7.5 depicts the following three examples:

   (i)   Fixed at both ends: $u(0) = u'(0) = 0,$ $u(\ell) = u'(\ell) = 0,$

  (ii)   Fixed at one end: $u(0) = u'(0) = 0,$ $u(\ell),$ $u'(\ell)$ arbitrary,

 (iii)   Loose at both ends: $u(0) = u(\ell) = 0,$ $u'(0),$ $u'(\ell)$ arbitrary.

$$\tag{7.50}$$

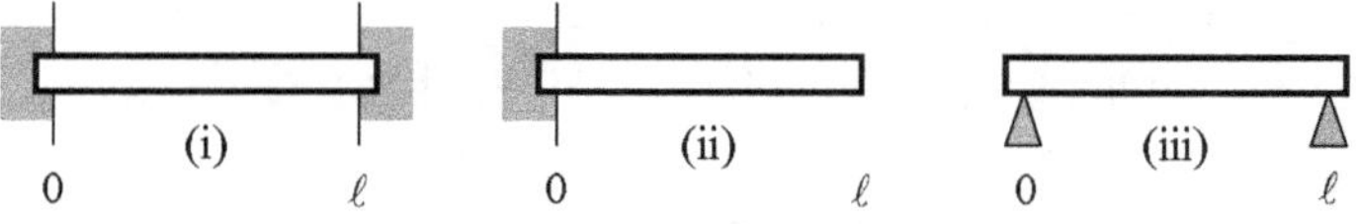

Fig. 7.5   Three examples for supporting a beam in the gravitational field: (i) fixed support at both ends; (ii) fixed support at one end, free at the other; (iii) loose support at both ends.

For case (i), there are four conditions to fix the four constants C_i. For the other two cases, however, it is not obvious how this should be done. To clarify the situation, we need to go back and look at how the Euler–Lagrange equation (7.48) is derived. The generic functional that describes (7.47) reads

$$J[u] = \int_0^\ell dx\, F(u'', u), \tag{7.51}$$

with

$$F = \frac{b}{2} u''^2 + \rho g u \tag{7.52}$$

for the present problem. Its (vanishing) variation is

$$0 = \delta J = J[u + \delta u] - J[u]$$

$$= \int_0^\ell dx \left(\frac{\partial F}{\partial u''} \delta u'' + \frac{\partial F}{\partial u} \delta u \right)$$

$$= \left[\frac{\partial F}{\partial u''} \delta u' \right]_\ell - \left[\frac{\partial F}{\partial u''} \delta u' \right]_0 - \left[\frac{d}{dx} \frac{\partial F}{\partial u''} \delta u \right]_\ell + \left[\frac{d}{dx} \frac{\partial F}{\partial u''} \delta u \right]_0$$

$$+ \int_0^\ell dx \left(\frac{d^2}{dx^2} \frac{\partial F}{\partial u''} + \frac{\partial F}{\partial u} \right) \delta u. \tag{7.53}$$

The last equality follows from integrating the $\delta u''$ term by parts twice resulting in the four boundary-value terms in the square brackets. To obtain (7.48), they must all vanish individually. The situations described in (7.50) translate to

$$\begin{aligned}
&\text{(i)} \quad \delta u(0) = \delta u'(0) = 0, &&\delta u(\ell) = \delta u'(\ell) = 0, \\
&\text{(ii)} \quad \delta u(0) = \delta u'(0) = 0, &&\delta u(\ell),\ \delta u'(\ell)\ \text{arbitrary}, \\
&\text{(iii)} \quad \delta u(0) = \delta u(\ell) = 0, &&\delta u'(0),\ \delta u'(\ell)\ \text{arbitrary}.
\end{aligned} \tag{7.54}$$

In the first case, all boundary-value terms in Eq. (7.53) vanish as a matter of course. For the other two cases, however, these terms will only vanish if the respective coefficients of the arbitrary variations δu or $\delta u'$ vanish. We thus require

$$\text{(ii)} \qquad\qquad \text{at } x = \ell: \quad
\begin{cases}
\dfrac{\partial F}{\partial u''} = b u'' = 0, \\[2mm]
\dfrac{d}{dx} \dfrac{\partial F}{\partial u''} = b u''' = 0,
\end{cases} \tag{7.55a}$$

$$\text{(iii)} \qquad \text{at } x = 0 \text{ and } x = \ell: \quad F_{u''} \equiv \frac{\partial F}{\partial u''} = b u'' = 0. \tag{7.55b}$$

In summary, therefore, the conditions are

$$
\begin{aligned}
\text{(i)} \quad & u(0) = 0, & u'(0) = 0, & \quad u(\ell) = 0, & u'(\ell) = 0, \\
\text{(ii)} \quad & u(0) = 0, & u'(0) = 0, & \quad u''(\ell) = 0, & u'''(\ell) = 0, \\
\text{(iii)} \quad & u(0) = 0, & u''(0) = 0, & \quad u(\ell) = 0, & u''(\ell) = 0.
\end{aligned}
\tag{7.56}
$$

The physical specification of arbitrary u thus is translated into the condition $u''' = 0$ and for arbitrary u', one has $u'' = 0$.

With these boundary conditions, one then immediately finds

$$
\begin{aligned}
\text{(i)} \quad & u(x) = -Ax^2 \left(x - \ell\right)^2, \\
\text{(ii)} \quad & u(x) = -Ax^2 \left(x^2 - 4\ell x + 6\ell^2\right), \\
\text{(iii)} \quad & u(x) = -Ax \left(x^3 - 2\ell x^2 + \ell^3\right)
\end{aligned}
\tag{7.57}
$$

as the static-equilibrium solutions for the three situations depicted in Fig. 7.5.

7.3 Fluid Dynamics

The description of mechanical phenomena associated with liquids or gases, generically called *fluids*, is called *fluid dynamics*, or (in older publications in particular) *hydrodynamics*. For fluids, individual point masses are replaced by the *mass density*,

$$
\rho(\mathbf{r}, t) = \frac{\Delta m}{\Delta V},
\tag{7.58}
$$

where Δm is the mass contained in the volume element ΔV at the position $\mathbf{r}$. It should be noted here that ΔV *cannot* be made arbitrarily small since for very small volumes the discrete nature of the mass distribution in terms of atoms and molecules becomes relevant. Rather, to determine a meaningful mass density for the present purpose, one must choose ΔV so that it contains a sufficiently large number of atoms or molecules approaching a continuous distribution, but it must be small enough so that there is no noticeable variation of macroscopic properties across the volume. Within the range of these specifications, $\rho(\mathbf{r}, t)$ then will not depend on the choice of ΔV. For example, for ΔV to contain about 10^7 atoms, a typical size for liquids is $\Delta V \approx (10^{-5}\,\mathrm{cm})^3$ and for gases $\Delta V \approx (10^{-4}\,\mathrm{cm})^3$.

By the same token, the individual velocities of particles are replaced by the *velocity field*

$$
\mathbf{v}(\mathbf{r}, t) = \begin{cases} \text{average velocity} \\ \text{of all particles} \\ \text{within } \Delta V \text{ at } \mathbf{r}. \end{cases}
\tag{7.59}
$$

This average value typically is several orders of magnitude smaller than the velocities of individual particles within ΔV. For example, for a room at normal temperature, the Maxwell distribution of particle velocities for gases shows that the actual velocities of individual air molecules are in the range of about 100–1000 m/s, but the macroscopically noticeable air movement, described by $\mathbf{v}(\mathbf{r}, t)$, is normally at most in the range of a few centimeters per second.

Individual particles in a fluid hitting a surface will impart an impulse on this surface. The time average of these impulses gives rise to an overall force ΔF felt across the surface's area ΔA. Throughout the medium, therefore, we can define a *pressure field*

$$P(\mathbf{r}, t) = \frac{\Delta F}{\Delta A} \tag{7.60}$$

by choosing sufficiently small — but not too small! — areas ΔA [comparable in size to $(\Delta V)^{2/3}$] located at $\mathbf{r}$. For a fluid at rest, the pressure at any point is the same in all directions, i.e., for any orientation of ΔA at a point, the pressure is the same.

Under the assumptions made here, the fields $\rho(\mathbf{r}, t)$, $\mathbf{v}(\mathbf{r}, t)$, and $P(\mathbf{r}, t)$ describe different aspects of the macroscopically relevant behavior of liquids or gases. If these three quantities provide a *complete* description of the state of a liquid or gas, it is called *ideal*. The notion of an ideal gas provides a good approximation of real gases as long as the densities are low, velocities are at an intermediate range, and the temperature is uniform across the entire (macroscopic) volume. By contrast, the only liquid that may be regarded as truly ideal is superfluid Helium. Nevertheless, in many circumstances the viscosity plays no important role; therefore, treating these liquids by the ideal-fluid equations will often result in reasonably good approximations of reality. Liquids of high viscosity, however, cannot be treated as ideal.

7.3.1   *Continuity equation*

As the particles of the medium move, they form a matter current. Since matter does not get lost during this motion, the current through the closed surface of a volume must be directly related to the mass density within the volume. Defining the *current density* as

$$\mathbf{j}(\mathbf{r}, t) = \rho(\mathbf{r}, t)\, \mathbf{v}(\mathbf{r}, t), \tag{7.61}$$

with units mass/area·time, one has

$$\oint_{\partial V} d\mathbf{A} \cdot \mathbf{j}(\mathbf{r}, t) = -\frac{d}{dt} \int_V d^3r\, \rho(\mathbf{r}, t) = -\int_V d^3r\, \frac{\partial \rho(\mathbf{r}, t)}{\partial t}, \tag{7.62}$$

where ∂V denotes the closed surface of the volume V, with the area element $d\mathbf{A}$ pointing outward. The left-hand side here describes the matter current (= mass/time) that flows outward through the surface ∂V enclosing the volume. This *net current* lowers the mass density inside the volume by the expression on the right. It is assumed here that the volume stays constant and remains at rest in the inertial system in which this equation is valid. Using Gauss's law, the area integral of the left-hand side can be transformed into a volume integral as well; one then has

$$\int_V d^3r \left(\operatorname{div}\mathbf{j}(\mathbf{r},t) + \frac{\partial \rho(\mathbf{r},t)}{\partial t} \right) = 0. \tag{7.63}$$

Since this is valid for arbitrary volumes, the integrand itself must vanish, providing the *continuity equation*

$$\boxed{\operatorname{div}\mathbf{j}(\mathbf{r},t) + \frac{\partial \rho(\mathbf{r},t)}{\partial t} = 0}. \tag{7.64}$$

This is the differential expression for the *conservation of mass* in a gaseous or liquid medium.

7.3.2   *Euler's equation of fluid dynamics*

A (macroscopic) force $\Delta\mathbf{F}$ on the mass $\Delta m = \rho\Delta V$ within a small volume element ΔV satisfies Newton's second axiom,

$$\Delta m \frac{d\mathbf{v}(\mathbf{r},t)}{dt} = \rho(\mathbf{r},t)\Delta V \frac{d\mathbf{v}(\mathbf{r},t)}{dt} = \Delta\mathbf{F} = \Delta\mathbf{F}_{\text{ext}} + \Delta\mathbf{F}_P, \tag{7.65}$$

where the last equality divides the force into an external part and an internal part due to the pressure in the medium. The external force on the fluid can be expressed in terms of a force density $\mathbf{f}(\mathbf{r},t)$ (= force/volume; for gravity, for example, the force density is $\mathbf{f} = \rho\mathbf{g}$). The external force then is written as

$$\Delta\mathbf{F}_{\text{ext}}(\mathbf{r},t) = \mathbf{f}(\mathbf{r},t)\Delta V. \tag{7.66}$$

To arrive at an expression for $\Delta\mathbf{F}_P$, let us consider a small rectangular volume $\Delta V = \Delta x \Delta y \Delta z$ at $\mathbf{r} = (x,y,z)$ and add the forces due to the pressure on the six sides of the box volume. Collecting only results up to order $\Delta x \Delta y \Delta z$ (i.e., neglecting all contributions containing nonlinear powers of Δx, Δy, or Δz), one finds (see Fig. 7.6)

$$\Delta\mathbf{F}_P(\mathbf{r},t) = -\oint_{\partial\Delta V} d\mathbf{A}\, P$$

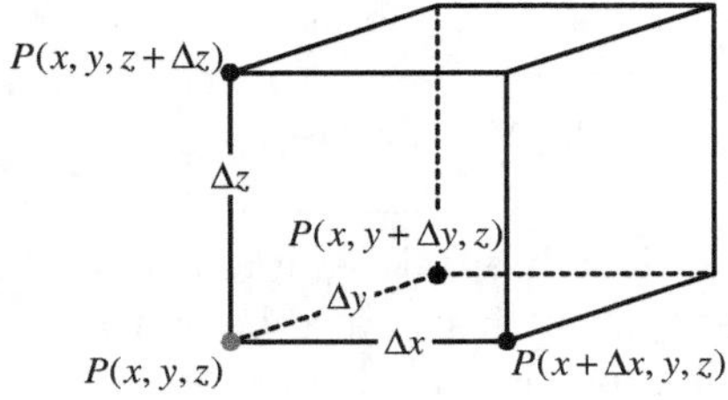

Fig. 7.6 Rectangular fluid volume employed in Eq. (7.67). Only the pressure values at the points indicated by solid dots enter this difference equation. Values at other points give rise to contributions that contain squared (or higher) powers in the cube dimensions Δx, Δy, or Δz.

$$
= -\mathbf{e}_x \Delta y \Delta z [P(x + \Delta x, y, z) - P(x, y, z)]
$$
$$
- \mathbf{e}_y \Delta x \Delta z [P(x, y + \Delta y, z) - P(x, y, z)]
$$
$$
- \mathbf{e}_z \Delta x \Delta y [P(x, y, z + \Delta z) - P(x, y, z)]
$$
$$
= -\Delta x \Delta y \Delta z \left[\mathbf{e}_x \frac{P(x + \Delta x, y, z) - P(x, y, z)}{\Delta x} \right.
$$
$$
+ \mathbf{e}_y \frac{P(x, y + \Delta y, z) - P(x, y, z)}{\Delta y}
$$
$$
\left. + \mathbf{e}_z \frac{P(x, y, z + \Delta z) - P(x, y, z)}{\Delta z} \right]
$$
$$
\approx - \left(\mathbf{e}_x \frac{\partial P}{\partial x} + \mathbf{e}_y \frac{\partial P}{\partial y} + \mathbf{e}_z \frac{\partial P}{\partial z} \right) \Delta x \Delta y \Delta z, \tag{7.67}
$$

where the overall minus sign takes care of the fact that $\Delta \mathbf{F}_P(\mathbf{r}, t)$ points inward into the volume (since it is the force acting on the volume) while, by convention, the surface-area element $d\mathbf{A}$ points outward. The last expression in the parentheses is the gradient of the pressure field,[113] i.e.,

$$
\Delta \mathbf{F}_P(\mathbf{r}, t) = -\boldsymbol{\nabla} P(\mathbf{r}, t) \, \Delta V. \tag{7.68}
$$

Finally, with the total differential of the velocity field given by

$$
d\mathbf{v}(\mathbf{r}, t) = \frac{\partial \mathbf{v}}{\partial t} dt + \frac{\partial \mathbf{v}}{\partial x} dx + \frac{\partial \mathbf{v}}{\partial y} dy + \frac{\partial \mathbf{v}}{\partial z} dz
$$
$$
= \frac{\partial \mathbf{v}}{\partial t} dt + \frac{\partial \mathbf{v}}{\partial x} \frac{dx}{dt} dt + \frac{\partial \mathbf{v}}{\partial y} \frac{dy}{dt} dt + \frac{\partial \mathbf{v}}{\partial z} \frac{dz}{dt} dt
$$
$$
= \frac{\partial \mathbf{v}}{\partial t} dt + (\mathbf{v} \cdot \boldsymbol{\nabla}) \mathbf{v} \, dt, \tag{7.69}
$$

---

[113]We emphasize that, in keeping with the definition (7.60), the last step in (7.67) that replaces the difference ratios be partial derivatives is to be interpreted as the mathematical *extrapolation* of (7.60) for vanishing volumes, i.e., it is not meant to imply that one actually looks at the physics inside an infinitesimal volume.

Newton's equation of motion can then be recast in the form

$$\rho(\mathbf{r}, t)\left[\frac{\partial \mathbf{v}(\mathbf{r}, t)}{\partial t} + \left[\mathbf{v}(\mathbf{r}, t) \cdot \boldsymbol{\nabla}\right] \mathbf{v}(\mathbf{r}, t)\right] = -\boldsymbol{\nabla} P(\mathbf{r}, t) + \mathbf{f}(\mathbf{r}, t) \; . \tag{7.70}$$

This is *Euler's equation of fluid dynamics*. Strictly speaking, it applies only to fluids without internal friction (viscosity), but it can also be employed to treat liquids with low viscosity in an approximate manner.

7.3.3   *Equation of state*

Euler's equation and the continuity equation together comprise four partial differential equations. To completely determine the five fields $\rho(\mathbf{r}, t)$, $\mathbf{v}(\mathbf{r}, t)$, and $P(\mathbf{r}, t)$, we require another equation. The simplest possibility for this additional equation is a *pressure-density relation* of the generic form

$$P = P\big(\rho(\mathbf{r}, t)\big). \tag{7.71}$$

This simple relation is only valid for isothermal and adiabatic processes. More complex situations require taking into account details of the thermodynamics of the process like heat transport, etc.

For the present purpose, we restrict the discussion to the so-called *polytropic equation of state*

$$P = const \cdot \rho^{\gamma}. \tag{7.72}$$

The relative change of density when changing the pressure is called the *compressibility* κ; for polytropic systems, it is given by

$$\kappa = \frac{1}{\rho}\frac{\partial \rho}{\partial P} = \frac{1}{\gamma P}. \tag{7.73}$$

Table 7.1 lists γ values for some physical systems that are described by the polytropic equation of state. For liquids, a change in pressure usually results only in a very small change in density ($\gamma \gg 1$). In the limit $\gamma = \infty$, the liquid becomes incompressible, i.e., $\rho = const \cdot P^{1/\gamma} = const$.

Note also that the γ values may depend on state variables other than pressure. For an ideal gas, for example, keeping the temperature constant provides a different value than keeping the entropy constant. The equation of state for an ideal gas is

$$PV = NkT \qquad \text{(ideal gas)}, \tag{7.74}$$

where N is the number of gas molecules (each with the same mass) and k the Boltzmann constant. For isothermal changes (temperature $T = const$),

Table 7.1   Values of the exponent γ of the polytropic equation of state (7.72) for some physical systems.

γ	System
∞	Incompressible liquid
$\gg 1$	Real liquid
1	Ideal gas, constant temperature (isothermal)
c_P/c_V	Ideal gas, constant entropy (adiabatic)
$5/3$	Fermi gas, $T \approx 0$, nonrelativistic
$4/3$	Fermi gas, $T \approx 0$, relativistic

this provides $P = const \cdot N/V = const \cdot \rho$ and hence $\gamma = 1$. The isothermal compressibility, therefore, is

$$\kappa_T = \frac{1}{P} \qquad (\text{ideal gas, } T = const). \tag{7.75}$$

For adiabatic changes (entropy $S = const$), one has $PV^{c_P/c_V} = const$, which already is of the form (7.72), and therefore

$$\kappa_S = \frac{c_V}{c_P\, P} \qquad (\text{ideal gas, } S = const), \tag{7.76}$$

where c_P and c_V are the specific heat values for constant pressure and constant volume, respectively.

7.3.4   Examples

The dynamics of ideal fluids thus are summarized in the five partial differential equations

$$
\begin{aligned}
\dot{\rho} + \boldsymbol{\nabla} \cdot (\rho \mathbf{v}) &= 0 & &\text{Continuity equation,} \\
\rho \dot{\mathbf{v}} + \rho(\mathbf{v} \cdot \boldsymbol{\nabla})\mathbf{v} &= -\boldsymbol{\nabla} P + \mathbf{f} & &\text{Euler equation,} \\
P &= P(\rho) & &\text{Equation of state}
\end{aligned}
\tag{7.77}
$$

for the five fields $\rho(\mathbf{r}, t)$, $\mathbf{v}(\mathbf{r}, t)$, and $P(\mathbf{r}, t)$. To apply them to a particular problem, they need to be supplemented by appropriate boundary values and initial conditions.

The nonlinearities of the fields appearing in these equations in general provide a formidable challenge when attempting to solve these equations. They may induce a very complex motion patterns (turbulence, etc.) that are very difficult to describe mathematically. It is possible, in particular, that small differences in the initial conditions may grow exponentially fast that make it impossible to obtain stable numerical solutions. For many applications, therefore, one cannot predict the dynamical behavior of the system for any length of time. Section 8.1 on chaotic systems will explore some of the resulting issues.

It should be emphasized that this complex behavior of fluid-dynamical systems is present even though the equations (7.77) neglect internal friction phenomena, i.e., they are already an idealization of the real situation. Taking into account the viscous behavior of real fluids leads to the *Navier–Stokes equation*[114]

$$\rho\dot{\mathbf{v}} + \rho(\mathbf{v} \cdot \boldsymbol{\nabla})\mathbf{v} = -\boldsymbol{\nabla}P + \mathbf{f} + \eta\boldsymbol{\nabla}^2\mathbf{v}, \tag{7.78}$$

where the positive quantity η is the *viscosity* of the liquid; it describes internal friction and it is highly temperature-dependent. The Navier–Stokes equations are even more complicated to solve than the Euler equations (obtained for $\eta = 0$) and very little is known about the general properties of their solutions.[115]

In the following, some simple applications of the ideal-fluid equations (7.77) are given.

Liquid at rest in the gravitational field

In the limiting case of *hydrostatics*, the velocity field vanishes and the density and pressure fields are time-independent,

$$\mathbf{v} = 0, \qquad \rho = \rho(\mathbf{r}), \qquad P = P(\mathbf{r}). \tag{7.79}$$

The continuity equation thus is satisfied trivially as a matter of course. The Euler equation reduces to

$$\boldsymbol{\nabla}P = \mathbf{f}, \tag{7.80}$$

describing the equilibrium situation of the pressure forces balancing the external forces.

Taking a fluid medium at rest in the homogeneous field of gravity,

$$\mathbf{f} = \rho\mathbf{g} = -\rho g\mathbf{e}_z, \tag{7.81}$$

the Euler equation becomes

$$\frac{dP(z)}{dz} = -\rho(z)g, \tag{7.82}$$

---

[114]This simplified form of the Navier–Stokes equation assumes $\boldsymbol{\nabla} \cdot \mathbf{v} = 0$ appropriate for an incompressible fluid. If this is not true, an additional term proportional to $\boldsymbol{\nabla}(\boldsymbol{\nabla} \cdot \mathbf{v})$ appears on the right-hand side.

[115]In fact, in 2000, a prize money of \$1 million was offered by the Clay Mathematics Institute for a "mathematical theory which will unlock the secrets hidden in the Navier–Stokes equations" (see http://www.claymath.org/ for details). To date, January 2021, the award remains unclaimed.

where the pressure and density fields only depend on z in view of the translation invariance in the x and y directions. To proceed further, let us consider equations of state for two cases: an incompressible liquid and an ideal gas.

For an incompressible fluid, the density $\rho = \rho_0 = const$ provides the solution

$$P(z) = P_0 - \rho_0 g z, \qquad (7.83)$$

where the integration constant is the pressure at $z = 0$. This equation describes the linearly increasing pressure with depth in a liquid. For a vertical column of fluid, with cross section A and height h, the pressure difference, $\Delta P = -\rho g(z - h - z) = \rho g h = Mg/A$, is directly related to the total mass M of the column.

For an ideal gas, Eq. (7.74) can be recast into

$$P = \rho \frac{kT}{m}, \qquad (7.84)$$

where m is the mass of an individual gas molecule. Equation (7.82) then reads

$$\frac{dP(z)}{dz} = -\frac{mg}{kT} P(z), \qquad (7.85)$$

which, assuming constant temperature, integrates to

$$P(z) = P_0 \, e^{-z/z_0}, \qquad \text{with} \quad z_0 = \frac{kT}{mg}, \qquad (7.86)$$

where P_0 is the pressure at $z = 0$. This is the *barometric equation* describing the change in pressure with height z. For Earth's atmosphere at $T \approx 300\,\mathrm{K}$, the value of $z_0 \approx 7.9\,\mathrm{km}$. To apply this formula presumes that $T = const$, which generally is not true for large height differentials in the atmosphere. To obtain reliable results, one needs a temperature profile $T = T(z)$ and integrate Eq. (7.85) numerically.

Rotating liquid in the gravitational field

Consider a cylindrical vessel at sealevel filled with an incompressible fluid rotating with constant angular velocity $\boldsymbol{\omega} = \omega \mathbf{e}_z$ about its symmetry axis in the constant field of gravity $\mathbf{g} = -g\mathbf{e}_z$. At static equilibrium, the volume elements of the fluid will no longer move with respect to each other and there is no longer any friction. The velocity field will then correspond to that of a rigid rotation,

$$\mathbf{v} = \boldsymbol{\omega} \times \mathbf{r}, \qquad (7.87)$$

and we may apply the Euler equations.

To calculate the shape of the surface of the rotating liquid, we employ cylindrical coordinates $r = \sqrt{x^2 + y^2}$, φ, and z. The relevant input equations then read

$$\mathbf{v} = \omega r \mathbf{e}_\varphi, \quad \mathbf{f} = -\rho_0 g \mathbf{e}_z, \quad P = P(r, z), \tag{7.88}$$

where ρ_0 is the constant density of the liquid. The continuity equation is satisfied as a matter of course. Using

$$(\mathbf{v} \cdot \boldsymbol{\nabla})\mathbf{v} = \omega r \frac{1}{r} \frac{\partial}{\partial \varphi} \omega r \mathbf{e}_\varphi = -\omega^2 r \mathbf{e}_r, \tag{7.89}$$

$$\boldsymbol{\nabla} P = \frac{\partial P}{\partial r} \mathbf{e}_r + \frac{\partial P}{\partial z} \mathbf{e}_z, \tag{7.90}$$

and $\partial \mathbf{v}/\partial t = 0$, the Euler equation has two components,

$$\frac{\partial P}{\partial r} = \rho_0 \omega^2 r \qquad \text{and} \qquad \frac{\partial P}{\partial z} = -\rho_0 g, \tag{7.91}$$

which are seen to be just the force densities associated with the centrifugal and gravitational forces acting on a fluid volume element. The pressure field then is

$$P(r, z) = \rho_0 \left(\frac{\omega^2 r^2}{2} - gz \right) + const. \tag{7.92}$$

[For $\omega = 0$, this corresponds to (7.83).] For surfaces of constant pressure (isobars), the first term must vanish, i.e.,

$$z = \frac{\omega^2}{2g} r^2 + const \qquad \text{(isobars)}, \tag{7.93}$$

providing the functional form of a *rotation paraboloid*. The net force resulting from adding the centrifugal and the gravitational forces is perpendicular to this surface everywhere, and hence this is the shape that is usually associated with the upper surface of the liquid as well.

Note, however, that this is only true if the pressure everywhere outside the surface is equal to the constant pressure within the surface. To see how large a deviation from this result is obtained if we assume that the surface is exposed to the atmosphere, we use the pressure variation described by the barometric equation (7.86), and ignore possible effects of the air-to-fluid interface. Measuring z from the lowest point of the fluid surface, the constant in (7.92) corresponds to P_0 in Eq. (7.86). The condition for pressure equilibrium at the surface then is

$$\rho_0 \left(\frac{\omega^2 r^2}{2} - gz \right) + P_0 = P_0 \, e^{-z/z_0} \approx P_0 \left(1 - \frac{z}{z_0} \right), \tag{7.94}$$

where the exponential on the right-hand side was linearized assuming $z/z_0 \ll 1$. This not only reasonable, since $z_0 \approx 8\,\mathrm{km}$ for Earth's atmosphere, but necessary — unless one wants to take into account terms of order z^2 beyond $\rho_0 g z$ for the gravitational potential as well. One then finds immediately

$$z = \frac{\omega^2}{2g^*}\, r^2, \qquad \text{where} \qquad g^* = g\left(1 - \frac{\rho_{\mathrm{air}}}{\rho_0}\right), \qquad (7.95)$$

with ρ_{air} being the density of air. This result still describes a rotation paraboloid, albeit with a slightly lower effective acceleration constant g^*. For water, for example, the ratio $\rho_{\mathrm{atm}}/\rho_0$ is of the order of 10^{-3}. Deviations from the paraboloidal shape enter only at order $(z/z_0)^2$. At this order, one must also consider contributions to the gravitational potential of order z^2.

Bernoulli's equation

Consider the stationary flow of an incompressible fluid in the field of constant gravity given by

$$\rho = \rho_0, \quad \mathbf{f} = -\rho_0 \boldsymbol{\nabla}\hat{U}, \quad \mathbf{v} = \mathbf{v}(\mathbf{r}), \quad P = P(\mathbf{r}), \qquad (7.96)$$

where the gravitational force density $\mathbf{f} = \rho_0 \mathbf{g} = -\rho_0 \boldsymbol{\nabla}\hat{U}$ is written in terms of the gradient of the gravitational potential per unit mass $\hat{U} = gz$.

The corresponding Euler equation

$$\rho_0(\mathbf{v} \cdot \boldsymbol{\nabla})\mathbf{v} = -\rho_0 \boldsymbol{\nabla}\hat{U} - \boldsymbol{\nabla}P \qquad (7.97)$$

can be rewritten equivalently as

$$\frac{\rho_0}{2}\boldsymbol{\nabla}\mathbf{v}^2 - \rho_0 \mathbf{v} \times (\boldsymbol{\nabla} \times \mathbf{v}) + \rho_0 \boldsymbol{\nabla}\hat{U} + \boldsymbol{\nabla}P = 0. \qquad (7.98)$$

The fluid elements will flow along stationary *streamlines* where every tangential path increment $d\mathbf{r}$ is parallel to $\mathbf{v}(\mathbf{r})$,

$$d\mathbf{r} \parallel \mathbf{v}(\mathbf{r}) \qquad \text{(along streamline).} \qquad (7.99)$$

Upon integration between two arbitrary points along the same streamline, we find from (7.98)

$$\int_1^2 \left[\frac{\rho_0}{2}\boldsymbol{\nabla}\mathbf{v}^2 + \rho_0 \boldsymbol{\nabla}\hat{U} + \boldsymbol{\nabla}P\right] \cdot d\mathbf{r} = \rho_0 \int_1^2 \left[\mathbf{v} \times (\boldsymbol{\nabla} \times \mathbf{v})\right] \cdot d\mathbf{r} = 0, \qquad (7.100)$$

or, using $\boldsymbol{\nabla}\Phi \cdot d\mathbf{r} = d\Phi$,

$$\int_1^2 d\left[\frac{\rho_0}{2}\mathbf{v}^2 + \rho_0 \hat{U} + P\right] = 0. \qquad (7.101)$$

Hence

$$\boxed{\;P + \frac{\rho_0}{2}\mathbf{v}^2 + \rho_0 \hat{U} = const \quad \text{along a streamline}\;}. \tag{7.102}$$

This is *Bernoulli's equation*; it says that the sum of the static pressure (P) and the kinetic-energy density $(\rho_0 \mathbf{v}^2/2$, also called the *dynamic pressure*) and the potential-energy density $(\rho_0 \hat{U}$, also called the *buoyancy term*) does not change along a streamline. Note that this does not necessarily mean that neighboring streamlines correspond to the same constant. The latter would only be true for an irrotational flow; this follows from Eq. (7.100) being zero for $\boldsymbol{\nabla} \times \mathbf{v} = 0$ irrespective of whether points 1 and 2 are on the same streamline or not.

This equation implies *Bernoulli's principle*: *There is less static pressure in regions of constricted flow cross sections since the flow rate is higher there.* This entails, for example, that the pressure differential due to the higher flow rate at the upper (convex) side of an airfoil as compared to the lower (concave) side provides a net upward pressure. (This should *not* be taken as a complete explanation for the lift force that lets airplanes stay airborne — the actual theory of lift is more complicated.)

Laminar flow

Stationary velocity fields that satisfy

$$\boldsymbol{\nabla} \times \mathbf{v}(\mathbf{r}) = 0 \qquad \text{(irrotational, stationary)} \tag{7.103}$$

are called *irrotational* or *laminar*. The associated laminar flow is also called a *potential flow* since there exists a *velocity potential* $\phi(\mathbf{r})$ with

$$\mathbf{v}(\mathbf{r}) = \boldsymbol{\nabla}\phi(\mathbf{r}). \tag{7.104}$$

For an incompressible fluid, with $\rho = const$, the continuity equation then reduces to *Laplace's equation*,

$$\boldsymbol{\nabla}^2 \phi = 0, \tag{7.105}$$

which is well-known from electrostatics. It may be used to calculate, for example, the laminar flow around an object. Putting an object in a homogeneous flow $\mathbf{v} = v_0 \mathbf{e}_z$, the appropriate boundary condition requires that the normal component $\mathbf{n} \cdot \mathbf{v}$ vanish at the surface of the object,

$$[\mathbf{n} \cdot \boldsymbol{\nabla}\phi]_{\text{surface}} = 0. \tag{7.106}$$

Far away from the object, the flow remains homogeneous, therefore

$$\phi \xrightarrow{\; r \to \infty \;} v_0 z. \tag{7.107}$$

The problem specified in this manner corresponds exactly to the electrostatic problem of placing a conducting body in an homogeneous electric field. For this example, the streamlines of fluid dynamics correspond to the equipotential lines of electrostatics.

7.3.5 Sound waves

The propagation of sound may also be derived from the nonlinear hydrodynamic equations (7.77) by *linearizing* them. We start from the equilibrium situation

$$\rho_0 = const, \qquad \mathbf{v}_0 = 0, \qquad P_0 = const, \tag{7.108}$$

which satisfies the equations as a matter of course, and assume small changes

$$\rho = \rho_0 + \delta\rho, \tag{7.109a}$$

$$\mathbf{v} = \delta\mathbf{v}, \tag{7.109b}$$

$$P = P_0 + \delta P. \tag{7.109c}$$

The equations are now solved by neglecting all but terms linear in the small changes. In sound waves, the density changes are so fast that no appreciable heat exchange takes place. Hence, the adiabatic compressibility (7.76) describes the relation between pressure and density changes,

$$\kappa_s = \frac{1}{\rho}\frac{\partial\rho}{\partial P} \quad\Rightarrow\quad \delta P = \frac{1}{\kappa_s\rho_0}\delta\rho, \tag{7.110}$$

which is the equation of state pertinent to the present situation. The linearized continuity and Euler equations then are

$$\frac{\partial}{\partial t}\delta\rho + \rho_0\boldsymbol{\nabla}\cdot\delta\mathbf{v} = 0, \tag{7.111a}$$

$$\rho_0\frac{\partial}{\partial t}\delta\mathbf{v} + \frac{1}{\kappa_s\rho_0}\boldsymbol{\nabla}\delta\rho = 0, \tag{7.111b}$$

respectively. Taking the time derivative of the first and the divergence of the second and subtracting the resulting equations yields

$$\left(\frac{\partial^2}{\partial t^2} - \frac{1}{\kappa_s\rho_0}\boldsymbol{\nabla}^2\right)\delta\rho = 0. \tag{7.112a}$$

Proceeding similarly by taking the gradient of the first and the time derivative of the second equation provides

$$\left(\frac{\partial^2}{\partial t^2} - \frac{1}{\kappa_s\rho_0}\boldsymbol{\nabla}^2\right)\delta\mathbf{v} = 0, \tag{7.112b}$$

where the operator identity $\nabla(\nabla \cdot \delta\mathbf{v}) = \nabla^2\delta\mathbf{v} + \nabla \times (\nabla \times \delta\mathbf{v})$ was used and a curl-free $\delta\mathbf{v}$ was assumed.

The wave equations (7.112) have the well-known *plane-wave solutions*

$$\delta\rho = \Delta\rho \sin(\mathbf{k} \cdot \mathbf{r} - \omega t), \tag{7.113a}$$

$$\delta\mathbf{v} = \Delta\mathbf{v} \sin(\mathbf{k} \cdot \mathbf{r} - \omega t), \tag{7.113b}$$

where ω and $k = |\mathbf{k}|$ are related by

$$\frac{\omega}{k} = \frac{1}{\sqrt{\rho_0 \kappa_s}} \equiv v_s. \tag{7.114}$$

The *speed of sound* v_s describes the propagation speed of the wave. It follows from any of the equations (7.111) that the constant amplitudes $\Delta\rho$ and $\Delta\mathbf{v}$ are related by

$$\frac{\Delta\mathbf{v}}{v_s} = \frac{\mathbf{k}}{k} \frac{\Delta\rho}{\rho_0}. \tag{7.115}$$

The waves considered here, therefore, are *longitudinal*,

$$\mathbf{k} \parallel \Delta\mathbf{v}, \tag{7.116}$$

propagating in the direction of the *wave vector* $\mathbf{k}$. Moreover, only one of the amplitudes $\Delta\rho$ or $|\Delta\mathbf{v}|$ can be chosen freely. This also determines the pressure wave [cf. (7.110)]

$$\delta P = v_s^2 \Delta\rho \sin(\mathbf{k} \cdot \mathbf{r} - \omega t). \tag{7.117}$$

Equations (7.113) through (7.117) specify *sound waves*.

The *dispersion relation* $\omega = v_s k$ is linear, i.e., wave packets propagate *non-dispersive*, without any distortion of their form. A nonlinear dispersion relation would imply different propagation times for different frequencies, which would make acoustic communication very difficult, if not impossible.

We can easily estimate the speed of sound in air via

$$v_s = \sqrt{\frac{\gamma P}{\rho_0}} = \sqrt{\frac{d+2}{d} \frac{kT}{m}} \approx \sqrt{\frac{7}{5} \frac{\text{eV}}{40} \frac{c^2}{29\,\text{GeV}}} \approx 330 \frac{\text{m}}{\text{s}}. \tag{7.118}$$

It was used here that $\gamma = c_P/c_V = (d+2)/d$, where $d = 5$ is the number of degrees of freedom of an air molecule (each having 3 degrees of freedom of translation and 2 for rotation), $P = \rho_0 kT/m$ from the ideal-gas law, with $kT \approx 1\,\text{eV}/40$ for room temperature and $m \approx 29\,\text{GeV}/c^2$ for the average mass of the N_2 and O_2 molecules weighted according to their natural abundance. The measured value is $v_s = 331\,\text{m/s}$.

The wave solutions discussed here are *plane waves*. A point-like source emits *spherical waves* $\sin(\mathbf{k} \cdot \mathbf{r} - \omega t)/r$ whose intensity falls off like $1/r^2$. Real sound waves experience damping and fall off even faster. To describe damping requires taking into account viscosity terms. Finally, we mention that sound waves also occur in solids, where in addition to longitudinal pressure waves also transversal shearing waves are possible.

Chapter 8

Beyond Classical Mechanics

In this part, we will explore two issues that go beyond the traditional range of validity of classical mechanics, namely the *chaotic behavior* of some classical systems and the resulting unpredictability of trajectories, and the non-Galilean transformations between inertial frames for objects moving at high speeds that follow from the *theory of special relativity*.

8.1 Chaotic Systems

The equations of classical mechanics are time-reversal invariant and deterministic. In principle, therefore, for any system that has evolved forward in time starting from some given initial conditions, we should be able to take the system's state at any point in time, trace it back to it origins, and thus completely recover the system at its initial state. The underlying assumption here is that the trajectories are stable against small variations of the initial conditions since experimentally we can never know the initial conditions with infinite precision. It had been realized quite early, most notably perhaps through the works of Henri Poincaré (1854–1912) and Aleksandr Lyapunov (1857–1918), that this assumption does not necessarily hold true for nonlinear systems. Indeed, the instability of some systems may be so large that, for all practical purposes, they 'forget' their initial conditions, thus making it impossible to recover them by reversing time.[116] Phase-space plots of the resulting trajectories exhibit highly irregular — seemingly chaotic — behavior.

---

[116]This sensitivity to initial conditions is often referred to as the "Butterfly Effect" after the title of a talk given by Edward Lorenz (MIT) at the December 1972 meeting of the American Association for the Advancement of Science in Washington, D.C., *"Predictability: Does the Flap of a Butterfly's Wings in Brazil set off a Tornado in Texas?"*

The early investigations, by Poincaré, Lyapunov, and others, were based on traditional 'pencil-and-paper' mathematics and as such their full implications became only apparent in the 1970s, and later, after it had become easily possible to perform numerical simulations of such *chaotic systems*, as they are called now. It is for this reason that the introduction to their dynamics is placed here, in the part called "Beyond Classical Mechanics", despite the fact that their basic treatment still is based on the equations of classical mechanics.

The study of deterministic chaos is a vast, still ongoing, and highly active area of research, requiring sophisticated mathematical tools and extensive computational simulations, and any attempt at a comprehensive treatment would be beyond the scope of the present course. The following presentation, therefore, is only intended to introduce some of the more elementary issues associated with the dynamics of chaotic systems. In particular, we will use the example of the periodically driven planar pendulum with damping[117] to discuss some of the common features of chaotic systems, in particular those for which the route to chaos goes through period-doubling cascades.[118]

8.1.1 *The road to chaos*

One of the necessary common features of chaotic systems is their *nonlinearity*. Linear systems — systems where the differential equations of motion depend only linearly on the unknown variable and its derivatives — never exhibit chaotic behavior. Linear harmonic oscillators, for example, never behave chaotically, even if combined with a (linear) damping force and if driven by a sinusoidal external force. The solutions of the resulting equation of motion for the corresponding one-dimensional problem given in Eq. (1.101),

$$\ddot{x} + 2\lambda\dot{x} + \omega_0^2 x = a_0 \cos\omega t, \qquad\qquad (1.101')$$

can be analyzed by conventional means in a straightforward manner, as discussed in detail in Secs. 1.3.3 and 1.3.4. The unknown displacement $x(t)$ and its derivatives appear here linearly on the left-hand side, which means,

---

[117]D. D'Humières, M. R. Beasley, B. A. Huberman, and A. Libchaber, "Chaotic States and Routes to Chaos in the Forced Pendulum," Phys. Rev. A **26**, 3483 (1982).

[118]A more detailed discussion of this case, including examples of numerical simulations, can be found in Taylor [7]. For more general aspects of the computational treatment of this case and other nonlinear systems, see also R. H. Landau, M. J. Páez, and C. C. Bordeianu, *Computational Physics*, 2nd eddition (Wiley-VCH Verlag, Weinheim, 2007).

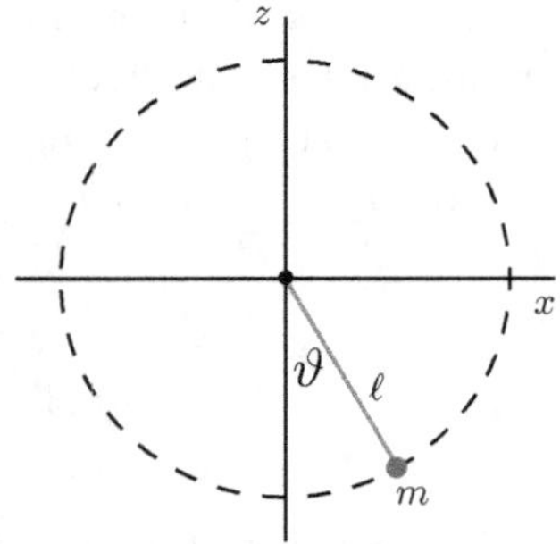

Fig. 8.1 Idealized mathematical pendulum, with bob of mass m at the end of a rigid, mass-less rod of length ℓ. Gravity (with strength g) acts along the negative z axis. The dashed circle indicates allowed trajectories. Depending on initial conditions and/or external driving force, over-the-top motion is possible.

in particular, that two solutions differ only by a solution of the homogeneous case. The general solution, therefore, is obtained as a particular solution of the inhomogeneous equation (1.101′) and the general solution of the homogeneous problem where the driving force is switched off. The linearity of the problem, in particular, allows one to express the latter as linear combinations of elementary solutions, which greatly simplifies the solution procedure.

A *nonlinear* problem very close to the linear harmonic oscillator case of Eq. (1.101′) is provided by the motion of the planar pendulum treated in Sec. 2.1.6.1. The corresponding undamped homogeneous equation of motion reads

$$\ddot{\vartheta} + \frac{g}{\ell} \sin \vartheta = 0, \tag{2.127′}$$

where ϑ is the displacement angle from the vertical, as depicted in Fig. 8.1, and the nonlinearity arises from the projection of gravity in tangential direction. The resulting sine term,

$$\sin \vartheta = \vartheta - \frac{1}{6}\vartheta^3 + \frac{1}{120}\vartheta^5 - \cdots, \tag{8.1}$$

contributes odd powers of ϑ to all orders. Allowing here for a damping term proportional to $\dot{\vartheta}$ and adding an external sinusoidal driving force $F(t) = f \cos \omega t$ with driving frequency ω along the tangent of the circular path, one obtains

$$\boxed{\ddot{\vartheta} + 2\lambda\dot{\vartheta} + \omega_0^2 \sin \vartheta = A\omega_0^2 \cos \omega t}, \tag{8.2}$$

where

$$\omega_0^2 = \frac{g}{\ell} \tag{8.3}$$

is the oscillator's squared natural frequency. The relationship of the strength λ of the damping term to ω_0 provides the underdamping ($\lambda^2 < \omega_0^2$), overdamping ($\lambda^2 > \omega_0^2$), and critical damping ($\lambda^2 = \omega_0^2$) cases discussed in Sec. 1.3.3 in the context of the linear equation of motion for the undriven, damped oscillator. The dimensionless strength A of the external driving term on the right-hand of Eq. (8.2) is easily seen to be related to the strength f of the driving force $F(t) = f \cos \omega t$ by

$$A = \frac{f}{mg}. \tag{8.4}$$

In other words, A provides a relative measure for how much the force f counteracts the weight mg of the pendulum bob. We may expect, therefore, a transition to qualitatively different trajectories around $A \approx 1$, when the force f balances the weight. For $A \gtrsim 1$, in particular, we expect — and indeed will find — that the driving force may be capable of pushing the bob over the top, at $\vartheta = \pi$. (The assumption here is that the damping parameter λ is not too strong. Therefore, to allow relatively unhindered angular motion, we will here, and in all that follows, only consider the underdamped case where $\lambda^2 < \omega_0^2$.)

For small-angle displacements, we may truncate the expansion (8.1) of the sine function at linear contributions, $\sin \vartheta \approx \vartheta$. Equation (8.2) then reverts back to a linear equation that is structurally identical to Eq. (1.101′), and all results discussed in Secs. 1.3.3 and 1.3.4 apply here as well. We then conclude that *for initial conditions with small angles, small angular velocities, and a weak driving force, $A \ll 1$,* the steady-state solution of (8.2), when all transients have died out, is given by

$$\vartheta(t) = A_r \cos(\omega t - \varphi), \quad \text{where} \quad A_r = \frac{A\omega_0^2}{\sqrt{(\omega_0^2 - \omega^2)^2 + (2\lambda\omega)^2}}, \tag{8.5}$$

similar to the particular solution (1.107) of Eq. (1.101′). The resonant amplitude A_r here exhibits the ususal enhancement when the driving frequency ω equals the natural frequency ω_0. This solution describes harmonic motion around $\vartheta = 0$ with the frequency ω of the external driving force, with a phase angle φ described by Eq. (1.106). For small initial angles and velocities and small force amplitude A, *all* solutions eventually settle on the steady-state form (8.5), independent of the particular details of the initial conditions. In the language of chaotic dynamics, a solution of this kind is called an *attractor* because it provides the stable limit for all such initial conditions.

Beyond the linear case

As the strength A of the driving force increases resulting in larger angular amplitudes, higher-order powers of ϑ need to be taken into account, with odd higher powers of the sinusoidal driving term entering the equation via the sine expansion (8.1). In view of

$$\cos^{2n+1}\xi = \sum_{i=0}^{n} C_i \cos\left((2i+1)\xi\right), \quad \text{with} \quad n = 0, 1, 2, \ldots, \tag{8.6}$$

where C_i are some real coefficients, the structure of the resulting steady-state solution, after all transients have died out, is given by

$$\vartheta(t) = a\cos(\omega t - \varphi_1) + b\cos 3(\omega t - \varphi_1) + c\cos 5(\omega t - \varphi_1) + \cdots, \tag{8.7}$$

which mixes odd higher-order *harmonics* 2ω, 3ω, etc. with the fundamental driving frequency ω, thus destroying the pure sinusoidal behavior (8.5) of the pure linear case.[119] However, because the magnitude relationships between the amplitudes,

$$|a| \gg |b| \gg |c| \gg \cdots, \tag{8.8}$$

is such that the term with fundamental frequency ω still remains dominant, the higher-order harmonics will only affect the amplitude of the fundamental-frequency contribution. In other words, the steady-state behavior (8.7) of the full solution can be written equivalently as

$$\vartheta(t) = a_1\,\theta_1(t)\,\cos(\omega t - \varphi_1), \tag{8.9}$$

where a_1 is some constant amplitude. The function $\theta_1(t)$ here arises from the higher-order harmonic contributions; it therefore is a *periodic* nonsinusoidal function with the same period $T = 2\pi/\omega$ as the driving force. The relative variation of $\theta_1(t)$ over the period is small and thus it modifies only the amplitude of the pure sinusoidal function $\cos(\omega t - \varphi)$ but does not contribute additional nodes to the solution. Figure 8.2 provides a schematic plot for this behavior.

In numerical simulations,[120] one finds that the number of transient cycles grows with increasing amplitude A, with more and more pronounced erratic behavior during the transition, but the motion eventually settles on an attractor solution with fundamental period ω as long as $A < A_1 = 1.0663$ (given to five significant figures). For $1 \lesssim A < A_1$, the pendulum bob

---

[119]In general, the phase φ_1 here will be different from the phase φ of the linear case.
[120]The road to chaos described here follows Taylor's numerical results [7] obtained with parameters $\omega_0 = 3\omega/2$ and $\lambda = \omega/2$.

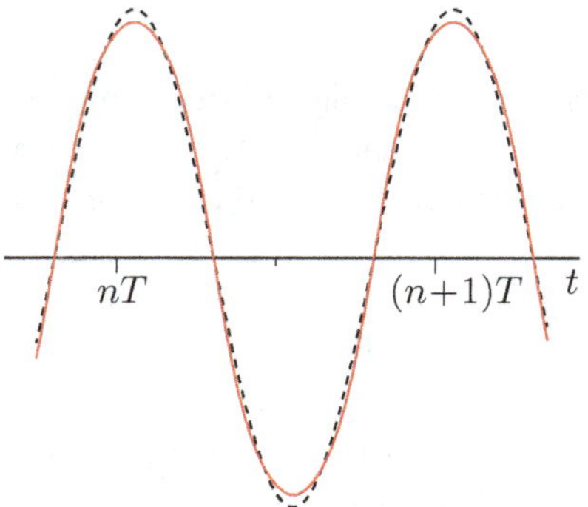

Fig. 8.2 Schematic, exaggerated plot of $\vartheta(t)$ (solid line) describing the behavior discussed in the context of Eq. (8.9) where the higher-order contributions contained in $\theta_1(t)$ distort the amplitude of the pure sinusoidal from $\cos(\omega t - \varphi_1)$ (dashed line). The time axis uses units of the period $T = 2\pi/\omega$; n is some arbitrary integer number large enough so that transients have died out.

can indeed be pushed over the top, with final attractor solutions centered around integer multiples of 2π. For 2π, for example, this means that the bob gets pushed over the top effectively once during the transient cycles and then settles oscillating with frequency ω about the equilibrium position $\vartheta = 0$. (For this example, it is possible that the over-the-top motion happens several times in different directions of ϑ during the transient cycles, with 2π being the effective net result.)

8.1.1.1   *Period doubling*

Beyond $A > A_1 = 1.0663$, a new phenomenon occurs. The steady-state solution is now no longer periodic with frequency ω. Instead, one finds that even though the nodes of the solution still exhibit the basic periodicity of the driving frequency ω, the amplitudes are now modulated by higher-order *subharmonics* with frequencies $\omega/2$. The period, therefore, has *doubled* and not too far above the threshold value A_1, we may write

$$\text{Period } 2T: \qquad \vartheta(t) = a_2\,\theta_2(t)\cos(\omega t - \varphi_2), \qquad (8.10)$$

where $\theta_2(t)$ is (nonsinusoidal) periodic function with period $2T$. Because the relative variation of $\theta_2(t)$ over its period is small, the dominant structure is still described by $\cos(\omega t - \varphi_2)$, but the crests and troughs of this functional behavior are modulated by frequency $\omega/2$, as shown in the schematic plot of Fig. 8.3. As the driving strength A is increased beyond the period-doubling threshold A_1, it takes more and more transient cycles with increasingly wild behavior before the regular attractor-solution behavior emerges.[121]

---

[121] In fact, depending on initial conditions, one may also observe attractors with period $3T$ as A approaches the value 1.0793. This shows that the damped driven pendulum is *not* a pure period-doubling problem, as discussed in connection with the logistic map in Sec. 8.1.3.

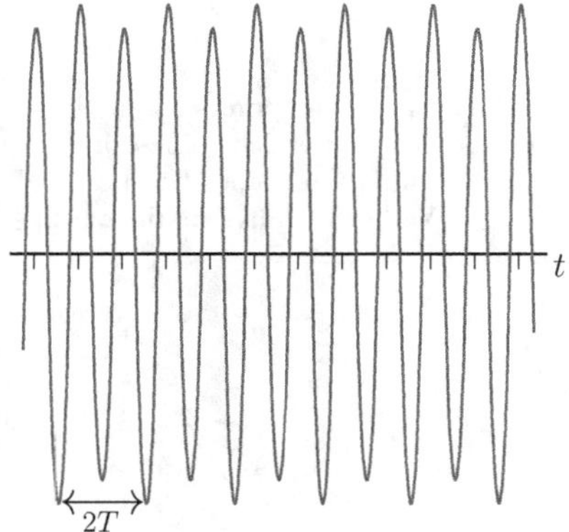

Fig. 8.3 Schematic, exaggerated plot of steady-state solution $\vartheta(t)$ for period doubling as described in the context of Eq. (8.10). Time axis given in units of the period $T = 2\pi/\omega$, with specific values depending on how long it takes for transients to die out. Crests and troughs of same height or depth, respectively, are separated by period $2T$, as indicated. The amplitude values of $\vartheta(t)$ are irrelevant for this schematic plot. The centroid value of ϑ for the t axis depends on the specific attractor.

At $A = A_2 = 1.0793$, the period is doubled again, resulting in attractor solutions with structure

$$\text{Period } 4T: \qquad \vartheta(t) = a_4\,\theta_4(t)\cos(\omega t - \varphi_4) \qquad (8.11)$$

not far above A_2. The function $\theta_4(t)$ now has period $4T$ (also simply called period four). Further period doubling occurs for increasing values of the control parameter A, with threshold values — called *bifurcation points* — leading to higher 2^n periodicity as indicated in Table 8.1. The spacing between adjacent bifurcation points becomes smaller and smaller, and it vanishes altogether at the critical value $A_c = 1.0829$. For values of A beyond this accumulation threshold, the fundamental periodicity due to the driving force is obliterated as well and solutions exhibit erratic behavior for all times, without any periodicity whatsoever. — *Chaos has set in.*

Period-doubling cascade, universality, and Feigenbaum δ

The so-called *period-doubling cascade* of bifurcation values in Table 8.1 provide the road to chaos for the damped driven pendulum of Eq. (8.2). Indeed, period-doubling behavior is found for many, many nonlinear physical systems and it is therefore oftentimes characterized as *universal*.[122] All physical systems that become chaotic via a cascade of period-doubling transitions share the universal feature that the ratio of adjacent control-parameter intervals between bifurcation points in the cascade has a con-

---

[122]It should be emphasized, however, that there are systems that exhibit chaotic behavior without going through a period-doubling cascade.

n	A_n	Periodicity 2^n
1	1.0663	2
2	1.0793	4
3	1.0821	8
4	1.0827	16
$\vdots$	$\vdots$	$\vdots$
A_c	1.0829	∞

Table 8.1 Bifurcation points A_n at which the periodicity is doubled. The doubling has an accumulation point at the critical value A_c. (Values with five significant figures taken from Ref. [7].)

stant limit,

$$\delta = \lim_{n\to\infty} \frac{A_n - A_{n-1}}{A_{n+1} - A_n} = 4.669201609\ldots , \qquad (8.12)$$

independent of the particular system, called the *Feigenbaum* δ, so named after its discoverer, Mitchell J. Feigenbaum (1944–2019), one of the pioneers of chaos theory.

8.1.2 *Lyapunov exponent*

In view of the fact that there are infinitely many bifurcation points in an interval $A \in [A_c - \varepsilon, A_c]$ however small we make ε (>0), it should be clear that in the chaotic region of physical parameters, no degree of precision for the initial condition in an experimental measurement or a numerical simulation will allow us to be able to make precise predictions for the long-term behavior of the solution, let alone allow us to recover the initial conditions when going back in time from a given final state.

The mathematical theory of stability describing the sensitivity of the time evolution of physical system to initial conditions is quite complex and beyond the scope of the present course. However, the basic idea due to Lyapunov[123] is to express the propagation of a (small) change $\boldsymbol{\delta}_0$ of initial conditions in terms of the exponential function[124]

$$|\boldsymbol{\delta}(t)| \approx e^{\lambda_{\mathrm{L}} t}\, |\boldsymbol{\delta}_0|, \qquad (8.13)$$

where $\boldsymbol{\delta}_0$ describes the initial separation vector in phase space. The *Lyapunov exponent* λ_{L} provides a measure for how two points close to each

---

[123] Alexandr Mikhailovich Lyapunov (1857–1918), Russian mathematician.

[124] We use the notation λ_{L} for the Lyapunov exponent since the more standard λ is used here already for the damping parameter. For the present problem, for small initial conditions and small driving force, the two are actually closely related, with $\lambda_{\mathrm{L}} \approx -\lambda$, as shown in Eq. (8.17).

other in phase space separate (for $\lambda_{\mathrm{L}} > 0$) or merge (for $\lambda_{\mathrm{L}} < 0$) as time progresses. In general, the exponent depends on the orientation of the initial separation vector $\boldsymbol{\delta}_0$ and one, therefore, obtains a spectrum of Lyapunov exponents commensurate with the dimension of phase space.[125] The largest of these values defined as

$$\lambda_{\mathrm{L}} = \lim_{t \to \infty} \lim_{|\boldsymbol{\delta}_0| \to 0|} \frac{1}{t} \ln \frac{|\boldsymbol{\delta}(t)|}{|\boldsymbol{\delta}_0|} \tag{8.14}$$

usually has the largest influence on the dynamical stability of the system, and this is the value usually quoted as *the* Lyapunov exponent.

In the present case, phase space is just two-dimensional, and one can extract the exponential behavior (8.13) in a fairly straightforward manner by considering the difference

$$\Delta\vartheta(t) = \vartheta_1(t) - \vartheta_2(2) \tag{8.15}$$

of two solutions ϑ_1 and ϑ_2 of the damped pendulum (8.2) with same driving strength A that differ only by small changes of initial conditions. For a weak driving force, $A \ll 1$, with small-angle and small-velocity initial conditions, we know the system basically behaves like a damped linear oscillator and we expect the difference to be very close to the regular behavior described by the corresponding solution (1.108). For two solutions that differ only by a small change δ_0 for the initial angle $\vartheta(0)$, but with the same initial angular velocity $\dot{\vartheta}(0)$ and strength A, the difference is determined only by the transient term in Eq. (1.108), and one easily finds

$$\Delta\vartheta(t) \approx \mathrm{e}^{-\lambda t} \, \delta_0 \, \frac{\omega_0}{\Omega} \, \cos(\Omega t - \varphi), \tag{8.16}$$

where the phase angle is given by $\tan\varphi = \lambda/\Omega$ and $\Omega = \sqrt{\omega_0^2 - \lambda^2}$ is the frequency of the underdamped case (1.96). The Lyapunov exponent for this case is negative, given by

$$\lambda_{\mathrm{L}} \approx \lim_{t \to \infty} \frac{1}{t} \ln \left| \frac{d\Delta\vartheta(t)}{d\delta_0} \right| = -\lambda, \tag{8.17}$$

and thus directly related to the damping parameter λ. Admittedly, this is a rather trivial result since we know already that transients die out exponentially, however, it is still quite instructive because it shows that the time limit of the $1/t$ factor kills the oscillatory parts of the expression (8.16).[126]

---

[125]In a two-dimensional $(q, \dot{q})$ phase space, for example, the speed of q separation (or merger) generally is different from the speed of $\dot{q}$ separation (or merger).

[126]The same limit mechanism is also at work in making the maximal Lyapunov exponent λ_{L} in Eq. (8.14) the only surviving contribution of all values in the Lyapunov spectrum.

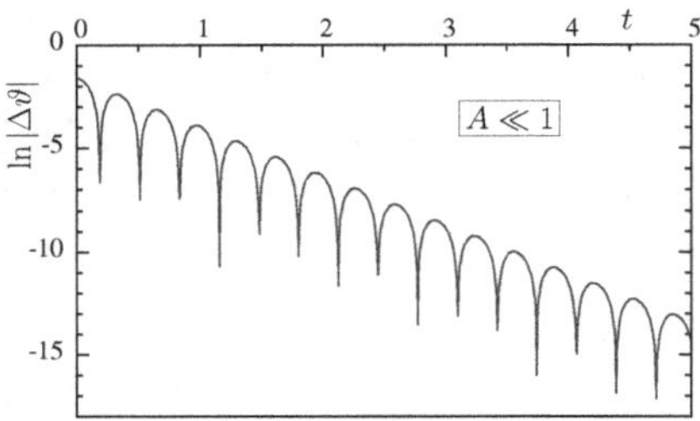

Fig. 8.4   Plot of $\ln|\Delta\vartheta(t)|$ of Eq. (8.16). The time axis is given in units of $T = 2\pi/\omega$. The downward spikes are plotting artifacts of $\ln|\Delta\vartheta| \to -\infty$ arising from where $\Delta\vartheta(t)$ vanishes. The Lyapunov exponent (8.17) can be extracted from the regular downward slope of the crest envelope.

Indeed, plotting $\ln|\Delta\vartheta|$ for this regime, one finds the behavior depicted in Fig. 8.4, with an oscillatory behavior of period π/Ω due to the nodes of the cosine factor in (8.16). To extract the Lyapunov exponent from this plot, one needs to determine the downward slope of the crest envelope. This graphical method will also work for larger values of A, beyond the linear regime, where closed-form expressions for the difference like (8.16) are no longer available. One finds[127] that as long as the strength stays within the bifurcation regime, $A < A_c$, the corresponding plots of $\ln|\Delta\vartheta|$ vs. t will always exhibit a downward slope of the crest amplitude for large enough t signifying a *negative* Lyapunov exponent and thus indicating asymptotic convergence to an attractor. However, the slope — and therefore the Lyapunov exponent — becomes smaller in magnitude as A increases. Increasing the strength parameter A beyond the critical value, $A > A_c$, the behavior changes. In this chaotic regime, the plots of $\ln|\Delta\vartheta|$ now have an upward slope and the spike pattern exhibits no discernible periodicity whatsoever. The associated *positive* Lyapunov exponent and the complete absence of attractors in this regime imply that even small changes in the initial values will eventually become magnified so much that it will be impossible in practical terms to correlate the motion's long-term behavior with initial conditions.

One might expect that chaos will reign for all values of A beyond A_c, but this is actually not true. One instead finds that for $A > A_c$, there are parameter intervals of A interspersed between chaotic regions where the motion is periodic again and not at all chaotic. To find the specific details

---

[127] For a thorough elementary discussion of these aspects of the road to chaos, see Taylor [7].

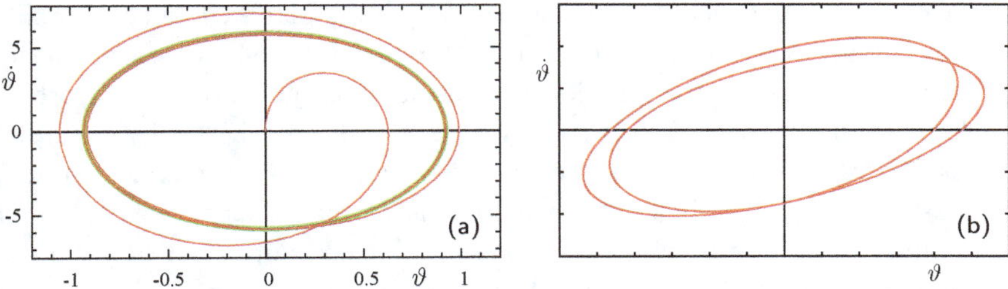

Fig. 8.5 (a) Clockwise phase-space trajectory for the linearized driven pendulum, similar to Eq. (1.108), with initial conditions $\vartheta_0 = 0$ and $\dot{\vartheta}_0 = 0$. The systems is seen to settle in the (shaded) stable elliptical phase-space trajectory once all transients have died out for large times. (b) Schematic picture of a phase-space attractor with period two. (Axes in arbitrary units.)

of such patterns for the present problem requires many, many numerical simulation runs with hours of computation time. However, in view of the inherent unpredictability of trajectories in the chaotic regime in general, much of these details would not be useful anyway beyond the realization that the motion is indeed unpredictably chaotic. But, as we shall see, some of the observed patterns are universal — like, for example, why periodic orbits occur within the chaotic regime — and may be explained in terms of simpler mechanisms that do not necessarily require lengthy detailed numerical simulations.

8.1.3   *Phase-space trajectories and Poincaré maps*

As discussed in Sec. 4.3 in connection with Liouville's Theorem, a point in phase space completely describes the state of a system at a given time. (Phase space, therefore, is also often referred to as *state space*.) For the present problem, phase space is just two-dimensional, comprised of angular displacement ϑ and angular velocity $\dot{\vartheta}$. Therefore, the time evolution of the driven damped pendulum can be depicted easily in two-dimensional $(\dot{\vartheta}, \vartheta)$ plots like the one in Fig. 8.5(a). The figure shows the rather simple example of the linearized forced damped pendulum, similar to Eq. (1.108). The details are not really relevant here, however, the important point is that the phase-space trajectory is seen to settle on the regular elliptical attractor after the initial transient behavior has died out.

Phase-space diagrams, therefore, offer a straightforward and simple way to discover whether motion is periodic or not. After sufficient time has elapsed, periodic orbits will always settle on fixed attractors like the one in Fig. 8.5(a). That fact that this is basically a distorted circle shows that the underlying motion has period one. Figure 8.5(b) shows a schematic picture of an attractor for a period-two motion comprised of two distorted circles that are connected tangentially at one point. Higher-order periodic orbits can also be easily identified in this manner. Chaotic motion, by contrast, will completely fill certain regions of phase space without any discernible repetitive behavior whatsoever. Such chaotic motion is quite impossible to analyze further with phase-space plots of trajectories obtained as continuous functions of time. However, as we shall see, more information can be obtained when considering iterated discrete-time maps.

In principle, phase-space diagrams offer a powerful tool for investigating the dynamics of the driven damped pendulum. This is relatively straightforward for the present problem because phase space is only two-dimensional and the resulting trajectories can be easily plotted and visually inspected for patterns of regularity. However, despite this relative simplicity, one still needs to numerically simulate the orbits for long times to find such patterns. While this may be easier than generating log plots of differences of solutions obtained for small perturbations of initial conditions as discussed in Sec. 8.1.2 on the Lyapunov exponent, it still requires lengthy numerical simulation runs to obtain a comprehensive picture.

Iterated maps

In practical terms, it is oftentimes easier to work with *iterated maps*, called *Poincaré maps*[128] that relate the phase space points at a time t_n to those at time $t_{n+1} = t_n + T$, where T is a characteristic time period. In other words, instead of following the system's phase-space trajectory for continuous times, one only records its state at discrete times. For the present problem, the characteristic time is the period $T = 2\pi/\omega$ of the driving force, and one then obtains the sequence of discrete phase-space mappings for the two-dimensional phase space $(\dot{\vartheta}, \vartheta)$ of the driven pendulum,

$$(\dot{\vartheta}_0, \vartheta_0) \to (\dot{\vartheta}_1, \vartheta_1) \to (\dot{\vartheta}_2, \vartheta_2) \to \cdots \to (\dot{\vartheta}_n, \vartheta_n) \to \cdots , \qquad (8.18)$$

where the index n here indicates discrete time $t_n = t_0 + nT$. Each discrete time step here is called an *orbit*.

---

[128]Named after Henri Poincaré (1854–1912), French mathematician. The maps are also called *return maps* or *successor maps* for obvious reasons. In higher-dimensional phase spaces, one usually considers *Poincaré sections* of two-dimensional cuts through phase space for easy visual interpretation.

Obviously, for a period-one motion, if the initial time t_0 is chosen large enough so that transient behavior is no longer relevant, this map consists of only *one* phase-space point that is revisited again and again after each orbit. And for a period-two motion, one obtains two points and, in general, for period N, one obtains N points.

For trajectories that do not exhibit periodicity, one finds that iterated points move away from each other, never to return to a previous point. As discussed before, such chaotic orbits completely fill regions of phase space when plotted as complete phase-space trajectories with continuous time, which defies easy interpretation. Iterated discrete-time maps, on the other hand, *do* reveal quite interesting patterns. One finds that in the chaotic region, iterations produce point patterns that *cannot* be described by continuous functions, instead they are *fractals*.

Fractals are among the most interesting and fascinating features of chaotic systems. One of the properties of a fractal is that when zooming in on a structure that seems continuous and smooth at a given scale, the magnification reveals more details that seem similar to the original structure. And this happens at *any* level of magnification. *Self-similarity*, therefore, is one of the defining properties of fractals.

Self-similar structures can indeed be found for the driven damped pendulum in Poincaré return maps for the chaotic regime of the driving strength A. The resulting phase-space point sets of fractal patterns can also be classified revealing that certain initial conditions always produce the same fractal patterns called *strange attractors*. In contrast to ordinary attractors which provide precise information about the long-term evolution of the system, a strange attractor only indicates the system will be somewhere on the phase-space points of the fractal attractor set, but because of imprecise information about any given point in view of its finite resolution, we do not know where it will be after the next iteration step. Therefore, rather than knowing where the system will be at any given time, we know where it will *not* be, namely *not* at any phase-space point not belonging to the strange attractor set.

Self-similarities can also be found when plotting the return values of ϑ_n vs. A or of $\dot{\vartheta}_n$ vs. A above the critical value.[129] The patterns produced for the latter example, in particular, look very similar in many respects to the patterns found for the logistic map in the subsequent Sec. 8.1.4, and we will defer the discussion until then.

---

[129] For numerical results, see Taylor [7], for example.

Discretizing differential equations

Even though discrete Poincaré return maps reveal more information about the dynamics in the chaotic regime than continuous phase-space orbits, to extract this information, one still needs to solve the differential equation (8.2) for hundreds of values of A and dozens if not hundreds of initial conditions and evolve the solution for hundreds or thousands of cycles, presenting a formidable numerical task. To reduce the effort, one would need to know the function F defined as

$$(\dot{\vartheta}_{n+1}, \vartheta_{n+1}) = F(\dot{\vartheta}_n, \vartheta_n) \tag{8.19}$$

that takes the system *directly* from time step n to $n + 1$. Clearly, this function must exist, but generally we do not know it. In practice, it is defined only operationally by solving the differential equation and explicitly evolving the system between times t_n and t_{n+1}. One way out is to try to *discretize* the underlying equation of motion by using finite-difference expressions for the derivatives occurring in the differential equations to obtain a simplified model of the problem. Now, to be clear, discretization is *exactly* what any numerical differential-equation solver does by choosing the required discrete time steps in a highly sophisticated adaptive manner, with time steps that are usually much, much smaller than the driving period T. To be much faster than such standard solver algorithms, one would need a discretized model that achieves the mapping (8.19) in *one single step*. However, such single-step discretization does not at all guarantee that the resulting coupled set of finite-difference equations for all phase-space variables provides a model that retains much or any of the dynamics of the original problem. Indeed, in most cases iterations will just produce nonsensical results, in particular, if higher-order derivatives are discretized in a naive manner. In general, therefore, successful discretization oftentimes is a bit of a black art.[130]

We will not dwell on this complicated problem any longer. Instead, we will discuss in the following section a very simple discretization model that reproduces all features of the bifurcation road to chaos found for the driven damped pendulum.

---

[130] For a discussion of the problems associated with discretizion in general and the present forced pendulum in particular, see R. Borrelli and C. Coleman, *Computers, Lies and the Fishing Season, College Math. Jour.* **25**, 401 (1994); J. Hubbard, *The Forced Damped Pendulum: Chaos, Complication and Control,* The American Mathematical Monthly **106**, 741 (1999).

8.1.4 The logistic map

The logistic map does not describe a mechanical system. Rather, it provides a semirealistic idealized model of how exponential population growth of biological systems is limited by prevailing conditions (food shortage, limited space, competing predator species, etc.). The model assumes that the exponential growth of a population $p(t)$,

$$p(t) = P_0 \, e^{gt} \quad \Rightarrow \quad dp(t) = g \, p(t), \tag{8.20}$$

is curtailed by modifying the growth parameter g according to

$$g \quad \to \quad G(t) = g\left(1 - \frac{p(t)}{P}\right). \tag{8.21}$$

This time-dependent modification depends linearly on the existing population $p(t)$ and introduces P as the allowed limit of the population. The differential growth is then determined by the quadratic equation

$$dp(t) = g \, p(t)\left(1 - \frac{p(t)}{P}\right). \tag{8.22}$$

If one now samples[131] this growth at regular times $t_n = t_0 + nT$, where t_0 is some initial time and $n = 0, 1, 2, \ldots$ enumerates the time increments in T, one obtains the discrete difference equation

$$p_{n+1} - p_n = g \, p_n\left(1 - \frac{p_n}{P}\right). \tag{8.23}$$

Defining a modified relative population $x(t)$ by

$$x(t) = \frac{g}{g+1}\frac{p(t)}{P}, \tag{8.24}$$

the corresponding discrete difference equation then reads

$$\boxed{x_{n+1} = rx_n(1 - x_n)}, \tag{8.25}$$

with a modified control parameter

$$r = g + 1 \tag{8.26}$$

limited to $r > 1$ since $g > 0$ for nontrivial results. The x values here are limited to $0 \le x \le 1$ since the iterates of both boundaries produce zero for all times corresponding to the population having become extinct. Overall the control parameter is limited to $1 < r \le 4$ since $f(x) = rx(1 - x)$ has

---

[131]Discrete sampling for this problem corresponds to counting the population at regular time intervals (say, once every year) and ignoring fluctuations during the interval, which is precisely what is being done for many real-world population records.

a maximum larger than 1 for $r > 4$ thus producing iterates outside of the allowed range.

Equation (8.25) is the *logistic map* that describes the time evolution of the system in discrete time steps.[132] It is perhaps one of the most famous examples of all systems exhibiting chaotic behavior. Despite its deceptively simple form, this *quadratic map* exhibits many of the universal features of more complex nonlinear systems in physics, biology, chemistry, and economics. Its properties have been investigated quite thoroughly, reported on in many publications and presented in many textbooks, and quite comprehensive discussions can also be found on the internet.[133] We will not repeat these discussions here.

For the present purpose, we will only consider the period-doubling feature of the logistic equation found as a function of the control parameter r. This feature is similar to the bifurcation cascades of the driven damped pendulum and it will allow us to understand better how regular periodic orbits can occur in the chaotic regime above the critical strength A_c.

Iterations of the logistic map (8.25) as a function of the control parameter r produce the bifurcation diagram shown in Fig. 8.6. Starting from arbitrary values of x and discarding the first few hundred transient iterates, the plot shows which x-values are visited for a given r-value upon iteration. A finite number of x-values then indicates periodicity.[134] For $r < 3$, there clearly is only one period. Period doubling starts at $r = 3$, doubles again at $r = 3.44949$, again at $r = 3.54409$, and so on, until the infinite bifurcation cascade reaches an accumulation point at the critical value $r_c = 3.56995$ where all periodicity is lost and chaos sets in. Qualitatively, this is exactly the same as the road to chaos found for the driven damped pendulum in terms of the bifurcation cascade of Table 8.1. The universal feature of all

---

[132]The continuous version of this equation goes back to P.-F. Verhulst, "Recherches mathématiques sur la loi d'acroissement de la population," Nouv. mém. de l'Academie Royale des Sci. et Belles-Lettres de Bruxelles **18**, 1 (1845); "Deuxième mémoire sur la loi d'acroissement de la population," Mém. de l'Academie Royale des Lettres et des Beaux-Arts de Belgique **20**, 1 (1847).

[133]See, for example, https://mathworld.wolfram.com/LogisticMap.html (accessed on Jan. 3, 2021), where further references can be found as well.

[134]Of course, since in practice the number of iterations is necessarily finite and the resolution of the bifurcation plot is finite as well, this statement needs to be taken with a grain of salt. What is meant is that in the periodic regime of r values, an arbitrarily large number of iterations will never give an indication of non-periodic behavior, whatever the level of magnification. In other words, if 2^n periodicity is found for more than 2^n iterations, this behavior will persist for more iterations and larger magnification. In the chaotic regime, by contrast, increasing the number of iterations will always tend to fill out completely at least certain ranges of x values, whatever the level of magnification.

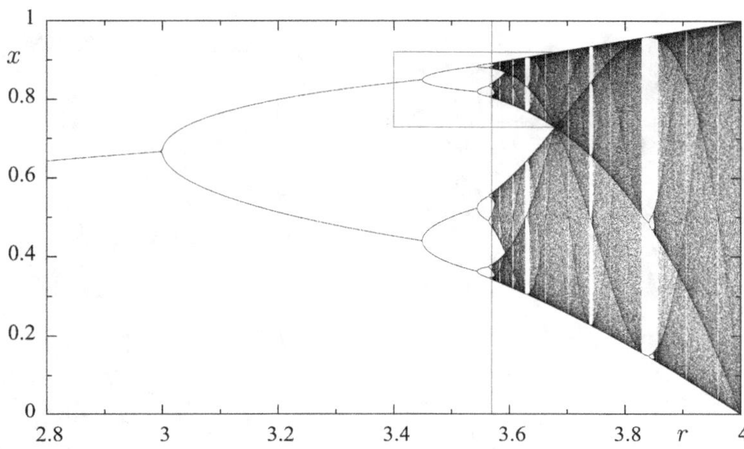

Fig. 8.6 Bifurcation diagram of the iterated logistic map $x_{n+1} = rx_n(1 - x_n)$ obtained with arbitrary starting values for x, discarding initial transient iterates and plotting the values of subsequent iterates of x vs. the control parameter r. The picture inside the rectangle on the top primary branch is a self-similar rescaled version of the entire iterated map. More self-similar copies spawned by every period-doubled branch at all scales can be found throughout the map. The critical value of the control parameter, $r_c \approx 3.56995$, where chaos starts, is indicated by the vertical line; it corresponds to the lowest r value for which self-similar copies of the 'collision' of the two primary branches at around $r \approx 3.68$ can be found. (The primary collision point is located at the right bottom corner of the inset rectangle.)

bifurcation cascades, the Feigenbaum δ of Eq. (8.12), is also found to be true here for the logistic map.

One of the most striking features of the bifurcation diagram 8.6 is that *self-similar* copies of the entire diagram occur at various values of $r > r_c$ throughout the diagram at smaller scales, thus revealing the *fractal nature* of the return map depicted in Fig. 8.6. Even at the necessarily limited resolution of this plot, one can see regular periodic orbits interspersed in the chaotic regime occur for all r values where fractal copies of the primary period-doubling regime ($3 < r < r_c$) start. Most of the corresponding intervals are very narrow and not even visible in the diagram (because of its necessarily finite size), but the self-similarity at all scales implies that there are infinitely many bands of r-value intervals that correspond to periodic orbits for $r > r_c$. Enlarged versions of two of the most prominent bands for period-six and period-three orbits are depicted in Fig. 8.7. As seen there, the period-six orbit actually results from *two* self-similar copies of the primary period-three orbit. The latter does not have the 2^n periodicity

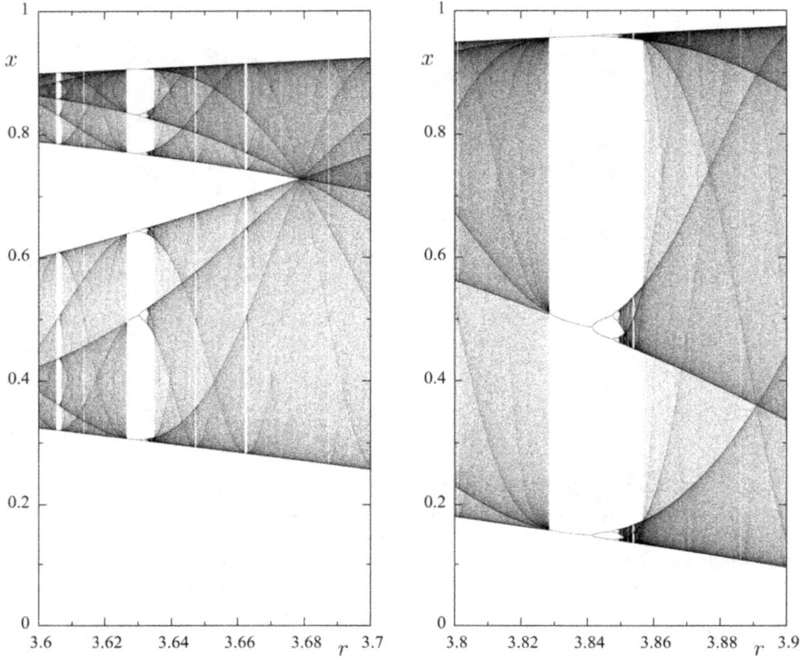

Fig. 8.7   Enlarged intervals in the chaotic regime with examples of iterates of period six (left figure, around $r \approx 3.63$) and period three (right figure, around $r \approx 3.83$). The period-six behavior on the left results from a self-similar double mapping of the period-three interval on the right to the first two primary period-doubled branches.

of the initial bifurcation cascade because it arises from the 'collision' of the two branches of the primary period-doubling cascade, as can be seen from Fig. 8.6.

It should be emphasized that even though the orbits of the driven damped pendulum do not correspond exactly to the quadratic map of the logistic equation, the inherent bifurcation features of this physical system are so dominant that the universality of this behavior — and indeed of *all* systems exhibiting period doubling — can indeed be understood by the findings for the logistic map. This explains, in particular, that the chaotic regime of the driven damped pendulum is governed by the fractal nature of strange attractors and the associated self-similarity of chaotic orbital dynamics.

8.1.5 *Nonlinear systems are normal*

Physical systems treated in classical mechanics courses usually are linear, however, most real-world systems are nonlinear or at least become nonlinear under certain conditions. A case in point is the pendulum discussed here for angular displacements so large that the linearization of the $\sin \vartheta$ term is no longer appropriate. Because it is not really unusual to consider initial conditions for this case with angular displacements close to the top, this exemplifies that nonlinearity is not an unusual feature of physical systems, but rather an integral part of the normal behavior spectrum of most physical systems. The fact that we are able to learn so much by studying linear systems is because physical systems if left to themselves will naturally go to their equilibrium configuration and thus the dominant first-order effects that perturb the system out of equilibrium are indeed linear, as discussed in Chapter 6. — At least that is the case if the equilibrium is sufficiently far away from the chaotic regime. However, for a physical system governed by period-doubling dynamics that lives, for example, very close to the left edge of the period-three band at $r \approx 3.83$ in Fig. 8.7, with corresponding regular trajectories, a perturbation considered small under normal circumstances might very well take it into the chaotic regime, with all its ensuing unpredictability.

This simple example shows that to understand the full complexity of the dynamics of classical mechanical systems, it is important to include the study of nonlinear systems. Needless to say, the cursory nature of the discussion in this chapter cannot do justice to the full complexity of the problem. It was merely meant as an introduction and encouragement for further studies of this important topic.

8.2 Special Relativity

In the 1860s, James Clerk Maxwell (1831–1879) and others put forward a unified theory of electric and magnetic phenomena — now called *classical electrodynamics* — that was shortly thereafter, in 1887, corroborated to conform with experimental reality by Heinrich Hertz (1857–1894). In the same year, 1887, the Michelson–Morley experiment established that the idea of a "luminiferous ether" that supposedly permeated the vacuum and acted as a carrier medium for the electromagnetic light waves was not supported by experimental reality.

From the point of view of mechanics as we have explored it so far, the most notable implication of Maxwell's theory and the subsequent experimental findings is the fact that light (in the vacuum of free space) travels with the *same* constant speed c in *any* inertial frame of reference[135] irrespective of the frames' relative uniform velocities. This is in profound contradiction to Newtonian mechanics and the underlying Galilei transformations (see Sec. 1.6).

In 1905, his *annus mirabilis*, Albert Einstein (1879–1955) resolved this contradiction in his paper on the electrodynamics of moving bodies[136] by realizing that this finding required a re-examination of the classical notion of simultaneity, thus leading to the notion of *spacetime*[137] and showing in the resulting *theory of special relativity* that the equations of classical mechanics are only valid in the limit of velocities that are small compared to the speed of light.

8.2.1 *The Einstein Principle of Relativity*

Let us consider a light ray traveling with speed c that is emitted from a point $\mathbf{r}_1 = (x_1, y_1, z_1)$ at time t_1 in an inertial frame S. If it is detected at some later time t_2 at the point $\mathbf{r}_2 = (x_2, y_2, z_2)$, we obviously have

$$c^2(t_2 - t_1)^2 = (x_2 - x_1)^2 + (y_2 - y_1)^2 + (z_2 - z_1)^2 \qquad (8.27)$$

which measures the squared relative (spatial) distance between the two *events* $(t_1, \mathbf{r}_1)$ and $(t_2, \mathbf{r}_2)$ in that frame (where we use the event notation introduced in Sec. 1.6.1). If we observe the *same* process from another inertial frame S', where the corresponding events are $(t_1', \mathbf{r}_1') = (t_1', x_1', y_1', z_1')$ and $(t_2', \mathbf{r}_2') = (t_2', x_2', y_2', z_2')$, we have

$$c^2(t_2' - t_1')^2 = (x_2' - x_1')^2 + (y_2' - y_1')^2 + (z_2' - z_1')^2, \qquad (8.28)$$

where it is found experimentally that light travels with the *same* speed c. This finding *cannot* be reconciled with the Galilei transformation of classical mechanics since they entail that the relative velocity of an observer needs to be added vectorially to the velocities of the particles of the observed system.

---

[135] Nowadays, the speed of light in vacuum is defined, without lack of generality, by the *exact* value $c = 299{,}792{,}458\,\mathrm{m/s}$; see Sec. 1.1.

[136] Albert Einstein, *"Zur Elektrodynamik bewegter Körper"*, Annalen der Physik **17**, 891 (1905).

[137] The basic notion of spacetime was introduced by Hermann Minkowski (1864–1909) who showed that Einstein's 1905 findings could be expressed very succinctly in a complex four-dimensional Euclidean space with an imaginary time component.

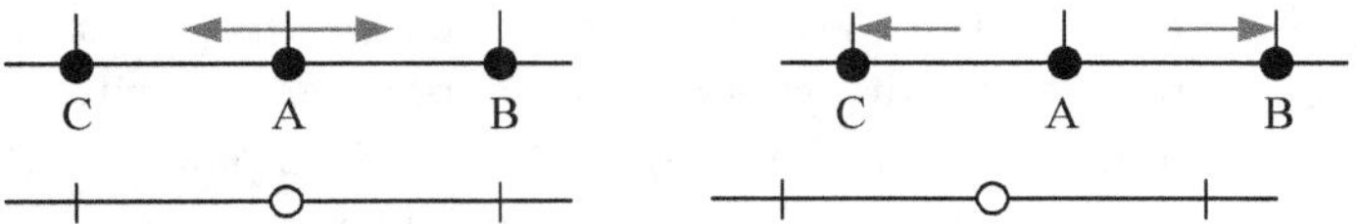

Fig. 8.8   Consider two inertial systems that move with respect to each other shown here as two parallel lines along the direction of relative motion. Shown on the left, an observer at A in the moving frame *simultaneously* sends two light beams to points B and C in his frame. Since they are equidistant to his position, from his point of view they will reach points B and C at the same time, as shown in the diagram on the right. An observer in the frame at rest, at the position indicated by the open circle, who was at the same point A when the signals were sent out, however, will *not* record them as reaching B and C simultaneously. Since the speed of light is the same in her frame, the observer at rest will record the signal to reach C before it reaches B since C is moving towards her and B is moving away, making the respective distances the light needs to travel different in her frame. The two events, occurring simultaneously in one frame ('his'), clearly are not simultaneous in the other ('hers').

Einstein elevated this experimental finding to the status of an axiomatic principle that replaces Galilei's principle (1.206):[138]

Relativity Principle (Einstein):

(1)   All inertial frames are equivalent.

(2)   The speed of light is the same in all inertial frames.

$$(8.29)$$

This has profound consequences for our understanding of the nature of time. Whereas in classical Newtonian mechanics time was an absolute that was the same everywhere and in all frames (apart from trivial time displacements), this experiment finds that this is not true. As is explained in Fig. 8.8, we find now that things that occur simultaneously in one frame, need not be simultaneous in other frames. Clearly, to accommodate Einstein's Relativity Principle, we need to find a new set of transformations between inertial frames that involve time in a non-trivial manner. They thus must be markedly different from Galilei transformations, yet at the same time, we expect that the Galilei transformations are recovered in the limit of small velocities since we know — again experimentally — that classical mechanics in all its ramifications provides an extremely good description of reality in this limit.

All descriptions of nature that adhere to Einstein's relativity principle are called 'relativistic'. Those that do not are called 'nonrelativistic', 'clas-

---

[138]Note that we could replace item (2) here equivalently with the statement "Maxwell's equations are valid in all inertial frames" which would make it conform with the wording of Galilei's Principle of Relativity (1.206).

sical', or 'Newtonian', etc. Clearly, *nature is relativistic.* In other words, the classical approach constitutes *always* an approximation with a limited range of applicability.[139] The term *special relativity* applies to the relativistic transformations between *inertial* frames. The relationships between arbitrary frames are the subject of *general relativity.*

8.2.1.1 *Invariant spacetime intervals*

To better understand the ramifications of Einstein's principle, let us introduce a squared 'distance' between two events as

$$s_{21}^2 = c^2(t_2 - t_1)^2 - (x_2 - x_1)^2 - (y_2 - y_1)^2 - (z_2 - z_1)^2, \qquad (8.30)$$

also referred to as the (squared) *spacetime interval* between two events. The experimental finding discussed above means that for events connected by a light ray the value

$$s_{21}^2 = 0 \qquad (8.31)$$

is an *invariant* that is valid in all frames. In fact, one finds that s_{12}^2 is an invariant whatever its value. To see this, consider the infinitesimal squared interval

$$ds^2 = c^2 dt^2 - dx^2 - dy^2 - dz^2 \qquad (8.32)$$

between two events that are separated by infinitesimals $(c\,dt, dx, dy, dz)$. The same interval viewed from different inertial systems must be related by

$$ds' = \lambda\, ds \qquad (8.33)$$

since they must be infinitesimals of the same order. The factor λ can only depend on the absolute value of the relative velocity between the two systems; it cannot depend on the direction of the relative velocity or on the space or time coordinates since this would violate the homogeneity of space and time.[140] By the same token, since there is nothing special about any one of the two systems, we must also have

$$ds = \lambda\, ds' \qquad (8.34)$$

---

[139]This means that if there is experimental evidence showing that a prediction based on classical mechanics is not correct, one of the first questions one must ask is whether the difference could be due to relativistic effects.

[140]This means that, by itself, there is nothing special about any particular point or any particular direction in space or time.

and thus

$$\lambda^2 = 1. \tag{8.35}$$

The only permitted values for λ then are the constants $\lambda = +1$ or $\lambda = -1$. The latter, however, is ruled out since the identity transformation, $ds' = ds$, clearly requires $\lambda = +1$. Therefore, since any transformation can be obtained continuously from the identity, this means that we must have $ds' = ds$ in all frames. We thus find that the infinitesimal distance (8.32) is an invariant which directly implies that the same is true as well for any finite distances s_{12}^2, as in (8.30), irrespective of what their values are. The actual values of s_{21}^2 can be positive, zero, or negative and we use the following notation for these values:

$$\boxed{\begin{aligned} s_{21}^2 &> 0: \quad \text{timelike,} \\ s_{21}^2 &= 0: \quad \text{lightlike,} \\ s_{21}^2 &< 0: \quad \text{spacelike.} \end{aligned}} \tag{8.36}$$

We therefore speak of two events as being separated by a timelike, a lightlike, or a spacelike distance. The experiments described above involving light signals between two events thus correspond to a lightlike separation (8.31).

An immediate consequence of the invariance of s_{21}^2 is that for two events to occur simultaneously, one must have $t_2 - t_1 = 0$ and therefore $s_{21}^2 < 0$. In other words, only for events with spacelike separation is it possible to find a reference frame where the events occur simultaneously. Similarly, only events with timelike separation can be transformed to a frame where they occur at the same point in space, i.e., $|\mathbf{r}_2 - \mathbf{r}_1| = 0$. Because of the invariance, the separation between spacelike and timelike intervals is absolute. The lightlike condition $c^2 t^2 - x^2 - y^2 - z^2 = 0$ defines a three-dimensional hypersurface — the *light cone* — in a four-dimensional coordinate system (ct, x, y, z) that separates time- and spacelike regions of this four-dimensional space (see Fig. 8.9).[141] A causal connection between two events, i.e., a connection in which the concepts of 'earlier' and 'later' are absolutes, is only possible for events within the timelike part of the light cone.

---

[141] In other words, the light cone associated with an event is the set of all events connected by light rays through that particular event.

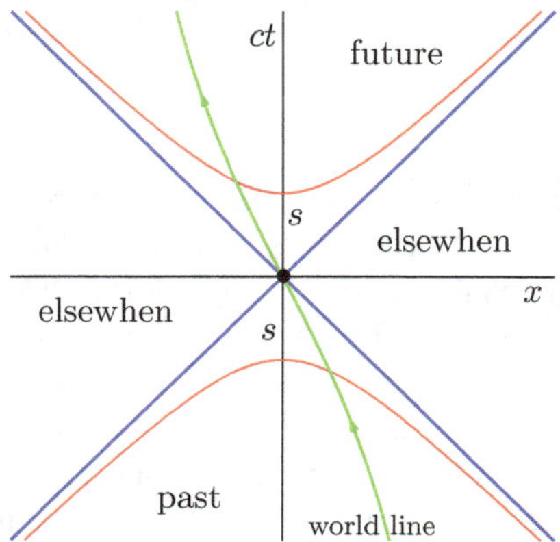

Fig. 8.9   Two-dimensional representation of spacetime, with time on the vertical and space on the horizontal axis (with x representing $\sqrt{\mathbf{r}^2}$), as it is seen from an event at the origin marked by the solid dot. The *light cone*, the three-dimensional locus of all points separated by a squared distance $c^2t^2 - \mathbf{r}^2 = 0$ from the origin, is indicated here schematically by the two slanted lines $ct = \pm x$ that intersect at the origin. It corresponds to all events that can be connected to the event at the origin by light signals. The event at the origin of the cone can only be causally influenced by events in the lower half of the *timelike* part of the cone (the 'past') and it itself can causally influence only events in the upper half (the 'future'). Events in the *spacelike* regions of space, $c^2t^2 - \mathbf{r}^2 < 0$, outside of the cone, are 'elsewhen', i.e, they do not have a definite causal connection since in these regions the concepts of 'simultaneous', 'earlier', and 'later' are frame-dependent. Physical particles move from the past to the future along timelike *world lines* x^μ that are associated with speeds always smaller than (or, in the case of massless particles, at most equal to) the speed of light c. The locus of all points that are a distance s from the origin are hyperboloids, $c^2t^2 - \mathbf{r}^2 = s^2$, centered around the time axis as indicated schematically by the two hyperbolas, with s being the distance of the origin from where the hyperboloids intersect with the time axis.

8.2.2   *Four-dimensional spacetime*

In the three-dimensional Euclidean space of nonrelativistic physics the invariant distance between two points is obtained by the sum of the squares of the component differences, i.e., $\Delta s^2 = \Delta x^2 + \Delta y^2 + \Delta z^2$. As a consequence, the Galilei transformations basically involve only the three space components and time is a mere parameter that could even be chosen the same in all frames without any problem. In relativity, on the other hand, we find that the squared invariant interval $ds^2 = c^2 dt^2 - dx^2 - dy^2 - dz^2$, in addition to the three space components, necessarily includes time in a non-trivial manner as part of the coordinates.

To accommodate this, we must now consider a space of four-component vectors, with time being added to the usual three space components, i.e.,

$$x^\mu = (ct, \mathbf{r}) \quad \longrightarrow \quad \begin{cases} x^0 = ct, \\ x^1 = x, \\ x^2 = y, \\ x^3 = z. \end{cases} \qquad (8.37)$$

The indices run through $\mu = (0, 1, 2, 3)$, where 0 denotes the *time component* and 1, 2, 3 enumerate the *space components*. Generally, a lower-case Greek letter μ, ν, σ, etc., will be used to indicate any of these four indices; lower-case Roman letters i, j, k, etc., will indicate a limitation to the three spatial components 1, 2, or 3. The fact that the index μ here is raised is significant and will be explained in the next section. Also note that in (8.37) we use the notation x^μ to refer to both the entire set of components — i.e., to the entire four-vector represented by components (ct, x, y, z) — and to the individual components. This *pars pro toto* notation is common albeit a bit sloppy, but it usually is clear from the context what is meant.

The *metric tensor* $G = \{g_{\mu\nu}\}$ of this space is fixed, by the definition of a metric,[142] by the condition that we must be able to write the invariant infinitesimal interval element as

$$\begin{aligned} ds^2 &= g_{\mu\nu} dx^\mu dx^\nu \\ &= dx^0 dx^0 - dx^1 dx^1 - dx^2 dx^2 - dx^3 dx^3, \end{aligned} \qquad (8.38)$$

where the $dx^\mu = (c\,dt, dx, dy, dz)$ are the infinitesimals of x^μ. We use here Einstein's summation convention where a summation over repeatedly appearing (dummy) indices is implied; here, the implied summation is over $\mu, \nu = 0, 1, 2, 3$. In view of the minus signs in front of the space-component contributions of ds^2, the space of four-vectors x^μ is not a Euclidean space.[143] It is the *four-dimensional Minkowski space* of what is called *relativistic space-time* whose metric tensor G is readily read off (8.38), i.e.,

$$G = \{g_{\mu\nu}\} = \begin{pmatrix} 1 & 0 & 0 & 0 \\ 0 & -1 & 0 & 0 \\ 0 & 0 & -1 & 0 \\ 0 & 0 & 0 & -1 \end{pmatrix}. \qquad (8.39)$$

---

[142]Loosely speaking, a *metric space* is one where one can define a distance between any two elements of the space, and the *metric tensor* (often simply called 'the metric') specifies how this distance is to be calculated. For a Euclidean space, for example, the usual metric tensor is the unit tensor.

[143]We mention without further explanation that it is possible to introduce a complex Euclidean time $x^4 = ict$ and, instead of (8.37), work with four-vectors (x^1, x^2, x^3, x^4) on a space with a Euclidean-type identity metric.

This matrix is its own inverse, $G^{-1} = G$, but we write $G^{-1} = \{g^{\mu\nu}\}$. In other words, the elements of the inverse metric tensor G^{-1}, by definition, are written as $g^{\mu\nu}$, with *raised* indices, just as the elements of the metric tensor G, by definition, are written as $g_{\mu\nu}$, with *lowered* indices.[144] Thus, for the non-zero diagonal elements, $g^{\mu\mu} = 1/g_{\mu\mu}$. We write

$$G^{-1}G = \{g^{\mu\sigma}g_{\sigma\nu}\} = \begin{pmatrix} 1 & 0 & 0 & 0 \\ 0 & 1 & 0 & 0 \\ 0 & 0 & 1 & 0 \\ 0 & 0 & 0 & 1 \end{pmatrix} = \mathbb{1} = \{\delta^{\mu}{}_{\nu}\}, \qquad (8.40)$$

where the Kronecker delta, $\delta^{\mu}{}_{\nu}$, denotes the elements of the four-dimensional unit matrix $\mathbb{1}$.

As we shall see, we can now exploit the invariance condition (8.38) to actually determine the correct transformations that replace the Galilei transformations in a relativistic context. Before we do so, let us first get a few notational issues out of the way.

8.2.2.1   *Contra- and covariant vector components*

The elements of this four-dimensional space are called *four-vectors* usually denoted by lower-case Roman letters a, b, x, p, etc. We introduce sets of *covariant* and *contravariant* basis vectors[145]

$$\{\hat{e}_0, \hat{e}_1, \hat{e}_2, \hat{e}_3\} \quad \text{and} \quad \{\hat{e}^0, \hat{e}^1, \hat{e}^2, \hat{e}^3\} \qquad (8.41)$$

that are reciprocal to each other,[146] with

$$\hat{e}^{\mu} \cdot \hat{e}_{\nu} = \delta^{\mu}{}_{\nu}, \qquad (8.42)$$

and they provide the metric tensor,

$$\hat{e}_{\mu} \cdot \hat{e}_{\nu} = g_{\mu\nu}, \qquad (8.43)$$

and its inverse,

$$\hat{e}^{\mu} \cdot \hat{e}^{\nu} = g^{\mu\nu}. \qquad (8.44)$$

The last three relations define the *scalar product* denoted by the dot "$\cdot$" (and thus also called the 'dot product'). An (abstract) four vector a can now be expanded according to

$$a = a^{\mu}\hat{e}_{\mu}, \quad \text{or equivalently} \quad a = a_{\nu}\hat{e}^{\nu}, \qquad (8.45)$$

---

[144]Note that since (*for the Minkowski space*) their elements are identical, one usually does not distinguish between the metric tensor G and its inverse G^{-1}, and simply refers to both $g_{\mu\nu}$ and $g^{\mu\nu}$ as 'the metric tensor'.

[145]The indices on the basis vectors number the vectors, *not* the components.

[146]One also says that the $\{\hat{e}^{\mu}, \mu = 0, 1, 2, 3\}$ are the *duals* of $\{\hat{e}_{\nu}, \nu = 0, 1, 2, 3\}$.

with unique expansion coefficients

$$a^\mu = \hat{e}^\mu \cdot a \quad \text{and} \quad a_\nu = \hat{e}_\nu \cdot a, \tag{8.46}$$

which are the projections of a along the respective unit vectors. Expansion coefficients a^μ and a_ν, respectively, are referred to as *contravariant* and *covariant* components of the four-vector a. Furthermore, we immediately find

$$a = a^\mu \hat{e}_\mu = a \cdot \hat{e}^\mu \hat{e}_\mu \quad \text{and} \quad a = a_\nu \hat{e}^\nu = a \cdot \hat{e}_\nu \hat{e}^\nu, \tag{8.47}$$

which means that the identity operation can be written in dyadic form,

$$\mathbf{1} = \hat{e}_\mu \hat{e}^\mu = \hat{e}^\nu \hat{e}_\nu. \tag{8.48}$$

One has

$$\hat{e}^\beta \cdot \mathbf{1} \cdot \hat{e}_\alpha = \delta^\beta_{\ \alpha}, \tag{8.49}$$

i.e., the matrix representation of the identity is the four-dimensional unit matrix, $\mathbb{1}$.

Because of

$$a_\mu = \hat{e}_\mu \cdot a = \hat{e}_\mu \cdot \hat{e}_\nu a^\nu \quad \text{and} \quad a^\mu = \hat{e}^\mu \cdot a = \hat{e}^\mu \cdot \hat{e}^\nu a_\nu, \tag{8.50}$$

we have

$$\boxed{a_\mu = g_{\mu\nu} a^\nu \quad \text{and} \quad a^\mu = g^{\mu\nu} a_\nu}, \tag{8.51}$$

i.e., the metric tensor tells us how to go back and forth between co- and contravariant components. One says that the indices are lowered or raised by the metric tensor. The relation between co- and contravariant components thus is

$$a_0 = a^0 \quad \text{and} \quad a_i = -a^i \quad \text{for} \quad i = 1, 2, 3. \tag{8.52}$$

As already alluded to above, in a somewhat sloppy but obvious notation, one writes[147]

$$a \to a^\mu = (a^0, a^1, a^2, a^3) \quad \text{and} \quad a \to a_\mu = (a_0, a_1, a_2, a_3) \tag{8.53}$$

if one needs to distinguish between the contra- and covariant representations of the abstract four-vector a. One may also write

$$a^\mu = (a^0, \mathbf{a}) \quad \text{and} \quad a_\mu = (a_0, -\mathbf{a}), \tag{8.54}$$

---

[147]This notation corresponds to the three-dimensional representation $\vec{a}$ of an abstract vector $\mathbf{a}$ discussed in Sec. 1.1.1.

where

$$\mathbf{a} = \begin{pmatrix} a^1 \\ a^2 \\ a^3 \end{pmatrix} \tag{8.55}$$

is an ordinary three-dimensional vector of $\mathbb{R}^3$ in Cartesian coordinates.[148]

The scalar product of two arbitrary four-vectors $a = (a^0, \mathbf{a})$ and $b = (b^0, \mathbf{b})$ is then found as

$$\boxed{a \cdot b = g_{\mu\nu} a^\mu b^\nu = a_\nu b^\nu = a^\mu b_\mu = a^0 b^0 - \mathbf{a} \cdot \mathbf{b}}. \tag{8.56}$$

The *length* of a four-vector a is given by the scalar product with itself,[149]

$$a^2 = a_\mu a^\mu = a_0^2 - \mathbf{a}^2. \tag{8.57}$$

Depending on whether $a^2 > 0$, $a^2 < 0$, or $a^2 = 0$, the vector is called *timelike, spacelike*, or a *null* vector, respectively. A null vector is also called a lightlike vector.

The length of a vector, a^2, is an invariant, of course, since it expresses an invariant distance between a and the origin in a particular frame. More generally, we will see below, in (8.75), that the scalar product $a \cdot b$ of any two four-vectors is an invariant.

(Finally, parenthetically, we mention that the formalism of co- and contravariant basis vectors is very similar to what one finds in Euclidean spaces with skewed — i.e., non-orthogonal — bases. For the special case of an orthogonal Euclidean basis, the metric tensor is the unit matrix and the co- and contravariant representations are the same; so there is no need to distinguish between them.)

8.2.3 *Lorentz transformations defined*

According to *Einstein's Relativity Principle*, physical laws must have the same form in all inertial frames — they must be *form invariant*[150] — and they must leave the distance between any two points invariant. For this to

---

[148] In the text, for simplicity, the notation $\mathbf{a} = (a^1, a^2, a^3)$ is usually employed throughout. Whether this designates a column or a row vector representation is usually obvious from the context.

[149] There is an unfortunate notational ambiguity in that a^2 may be read as the y-components of $\mathbf{a}$ and as the square of the four vector a, but the meaning should be clear from the context.

[150] One also says, equivalently, that physical laws must by *covariant*. This is *not* to be confused with the covariant components of a four-vector. There is no mathematical relationship between these two notions of covariance.

be true, the physically relevant transformations that relate different inertial frames to each other must meet certain requirements.

The simplest type of motion is force-free motion whose trajectories, in any inertial system, are straight lines. In parametric form they can be written as $x^\nu = A^\nu + r\,B^\nu$, where r is some (real) parameter and A (for $r = 0$) and $A + B$ (for $r = 1$) are two four-vectors on the line. The most general transformation of a four-vector x^ν that preserves this parametric representation is a *linear* mapping of the form

$$\boxed{x'^\mu = \Lambda^\mu{}_\nu x^\nu + a^\mu}\,, \tag{8.58}$$

with a 4×4 transformation matrix $\{\Lambda^\mu{}_\nu\}$ and a constant translation vector a^μ. The corresponding inverse mapping is given by

$$x^\mu = \left(\Lambda^{-1}\right)^\mu{}_\nu x'^\nu - a'^\mu, \tag{8.59}$$

where

$$a'^\mu = \left(\Lambda^{-1}\right)^\mu{}_\nu a^\nu. \tag{8.60}$$

The assumption here is, of course, that the inverse Λ^{-1} exists. This is obvious on physical grounds, but will be shown explicitly in Eq. (8.70) below.

In the four-dimensional Minkowski space, the infinitesimal line element ds describing the distance between two points x^μ and $x^\mu + dx^\mu$, separated by an infinitesimal increment $dx^\mu = (c\,dt, d\mathbf{r})$, is given by Eq. (8.38). As we have seen, Einstein's relativity postulate implies that the physically relevant transformations that relate different *inertial* spacetime coordinate systems must leave this scalar product invariant, i.e.,

$$\boxed{x^\mu \longmapsto x'^\mu: \qquad dx'_\mu dx'^\mu = dx_\mu dx^\mu = ds^2 \qquad \text{invariant}}\,. \tag{8.61}$$

Linear transformations of the form (8.58) that satisfy this postulate are the *Lorentz transformations*.

We can now exploit the postulate (8.61) to find the transformation matrix $\Lambda^\mu{}_\nu$. With

$$dx'^\mu = \Lambda^\mu{}_\nu dx^\nu, \tag{8.62}$$

we have

$$ds^2 = g_{\rho\mu} dx'^\rho dx'^\mu = g_{\rho\mu} \Lambda^\rho{}_\sigma \Lambda^\mu{}_\nu dx^\sigma dx^\nu = g_{\sigma\nu} dx^\sigma dx^\nu, \tag{8.63}$$

and hence, in view of the independence of the dx^μ,

$$\boxed{g_{\rho\mu} \Lambda^\rho{}_\sigma \Lambda^\mu{}_\nu = g_{\sigma\nu}, \qquad \text{or} \qquad \Lambda^{\mathrm{T}} G \Lambda = G}\,, \tag{8.64}$$

where the transformation matrix Λ is defined by

$$\Lambda = \{\Lambda^\mu{}_\nu\} = \left(\begin{array}{c|c} \Lambda^0{}_0 & \Lambda^0{}_j \\ \hline \Lambda^i{}_0 & \Lambda^i{}_j \end{array}\right). \tag{8.65}$$

The matrix Λ is written in blocks here to better exhibit the relation between temporal and spatial components.

Equations (8.58) and (8.64) define the *inhomogeneous Lorentz transformation*, also called a *Poincaré transformation*, characterized by (Λ, a). If there are no translations, i.e., $a^\mu = 0$, one obtains a *homogeneous Lorentz transformation* $(\Lambda, 0)$, oftentimes simply called a *Lorentz transformation*. The pure *translations* are given by $(\mathbb{1}, a)$. Note that by trivially extending every vector by a fifth component of value 1, one can write the Poincaré transformations in terms of 5×5 matrices,

$$\begin{pmatrix} x'^\mu \\ 1 \end{pmatrix} = \begin{pmatrix} \Lambda^\mu{}_\nu & a^\mu \\ 0 & 1 \end{pmatrix} \begin{pmatrix} x^\nu \\ 1 \end{pmatrix} = \begin{pmatrix} g^\mu{}_\rho & a^\mu \\ 0 & 1 \end{pmatrix} \begin{pmatrix} \Lambda^\rho{}_\nu & 0 \\ 0 & 1 \end{pmatrix} \begin{pmatrix} x^\nu \\ 1 \end{pmatrix}. \tag{8.66}$$

The last step shows that this may always be broken down into a pure Lorentz transformation $(\Lambda, 0)$ followed by a pure translation $(\mathbb{1}, a)$, i.e., $(\Lambda, a) = (\mathbb{1}, a)(\Lambda, 0)$. The study of the full Poincaré transformation can thus be separated into studying the homogeneous Lorentz transformations (henceforth simply called 'Lorentz transformations') and the translations independently.

The defining equation (8.64), in particular, means that the metric tensor is invariant under Lorentz transformations, i.e., it is always given by (8.39) in any frame. Indeed, Eq. (8.61) is completely equivalent to requiring that the transformations between inertial frames be such that the spacetime metric does not change.

Mathematically, the parameters $(\Lambda^\mu{}_\nu,\ a^\mu)$ of the Lorentz transformation may be complex. Physical Lorentz transformations, however, relate one physical (real) four-vector x to another physical (real) four-vector x' and thus have only real parameters. For our purposes, therefore, we only need real Lorentz transformations.[151]

The Lorentz transformations, $(\Lambda, 0)$, form a group, called $O(1, 3)$, where 1 and 3 stand for the number of positive and negative entries, respectively, in the diagonal metric (8.39). If one includes translations in space and time,

---

[151]In other words, we consider only Lorentz transformations on $\mathbb{R}^{1+3}$, where $1+3$ stands for the one time and three space components. However, complex Lorentz transformations need to be considered in the proof of the *PCT* Theorem.

(Λ, a), one obtains the Poincaré group $P(1,3)$. Below, we shall verify the group properties for $O(1,3)$.

Finally, we emphasize that the Lorentz transformation is defined for contravariant vectors in terms of elements $\Lambda^\mu{}_\nu$, with a contravariant row index μ and a covariant column index ν. Other index combinations, $\Lambda^{\mu\nu}$, $\Lambda_{\mu\nu}$ and $\Lambda_\mu{}^\nu$, are obtained by raising or lowering indices with the help of the metric tensor, i.e.,

$$\{\Lambda^{\mu\nu}\} = \{\Lambda^\mu{}_\rho g^{\rho\nu}\} = \left(\begin{array}{c|c} \Lambda^0{}_0 & -\Lambda^0{}_j \\ \hline \Lambda^i{}_0 & -\Lambda^i{}_j \end{array} \right), \tag{8.67a}$$

$$\{\Lambda_{\mu\nu}\} = \{g_{\mu\rho}\Lambda^\rho{}_\nu\} = \left(\begin{array}{c|c} \Lambda^0{}_0 & \Lambda^0{}_j \\ \hline -\Lambda^i{}_0 & -\Lambda^i{}_j \end{array} \right), \tag{8.67b}$$

$$\{\Lambda_\mu{}^\nu\} = \{g_{\mu\sigma}\Lambda^\sigma{}_\rho g^{\rho\nu}\} = \left(\begin{array}{c|c} \Lambda^0{}_0 & -\Lambda^0{}_j \\ \hline -\Lambda^i{}_0 & \Lambda^i{}_j \end{array} \right), \tag{8.67c}$$

where $\mu = (0, i = 1, 2, 3)$ and $\nu = (0, j = 1, 2, 3)$. The Lorentz transformation for covariant components is then immediately found as

$$\begin{aligned} x'_\mu = g_{\mu\nu}x'^\nu &= g_{\mu\nu}\left(\Lambda^\nu{}_\sigma x^\sigma + a^\nu\right) \\ &= g_{\mu\nu}\Lambda^\nu{}_\sigma g^{\sigma\rho}x_\sigma + a_\mu \\ &= \Lambda_\mu{}^\rho x_\rho + a_\mu, \end{aligned} \tag{8.68}$$

with $\Lambda_\mu{}^\rho$ given by Eq. (8.67c).

To show that the inverse Λ^{-1} exists, let us raise the index σ in (8.64), i.e.,

$$g^{\kappa\sigma}\Lambda_{\mu\sigma}\Lambda^\mu{}_\nu = \Lambda_\mu{}^\kappa\Lambda^\mu{}_\nu = g^{\kappa\sigma}g_{\sigma\nu} = g^\kappa{}_\nu = \delta^\kappa{}_\nu, \tag{8.69}$$

which provides the inverse Lorentz transformation,

$$(\Lambda^{-1})^\kappa{}_\mu = \Lambda_\mu{}^\kappa. \tag{8.70}$$

By construction, therefore, the inverse exists.[152] With

$$\Lambda = \left(\begin{array}{c|ccc} \Lambda^0{}_0 & \Lambda^0{}_1 & \Lambda^0{}_2 & \Lambda^0{}_3 \\ \hline \Lambda^1{}_0 & \Lambda^1{}_1 & \Lambda^1{}_2 & \Lambda^1{}_3 \\ \Lambda^2{}_0 & \Lambda^2{}_1 & \Lambda^2{}_2 & \Lambda^2{}_3 \\ \Lambda^3{}_0 & \Lambda^3{}_1 & \Lambda^3{}_2 & \Lambda^3{}_3 \end{array}\right), \tag{8.71}$$

its explicit form is given by

$$\Lambda^{-1} = \left(\begin{array}{c|ccc} \Lambda^0{}_0 & -\Lambda^1{}_0 & -\Lambda^2{}_0 & -\Lambda^3{}_0 \\ \hline -\Lambda^0{}_1 & \Lambda^1{}_1 & \Lambda^2{}_1 & \Lambda^3{}_1 \\ -\Lambda^0{}_2 & \Lambda^1{}_2 & \Lambda^2{}_2 & \Lambda^3{}_2 \\ -\Lambda^0{}_3 & \Lambda^1{}_3 & \Lambda^2{}_3 & \Lambda^3{}_3 \end{array}\right), \tag{8.72}$$

i.e., the inversion corresponds to transposing the entire matrix and changing the signs of the off-diagonal time-space and space-time elements. We can write

$$\Lambda^{-1} = P\Lambda^{\mathrm{T}}P, \tag{8.73}$$

where P is the space-inversion operator defined in Eq. (8.89) below (which by itself is also a Lorentz transformation; see footnote 153).

8.2.3.1 *Lorentz scalars, etc.*

It should be obvious that not just (8.61), but any scalar product $a \cdot b$ is invariant under a homogeneous Lorentz transformation. Indeed, with

$$a'^{\mu} = \Lambda^{\mu}{}_{\nu} a^{\nu} \qquad \text{and} \qquad b'^{\mu} = \Lambda^{\mu}{}_{\nu} b^{\nu}, \tag{8.74}$$

we have

$$a' \cdot b' = a'_{\mu} b'^{\mu} = a'^{\rho} g_{\rho\mu} b'^{\mu} = a^{\sigma} \Lambda^{\rho}{}_{\sigma} g_{\rho\mu} \Lambda^{\mu}{}_{\nu} b^{\nu} = a^{\sigma} g_{\sigma\nu} b^{\nu} = a_{\nu} b^{\nu} = a \cdot b, \tag{8.75}$$

and one says that $a \cdot b$ is a *Lorentz scalar*. This is completely analogous to the invariance of the scalar product $\mathbf{a} \cdot \mathbf{b}$ of three-vectors of $\mathbb{R}^3$ under ordinary three-dimensional rotations.

Note that the *four-gradient*,

$$\partial_{\mu} \equiv \frac{\partial}{\partial x^{\mu}} = \left(\frac{1}{c}\frac{\partial}{\partial t}, \boldsymbol{\nabla}\right) = \left(\frac{1}{c}\frac{\partial}{\partial t}, \frac{\partial}{\partial x}, \frac{\partial}{\partial y}, \frac{\partial}{\partial z}\right), \tag{8.76}$$

---

[152]Equation (8.85) provides another formal reason why this must be the case.

transforms like a covariant vector. This follows directly from

$$\partial_\mu' = \frac{\partial}{\partial x'^\mu} = \frac{\partial x^\rho}{\partial x'^\mu}\frac{\partial}{\partial x^\rho} = (\Lambda^{-1})^\rho{}_\mu \frac{\partial}{\partial x^\rho} = \Lambda_\mu{}^\rho \frac{\partial}{\partial x^\rho}, \qquad (8.77)$$

where Eqs. (8.59) and (8.70) were used. The *four-divergence* of a contravariant vector $A^\mu = (A^0, \mathbf{A})$ then reads

$$\partial_\mu A^\mu = \frac{1}{c}\frac{\partial}{\partial t}A^0 + \boldsymbol{\nabla} \cdot \mathbf{A}. \qquad (8.78)$$

This is a Lorentz scalar whose spatial part, $\boldsymbol{\nabla} \cdot \mathbf{A}$, is just the usual three-dimensional divergence. The associated contravariant vector is

$$\partial^\mu \equiv \frac{\partial}{\partial x_\mu} = \left(\frac{1}{c}\frac{\partial}{\partial t}, -\boldsymbol{\nabla}\right). \qquad (8.79)$$

This means, in particular, that the d'Alembert operator,

$$\Box = \partial_\mu \partial^\mu = \frac{1}{c^2}\frac{\partial^2}{\partial t^2} - \frac{\partial^2}{\partial x^2} - \frac{\partial^2}{\partial y^2} - \frac{\partial^2}{\partial z^2} = \frac{1}{c^2}\frac{\partial^2}{\partial t^2} - \boldsymbol{\nabla}^2, \qquad (8.80)$$

is a Lorentz scalar.

8.2.4 *Homogeneous Lorentz group $O(1,3)$*

The set of all Λ's that satisfy (8.64) form the *homogeneous Lorentz group* $O(1,3)$, also called the *full Lorentz group*. The four defining properties of a group are easily verified:

(1) The concatenation $\Lambda_2\Lambda_1$ of two Lorentz transformations Λ_1 and Λ_2 corresponds to another Lorentz transformation, i.e.,

$$(\Lambda_2\Lambda_1)^\mathrm{T} G\Lambda_2\Lambda_1 = \Lambda_1^\mathrm{T}\Lambda_2^\mathrm{T} G\Lambda_2\Lambda_1 = \Lambda_1^\mathrm{T} G\Lambda_1 = G. \qquad (8.81)$$

(2) From the properties of ordinary matrix multiplication, it follows that

$$\Lambda_3(\Lambda_2\Lambda_1) = (\Lambda_3\Lambda_2)\Lambda_1, \qquad (8.82)$$

and therefore associativity holds true trivially.

(3) There exists an identity transformation,

$$\mathbb{1}^\mu{}_\nu = \delta^\mu{}_\nu. \qquad (8.83)$$

(4) In view of

$$\det\left(\Lambda^\mathrm{T} G\Lambda\right) = \det G \quad \Rightarrow \quad \left(\det\Lambda\right)^2 = 1, \qquad (8.84)$$

one has

$$\boxed{\det\Lambda = \pm 1}, \qquad (8.85)$$

and since this is not zero, there exists an inverse Λ^{-1}. Moreover, multiplying Eq. (8.64) with $(\Lambda^{\mathrm{T}})^{-1}$ from the left and Λ^{-1} from the right, we see that Λ^{-1} is also a Lorentz transformation, i.e.,

$$\text{LHS:} \qquad (\Lambda^{\mathrm{T}})^{-1}\Lambda^{\mathrm{T}}G\Lambda\Lambda^{-1} = G, \tag{8.86a}$$

$$\text{RHS:} \qquad (\Lambda^{\mathrm{T}})^{-1}G\Lambda^{-1} = (\Lambda^{-1})^{\mathrm{T}}G\Lambda^{-1}, \tag{8.86b}$$

$$\text{and therefore} \qquad (\Lambda^{-1})^{\mathrm{T}}G\Lambda^{-1} = G. \tag{8.86c}$$

According to (8.85), the real Lorentz transformations can be separated into those with determinant $+1$ or -1, similar to what one finds for rotations in $\mathbb{R}^3$. In addition, they can be classified by the sign of the time-time component Λ_{00}. From

$$G_{00} = (\Lambda^{\mathrm{T}}G\Lambda)_{00} = \Lambda_{00}^2 - \sum_{i=1}^{3}\Lambda_{0i}^2 = 1 \quad \Longrightarrow \quad \Lambda_{00}^2 = 1 + \sum_{i=1}^{3}\Lambda_{0i}^2 \geq 1, \tag{8.87}$$

it follows that, for *real* transformations,

$$\boxed{\Lambda_{00} \geq +1 \qquad \text{or} \qquad \Lambda_{00} \leq -1}. \tag{8.88}$$

The transformations with $\Lambda_{00} \geq 1$ are called *orthochronous* since they preserve the direction of time. By contrast, *non-orthochronous* transformations with $\Lambda_{00} \leq -1$ reverse the flow of time.

Since there exists no continuous transformation that changes the sign of the determinant and/or changes the sign of Λ_{00}, the real Lorentz transformations can be grouped into four *disconnected* classes characterized by the respective signs. The discontinuous operations that connect the four classes are *time-reversal* T and *space-inversion* (or *parity*) P,

$$T = \{T^{\mu}{}_{\nu}\} = \left(\begin{array}{c|c} -1 & 0 \\ \hline 0 & \mathbf{1} \end{array}\right), \quad P = \{P^{\mu}{}_{\nu}\} = \left(\begin{array}{c|c} 1 & 0 \\ \hline 0 & -\mathbf{1} \end{array}\right), \tag{8.89}$$

where $\mathbf{1}$ is the 3×3 unit matrix.[153] The product $TP = PT = -\mathbb{1}$, in particular, changes the sign of Λ_{00} but not of $\det \Lambda$.

The four classes are summarized in Table 8.2. As indicated, only the orthochronous proper class forms a subgroup of its own, called the *restricted*

---

[153] We emphasize that T and P *are* Lorentz transformations, i.e., $T^{\mathrm{T}}GT = G$ and $P^{\mathrm{T}}GP = G$. By contrast $G = \{g_{\mu\nu}\}$ is *not* a Lorentz transformation, since by the definition (8.58) adopted here for contravariant four-vectors, a Lorentz transformation has a contravariant row and covariant column index. The entity $\{g^{\mu}{}_{\nu}\} = G^{-1}G = \mathbb{1}$ is simply the identity transformation.

Table 8.2   The four classes of homogeneous real Lorentz transformations. The first two classes, with $\det \Lambda = +1$, together form the proper Lorentz group $SO(1,3)$.

Class	Λ_{00}	$\det \Lambda$	Relation to restricted group
Orthochronous, proper: *restricted group*	$\geq +1$	$+1$	$\mathbb{1}$
non-orthochronous, proper	≤ -1	$+1$	$TP = -\mathbb{1}$
Orthochronous, improper	$\geq +1$	-1	P
non-orthochronous, improper	≤ -1	-1	T

group, but none of the other classes form a subgroup since they all lack the identity element. However, the union of the restricted group with any one of the other classes does yield a subgroup. The union of the first two classes, both with $\det \Lambda = +1$, in particular, is called the *proper Lorentz group* $SO(1,3)$. Members of each non-group class can be obtained from members of the restricted group by the discrete operation indicated in the last column. Therefore, to study the homogeneous Lorentz group, it is sufficient to study the restricted group and the two discrete mappings P and T.

8.2.5   *Constructing the Lorentz transformation*

All elements of the restricted group can be obtained by a continuous deformation of the identity. Any element Λ, in particular, that differs only infinitesimally from the identity can be written in the linearized form

$$\Lambda^{\mu}{}_{\nu} = g^{\mu}{}_{\nu} + \varepsilon\, \lambda^{\mu}{}_{\nu}, \qquad \text{or} \qquad \Lambda = \mathbb{1} + \varepsilon\, \lambda, \tag{8.90}$$

where $\mathbb{1}$ is the 4×4 unit matrix and ε is an infinitesimally small parameter. The matrix λ is restricted by (8.64),

$$(\mathbb{1} + \varepsilon\, \lambda)^{\mathrm{T}}\, G\, (\mathbb{1} + \varepsilon\, \lambda) = G \qquad \Longrightarrow \qquad \lambda = -G^{-1}\lambda^{\mathrm{T}}G. \tag{8.91}$$

In detail, this reads

$$\left(\begin{array}{c|ccc} \lambda^0{}_0 & \lambda^0{}_1 & \lambda^0{}_2 & \lambda^0{}_3 \\ \hline \lambda^1{}_0 & \lambda^1{}_1 & \lambda^1{}_2 & \lambda^1{}_3 \\ \lambda^2{}_0 & \lambda^2{}_1 & \lambda^2{}_2 & \lambda^2{}_3 \\ \lambda^3{}_0 & \lambda^3{}_1 & \lambda^3{}_2 & \lambda^3{}_3 \end{array}\right) = \left(\begin{array}{c|ccc} -\lambda^0{}_0 & \lambda^1{}_0 & \lambda^2{}_0 & \lambda^3{}_0 \\ \hline \lambda^0{}_1 & -\lambda^1{}_1 & -\lambda^2{}_1 & -\lambda^3{}_1 \\ \lambda^0{}_2 & -\lambda^1{}_2 & -\lambda^2{}_2 & -\lambda^3{}_2 \\ \lambda^0{}_3 & -\lambda^1{}_3 & -\lambda^2{}_3 & -\lambda^3{}_3 \end{array}\right). \tag{8.92}$$

We can conclude then that

$$\lambda^\mu{}_\mu = 0, \qquad \lambda^0{}_i = \lambda^i{}_0, \qquad \text{and} \qquad \lambda^i{}_j = -\lambda^j{}_i, \tag{8.93}$$

i.e., all diagonal elements vanish, the space-time (or time-space) components are symmetric and the space-space components antisymmetric. The general form of λ, therefore, is

$$\lambda = \begin{pmatrix} 0 & \lambda^0{}_1 & \lambda^0{}_2 & \lambda^0{}_3 \\ \lambda^0{}_1 & 0 & \lambda^1{}_2 & \lambda^1{}_3 \\ \lambda^0{}_2 & -\lambda^1{}_2 & 0 & \lambda^2{}_3 \\ \lambda^0{}_3 & -\lambda^1{}_3 & -\lambda^2{}_3 & 0 \end{pmatrix}, \tag{8.94}$$

with six independent elements.

Any matrix of that structure can be written as a linear combination of six independent traceless matrices; three of them must be symmetric, with only nonvanishing space-time components, and three must be antisymmetric, with only nonvanishing space-space components. These matrices form a six-dimensional vector space. A standard choice for these six matrices is[154]

$$\mathbb{K}_x = i \begin{pmatrix} 0 & 1 & 0 & 0 \\ 1 & 0 & 0 & 0 \\ 0 & 0 & 0 & 0 \\ 0 & 0 & 0 & 0 \end{pmatrix}, \qquad \mathbb{J}_x = i \begin{pmatrix} 0 & 0 & 0 & 0 \\ 0 & 0 & 0 & 0 \\ 0 & 0 & 0 & -1 \\ 0 & 0 & 1 & 0 \end{pmatrix},$$

$$\mathbb{K}_y = i \begin{pmatrix} 0 & 0 & 1 & 0 \\ 0 & 0 & 0 & 0 \\ 1 & 0 & 0 & 0 \\ 0 & 0 & 0 & 0 \end{pmatrix}, \qquad \mathbb{J}_y = i \begin{pmatrix} 0 & 0 & 0 & 0 \\ 0 & 0 & 0 & 1 \\ 0 & 0 & 0 & 0 \\ 0 & -1 & 0 & 0 \end{pmatrix}, \tag{8.95}$$

$$\mathbb{K}_z = i \begin{pmatrix} 0 & 0 & 0 & 1 \\ 0 & 0 & 0 & 0 \\ 0 & 0 & 0 & 0 \\ 1 & 0 & 0 & 0 \end{pmatrix}, \qquad \mathbb{J}_z = i \begin{pmatrix} 0 & 0 & 0 & 0 \\ 0 & 0 & -1 & 0 \\ 0 & 1 & 0 & 0 \\ 0 & 0 & 0 & 0 \end{pmatrix}.$$

---

[154]One could also work with matrices without the imaginary factors i. We follow here the convention of relativistic quantum mechanics.

The matrices $\mathbb{K}_k$ are anti-Hermitian and the $\mathbb{J}_k$ are Hermitian. The latter are trivial four-dimensional extensions of the usual Hermitian generators J_k of three-dimensional rotations [see Sec. 5.1.1.3, in particular, Eq. (5.32)], i.e.,

$$\mathbb{J}_k = \left(\begin{array}{c|c} 0 & 0 \\ \hline 0 & \mathsf{J}_k \end{array}\right). \tag{8.96}$$

The $\mathbb{J}_k$ therefore will be referred to also within the present context as the *generators of rotations*.[155] For reasons that will become obvious below, the $\mathbb{K}_k$ are called the *generators of boosts*.

The six generators can be grouped as the components of two *vector operators* $\mathbf{J}$ and $\mathbb{K}$ according to

$$\mathbf{J} = (\mathbb{J}_x, \mathbb{J}_y, \mathbb{J}_z), \quad \mathbb{K} = (\mathbb{K}_x, \mathbb{K}_y, \mathbb{K}_z). \tag{8.97}$$

(For the proof, one must show that $\mathbf{J}$ and $\mathbb{K}$ behave like vectors under rotations, which is established by considering their commutator relations; see Sec. 8.2.5.1.) We can now write the infinitesimal transformation (8.90) in the more explicit form

$$\Lambda = \mathbb{1} - i\varepsilon(\boldsymbol{\xi} \cdot \mathbb{K} + \boldsymbol{\omega} \cdot \mathbf{J}), \tag{8.98}$$

where the six real components of $\boldsymbol{\xi} = (\xi_1, \xi_2, \xi_3)$ and $\boldsymbol{\omega} = (\omega_1, \omega_2, \omega_3)$ correspond to the six independent parameters of λ. Taking $\varepsilon = 1/N$, where N is large, this can be exponentiated in the usual manner by considering a succession of N infinitesimal transformations and performing the limit $N \to \infty$, resulting in

$$\boxed{\Lambda = \Lambda(\boldsymbol{\xi}, \boldsymbol{\omega}) = \mathrm{e}^{-i(\boldsymbol{\xi} \cdot \mathbb{K} + \boldsymbol{\omega} \cdot \mathbf{J})}}. \tag{8.99}$$

This is the *exponential representation* of the restricted Lorentz group in terms of the generators of boosts and rotations. Note that it is *not unitary* in general since the boost generators $\mathbb{K}_k$ are *anti*-Hermitian. Special cases are the pure boosts and pure rotations

$$\Lambda_{\mathsf{B}}(\boldsymbol{\xi}) = \Lambda(\boldsymbol{\xi}, 0) = \mathrm{e}^{-i\boldsymbol{\xi} \cdot \mathbb{K}} \tag{8.100}$$

and

$$\Lambda_{\mathsf{R}}(\boldsymbol{\omega}) = \Lambda(0, \boldsymbol{\omega}) = \mathrm{e}^{-i\boldsymbol{\omega} \cdot \mathbf{J}}, \tag{8.101}$$

respectively.

---

[155] For a more explicit connection to rotations, see the discussion below related to Eq. (8.103.)

8.2.5.1 *Another Lie-algebra interlude*

In Sec. 5.1.1.3, we had introduced the fact that continuous groups like $SO(3)$ can be understood in terms of the structure of the underlying Lie algebras. As is shown by the preceding considerations employing infinitesimal transformations, the restricted Lorentz group is another example of a Lie group and we introduce here the underlying Lie algebra.

The following commutation relations among all the components $\mathbb{J}_k$ and $\mathbb{K}_l$ are easily established,

$$[\mathbb{J}_k, \mathbb{J}_l] = i \sum_{m=1}^{3} \varepsilon_{klm} \mathbb{J}_m, \qquad (8.102a)$$

$$[\mathbb{J}_k, \mathbb{K}_l] = i \sum_{m=1}^{3} \varepsilon_{klm} \mathbb{K}_m, \qquad (8.102b)$$

$$[\mathbb{K}_k, \mathbb{K}_l] = -i \sum_{m=1}^{3} \varepsilon_{klm} \mathbb{J}_m, \qquad (8.102c)$$

where again $[A, B] = AB - BA$ denotes the *commutator* of the generators. The first equation here is the *commutator Lie algebra of the angular momenta* that we encountered already in Eq. (5.42). This is now the four-dimensional representation relevant for the Minkowski space that follows by a trivial extension of the three-dimensional $\mathbb{J}_k$'s by filling up the additional time components by zeros. The three components $\mathbb{J}_k$, therefore, form a *vector operator* $\mathbf{J} = (\mathbb{J}_x, \mathbb{J}_y, \mathbb{J}_z)$. Furthermore, the commutation relations of the $\mathbb{K}_l$ with the $\mathbb{J}_k$ is the defining commutation relation for a vector operator and show that $\mathbf{K} = (\mathbb{K}_x, \mathbb{K}_y, \mathbb{K}_z)$ is indeed also a vector operator, as anticipated already above. Finally, the commutators among the components $\mathbb{K}_k$, Eq. (8.102c), show that the commutator algebra of the $\mathbb{J}_k$ and $\mathbb{K}_l$ is *closed*. The three sets of commutation relations (8.102) of the six generators thus provide a complete definition of the algebra relevant for the restricted Lorentz group.

The commutation relation (8.102a) says that any sequence of pure rotations about arbitrary axes can be replaced by a single rotation [and that $SO(3)$ is a subgroup of the proper Lorentz transformations]. Equation (8.102c), in contrast, means that a sequence of two boosts along different axes cannot be reduced to a single boost, but contains also a rotation. (This gives rise to the Thomas precession discussed in Sec. 8.2.7.3 below.) The rotations, therefore, form a closed algebra by themselves, but the boosts do not.

8.2.6 Rotations and boosts

Putting $\boldsymbol{\xi} = 0$ and $\boldsymbol{\omega} = \theta\hat{\mathbf{n}}$, where $\hat{\mathbf{n}}$ is a unit vector, the resulting Lorentz transformation is immediately found to have the structure

$$\Lambda_{\mathsf{R}} = \mathrm{e}^{-i\theta\hat{\mathbf{n}}\cdot\mathbf{J}} = \left(\begin{array}{c|c} 1 & 0 \\ \hline 0 & \mathsf{R}(\hat{\mathbf{n}}, \theta) \end{array}\right). \tag{8.103}$$

For this special case, the condition (8.64) translates into the usual orthogonality condition

$$\mathsf{R}^{\mathrm{T}}\mathsf{R} = \mathsf{R}\mathsf{R}^{\mathrm{T}} = 1 \tag{8.104}$$

for ordinary 3×3 rotation matrices R of $\mathbb{R}^3$ that form the orthogonal group $SO(3)$. The Lorentz transformation (8.103), therefore, describes a *pure active rotation* of the space coordinates by an angle θ about the axis $\hat{\mathbf{n}}$; it leaves the time component and one spatial direction (the rotation axis) unchanged. An active rotation about the z-axis, for example, is given by (see Sec. 5.1.1)

$$\Lambda_{\mathsf{R}_z} = \left(\begin{array}{c|c} 1 & 0 \\ \hline 0 & \mathsf{R}(\mathbf{e}_z, \theta) \end{array}\right) = \left(\begin{array}{c|ccc} 1 & 0 & 0 & 0 \\ \hline 0 & \cos\theta & -\sin\theta & 0 \\ 0 & \sin\theta & \cos\theta & 0 \\ 0 & 0 & 0 & 1 \end{array}\right). \tag{8.105}$$

Apart from the trivial extension to four dimensions, therefore, the rotations do not introduce any new features that are not known already from three dimensions. One still finds, for example, that any sequence of pure rotations corresponds to one combined rotation. This is due to the fact that the pure rotation group $SO(3)$ is a subgroup of the restricted Lorentz group.

Next, let us consider a pure boost along the x axis, i.e., $\boldsymbol{\xi} = (\xi, 0, 0)$ and $\boldsymbol{\omega} = 0$, and therefore

$$\Lambda_{\mathsf{B}_x} = \mathrm{e}^{-i\xi\mathbb{K}_x} = \sum_{n=0}^{\infty} \frac{(-i\xi\mathbb{K}_x)^n}{n!}. \tag{8.106}$$

Note that

$$(-i\mathbb{K}_x)^0 = \left(\begin{array}{c|ccc} 1 & 0 & 0 & 0 \\ \hline 0 & 1 & 0 & 0 \\ 0 & 0 & 1 & 0 \\ 0 & 0 & 0 & 1 \end{array}\right) \tag{8.107}$$

and

$$(-i\mathbb{K}_x)^2 = \begin{pmatrix} 1 & 0 & 0 & 0 \\ 0 & 1 & 0 & 0 \\ 0 & 0 & 0 & 0 \\ 0 & 0 & 0 & 0 \end{pmatrix} \implies (-i\mathbb{K}_x)^{2k} = \begin{pmatrix} 1 & 0 & 0 & 0 \\ 0 & 1 & 0 & 0 \\ 0 & 0 & 0 & 0 \\ 0 & 0 & 0 & 0 \end{pmatrix} \tag{8.108}$$

for $k > 0$. With

$$(-i\mathbb{K}_x)^{2k+1} = -i\mathbb{K}_x = \begin{pmatrix} 0 & 1 & 0 & 0 \\ 1 & 0 & 0 & 0 \\ 0 & 0 & 0 & 0 \\ 0 & 0 & 0 & 0 \end{pmatrix}, \tag{8.109}$$

the series then yields

$$\Lambda_{\mathsf{B}_x} = \sum_{n=0}^{\infty} \frac{(-i\xi\mathbb{K}_x)^n}{n!} = \sum_{k=0}^{\infty} \left[\frac{\xi^{2k}}{2k!} (-i\mathbb{K}_x)^{2k} + \frac{\xi^{2k+1}}{(2k+1)!} (-i\mathbb{K}_x)^{2k+1} \right]$$

$$= \begin{pmatrix} \cosh\xi & \sinh\xi & 0 & 0 \\ \sinh\xi & \cosh\xi & 0 & 0 \\ 0 & 0 & 1 & 0 \\ 0 & 0 & 0 & 1 \end{pmatrix}, \tag{8.110}$$

where the power series expansions for the hyperbolic sine and cosine were used. The physical significance of this result is seen when we use the familiar relation $\cosh^2\xi - \sinh^2\xi = 1$ and write

$$\sinh\xi = \frac{\beta}{\sqrt{1-\beta^2}} \quad \text{and} \quad \cosh\xi = \frac{1}{\sqrt{1-\beta^2}}, \quad \text{or} \quad \beta = \tanh\xi, \tag{8.111}$$

in terms of a real parameter β restricted by $0 \leq |\beta| \leq 1$. This transformation affects only the time and x components of the position vector; the y and z components remain unchanged, i.e.,

$$\Lambda_{\mathsf{B}_x}: \quad \boxed{\begin{aligned} ct' &= \gamma(ct + \beta\, x), \\ x' &= \gamma(x + \beta\, ct), \\ y' &= y, \\ z' &= z, \end{aligned}} \tag{8.112}$$

where we have introduced the *Lorentz factor*

$$\gamma = \frac{1}{\sqrt{1-\beta^2}}. \tag{8.113}$$

Now, consider a particle at rest in the unprimed system (i.e., $dx = 0$) that is boosted to a velocity v along the x-axis; one has

$$v = \frac{dx'}{dt'} = \sqrt{1 - \beta^2}\frac{dx'}{dt} = \beta c \quad \Rightarrow \quad \beta = \frac{v}{c}. \tag{8.114}$$

The *active* boost Λ_{B_x}, therefore, propels a particle from rest along the x-axis with a velocity

$$v = \beta c = c\tanh\xi. \tag{8.115}$$

We see here, in particular, that for $\beta \ll 1$ the boost transformation (8.112) reduces to

$$v \ll c: \qquad \begin{aligned} t' &= t, \\ x' &= x + v\,t, \\ y' &= y, \\ z' &= z, \end{aligned} \tag{8.116}$$

which is the corresponding Galilei transformation. As expected, therefore, the Lorentz boost does indeed reduce to a Galilei transformation in the limit of small velocities.

Rapidity

The quantity ξ is called the *rapidity*. Its significance is that while relativistic velocities do not add, rapidities do. To see this consider a particle with initial velocity $u = dx/dt$ along the x-axis that is boosted by a velocity v along the same axis. Putting $\beta = v/c$, we have from (8.112)

$$dt' = \frac{c\,dt + \beta dx}{c\sqrt{1 - \beta^2}} = dt\frac{1 + \frac{vu}{c^2}}{\sqrt{1 - \beta^2}}, \tag{8.117}$$

and thus

$$u' \equiv \frac{dx'}{dt'} = \frac{\sqrt{1 - \beta^2}}{1 + \frac{vu}{c^2}}\frac{dx'}{dt} = \frac{u + v}{1 + \frac{vu}{c^2}}, \tag{8.118}$$

which describes how velocities are to be added in a relativistic context.[156] Employing an obvious notation for the rapidities associated with the three velocities u', u, and v, this reads

$$\tanh\xi_{u'} = \frac{\tanh\xi_v + \tanh\xi_u}{1 + \tanh\xi_v \tanh\xi_u} = \tanh(\xi_u + \xi_v), \tag{8.119}$$

where a standard addition theorem for the hyperbolic tangent was used in the last equality. We thus find

$$\xi_{u'} = \xi_u + \xi_v, \tag{8.120}$$

i.e., for boosts along the same axis, the rapidities simply add, as stipulated.

---

[156]In the nonrelativistic limit, for $vu/c^2 \ll 1$, we recover the usual addition of velocities for Galilei transformations, $u' = u + v$.

8.2.6.1   *Boost along arbitrary direction*

To describe boosts in an arbitrary direction given by a velocity $\mathbf{v}$, it is convenient to introduce a dimensionless 'velocity' vector and its magnitude by

$$\boldsymbol{\beta} = \frac{\mathbf{v}}{c} \quad \text{and} \quad \beta = \frac{|\mathbf{v}|}{c}, \tag{8.121}$$

respectively. We also need the explicit representation of this vector as column and row three-vectors in terms of its Cartesian components β_i, i.e.,

$$\boldsymbol{\beta} \;\longrightarrow\; \vec{\beta} = \begin{pmatrix} \beta_1 \\ \beta_2 \\ \beta_3 \end{pmatrix} = \frac{1}{c}\begin{pmatrix} v_x \\ v_y \\ v_z \end{pmatrix} \quad \text{and} \quad \vec{\beta}^{\mathrm{T}} = (\beta_1,\, \beta_2,\, \beta_3)\,. \tag{8.122}$$

The *active* boost by a velocity $\mathbf{v}$ in an arbitrary direction can now be written as

$$\Lambda_{\mathsf{B}}(\mathbf{v}) = \begin{pmatrix} \gamma & \gamma\vec{\beta}^{\mathrm{T}} \\[2mm] \gamma\vec{\beta} & \mathbf{1} + \alpha\vec{\beta}\vec{\beta}^{\mathrm{T}} \end{pmatrix}, \tag{8.123}$$

where $\mathbf{1}$ is the three-dimensional unit matrix and $\vec{\beta}\vec{\beta}^{\mathrm{T}}$ is a $3{\times}3$ dyad matrix with elements $\beta_i\beta_j$. The factor α is

$$\alpha = \frac{\gamma - 1}{\beta^2} = \frac{\gamma^2}{\gamma + 1} \tag{8.124}$$

and the Lorentz factor γ is given in Eq. (8.113). More explicitly, one has

$$\Lambda_{\mathsf{B}}(\mathbf{v}) = \begin{pmatrix} \gamma & \gamma\beta_1 & \gamma\beta_2 & \gamma\beta_3 \\[2mm] \gamma\beta_1 & 1 + \alpha\beta_1^2 & \alpha\beta_1\beta_2 & \alpha\beta_1\beta_3 \\[2mm] \gamma\beta_2 & \alpha\beta_1\beta_2 & 1 + \alpha\beta_2^2 & \alpha\beta_2\beta_3 \\[2mm] \gamma\beta_3 & \alpha\beta_1\beta_3 & \alpha\beta_2\beta_3 & 1 + \alpha\beta_3^2 \end{pmatrix}. \tag{8.125}$$

As one reads off these expressions, the *pure boosts are symmetric,*

$$\Lambda_{\mathsf{B}}(\mathbf{v}) = \Lambda_{\mathsf{B}}^{\mathrm{T}}(\mathbf{v}). \tag{8.126}$$

In fact, any Lorentz transformation that is symmetric is a pure boost. Furthermore, the inverse boost follows simply by $\mathbf{v} \to -\mathbf{v}$, i.e., $\vec{\beta} \to -\vec{\beta}$, and

$$\Lambda_{\mathsf{B}}^{-1}(\mathbf{v}) = \Lambda_{\mathsf{B}}(-\mathbf{v}) = P\Lambda_{\mathsf{B}}(\mathbf{v})P, \tag{8.127}$$

where P is the space-inversion operator in (8.89).

It is easy to see that rotations and boosts in general do *not* commute. To this end, consider

$$\Lambda_\mathsf{R}\Lambda_\mathsf{B} = \begin{pmatrix} 1 & 0 \\ 0 & \mathsf{R} \end{pmatrix} \begin{pmatrix} \gamma & \gamma\vec{\beta}^\mathsf{T} \\ \gamma\vec{\beta} & \mathbf{1}+\alpha\vec{\beta}\vec{\beta}^\mathsf{T} \end{pmatrix} = \begin{pmatrix} \gamma & \gamma\vec{\beta}^\mathsf{T} \\ \gamma(\mathsf{R}\vec{\beta}) & \mathsf{R}+\alpha(\mathsf{R}\vec{\beta})\vec{\beta}^\mathsf{T} \end{pmatrix} \tag{8.128}$$

and

$$\Lambda_\mathsf{B}\Lambda_\mathsf{R} = \begin{pmatrix} \gamma & \gamma\vec{\beta}^\mathsf{T} \\ \gamma\vec{\beta} & \mathbf{1}+\alpha\vec{\beta}\vec{\beta}^\mathsf{T} \end{pmatrix} \begin{pmatrix} 1 & 0 \\ 0 & \mathsf{R} \end{pmatrix} = \begin{pmatrix} \gamma & \gamma(\vec{\beta}^\mathsf{T}\mathsf{R}) \\ \gamma\vec{\beta} & \mathsf{R}+\alpha\vec{\beta}(\vec{\beta}^\mathsf{T}\mathsf{R}) \end{pmatrix}. \tag{8.129}$$

Clearly, the two resulting matrices are only identical if the rotation R leaves the boost velocity unchanged, i.e.,

$$\mathsf{R}\vec{\beta} = \vec{\beta}, \tag{8.130}$$

which is equivalent to $\vec{\beta}^\mathsf{T}\mathsf{R} = \vec{\beta}^\mathsf{T}$. According to *Euler's Theorem* (5.116) for rotations, this is only true if the boost direction and the rotation axis coincide, i.e.,

$$\hat{\mathbf{n}} = \frac{\mathbf{v}}{|\mathbf{v}|}, \tag{8.131}$$

since the rotation axis is the only invariant vector under a rotation.[157]

In summary, this discussion shows that the six parameters of the Lorentz transformation may be identified with one rotation angle and a rotation axis according to

$$\boldsymbol{\omega} = \theta\hat{\mathbf{n}}, \tag{8.132}$$

and with three velocity components,

$$\mathbf{v} = c\boldsymbol{\beta} = c \begin{pmatrix} \tanh\xi_x \\ \tanh\xi_y \\ \tanh\xi_z \end{pmatrix}, \tag{8.133}$$

along an arbitrary boost direction $\hat{\mathbf{v}} = \mathbf{v}/|\mathbf{v}|$. It is the boosts, in particular, that give rise to the familiar notions of length contraction and time dilation, etc., associated with relativistic kinematics (see subsequent section).

The present discussion concerned *active* transformations that rotate and/or boost a four-vector within a given frame. For *passive* transformations, that describe a given four-vector in different frames, one needs to change the sign of the transformation parameters $\boldsymbol{\omega}$ and $\mathbf{v}$. For boosts, this is equivalent to changing the signs of the rapidities.

---

[157]In quantum mechanics, this gives rise to the *helicity formalism* which uses the direction of a particle's momentum as the quantization axis for the particle's spin.

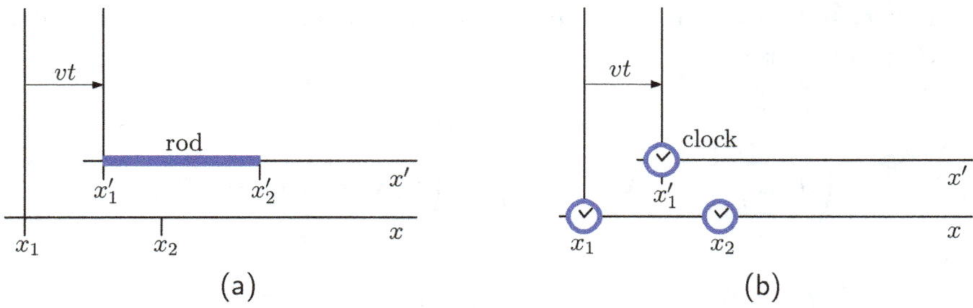

Fig. 8.10   (a) Length contraction: To measure the apparent length $L = x_2 - x_1$ of a moving rod with proper length $L_0 = x_2' - x_1'$, the two positions x_2 and x_1 of the end points of the rod as they appear in the lab frame must be measured at the same time. (b) Time dilation: The clock in the moving frame defines the origin of that frame. The two frames are synchronized as the clock passes the origin of the lab frame (i.e., $x_1' = x_1 = 0$ and $t_1' = t_1 = 0$) and the respective clock readings between moving frame and lab frame are compared for later times t_2 as the clock moves past $x_2 = vt_2$.

8.2.7   Kinematics

Four-dimensional spacetime provides some interesting effects that are purely kinematical, resulting either from spacetime topology itself or from Lorentz transformations between frames.

8.2.7.1   Length contraction, time dilation

Consider a rod at rest in an inertial frame. Its length L_0 can be established by comparing it with some fixed reference ruler (also at rest). The relativity principle implies that this procedure yields the same value L_0 in any inertial frame. In other words, the length L_0 — called the rod's *proper length* — is an invariant under Lorentz transformations, i.e., it is a Lorentz scalar.

Now, let us align the rod with the x-axis of the (inertial) lab frame and move it along this axis with uniform velocity $v = \beta c$, as shown in Fig. 8.10(a). Let x' be the corresponding rest-frame axis. In other words, in its own rest frame, the rod's end points are at positions x_2' and x_1' with $\Delta x' = x_2' - x_1' = L_0$. To establish the apparent length L in the (unprimed) lab frame, we need to place two observers so that the front and back ends of the rod at x_2 and x_1 pass the observers *at the same time* $t_1 = t_2$, making the length measured in the lab $L = x_2 - x_1$. Without lack of generality, we may choose the Lorentz frames such that the two positions x_1' and x_1 coincide when clocks at these positions in both frames show $t_1' = 0$ and $t_1 = 0$ and we may mark them as the respective origins of the two frames, i.e., $x_1' = 0$ and $x_1 = 0$. The corresponding Lorentz transformation (8.112) between

(x'_1, t'_1) and (x_1, t_1) then is trivial. The values of the positions $x'_2 = L_0$ and $x_2 = L$ now simply correspond to the respective lengths observed in the two frames, and with $t_1 = t_2 = 0$, the boost transformation[158] (8.112) for x'_2 reads

$$L_0 = x'_2 = \frac{x_2 - \beta t_2}{\sqrt{1 - \beta^2}} = \frac{L}{\sqrt{1 - \beta^2}}. \tag{8.134}$$

Hence, a moving rod that has proper length L_0 in its own rest frame will appear to have a *reduced* length L to the observer at rest in the lab, i.e.,

$$\boxed{L = L_0 \sqrt{1 - \beta^2}}. \tag{8.135}$$

This length contraction happens for objects of any shape along the direction of their motion. A sphere moving at relativistic speed, for example, will appear as a flat pancake to an observer at rest.[159] In general, we may write

$$L_\parallel = L_{0\parallel} \sqrt{1 - \frac{\mathbf{v}^2}{c^2}} \quad \text{and} \quad L_\perp = L_{0\perp}, \tag{8.136}$$

where $\mathbf{v}$ is a boost velocity in an arbitrary direction and $L_\parallel$ and $L_\perp$, etc., are the lengths aligned with and perpendicular to $\mathbf{v}$, respectively.

Next, consider a clock moving with velocity $v = \beta c$ through the lab along x, as shown in Fig. 8.10(b). The position of the moving clock may mark the origin of the moving frame. We then compare the reading of the moving clock with clock readings in the lab frame as the moving clock passes positions x_1 and x_2. As before, we may mark $x_1 = 0$ as the origin of the lab frame as the moving clock at $x'_1 = 0$ passes that point at $t'_1 = 0$ and also synchronize the clocks of the two frames by putting $t_1 = t'_1 = 0$. The boost transformation between the time t'_2 shown by the clock in the moving frame as it passes x_2 and the time t_2 shown by the clock at x_2 in the lab frame is given by

$$t'_2 = \left(t_2 - \frac{\beta}{c} x_2 \right) \frac{1}{\sqrt{1 - \beta^2}}. \tag{8.137}$$

---

[158] We are dealing with *passive transformations* here. Hence, compared to (8.112), we have $\beta \to -\beta$.

[159] Actually, the situation is a bit more complicated. While the flattening would indeed be measurable in the lab, an actual observer would still *see* a spherical object (assuming it is far enough away) because the light rays reaching his eyes at the same time emanated from the various parts of the moving object at different times which, as one can show by detailed calculations, preserves the sphericity of the image shape in the observer's eyes. However, if the sphere is not approaching the observer head-on, it would appear to be *rotated*.

Since the clock does not move within the moving frame, the position x_2 is simply $x_2 = vt_2$ and thus

$$t_2' = t_2\sqrt{1 - \beta^2}. \tag{8.138}$$

In a more generic notation, where $\Delta t_0 = t_2' - t_1'$ and $\Delta t = t_2 - t_1$ are the respective elapsed time intervals in both frames and $\beta^2 = \mathbf{v}^2/c^2$ is the squared boost velocity along any direction, we may write

$$\boxed{\Delta t = \frac{\Delta t_0}{\sqrt{1 - \beta^2}}}. \tag{8.139}$$

To the observer at rest, therefore, the clock in the moving frame appears to run *slower*. This *time dilation* effect is present whether one measures it or not and it has real consequences, for example, when measuring the decay time of particles moving at relativistic speeds close to c.

8.2.7.2 *World line, four-velocity, four-momentum, etc.*

The *four-velocity* u is defined as the rate of change of the spacetime trajectory x — called the particle's *world line* (see Fig. 8.9) — along an invariant spacetime distance s, i.e.,

$$\boxed{u^\mu = c\frac{dx^\mu}{ds} = \frac{dx^\mu}{d\tau} = (u^0, \mathbf{u})}, \tag{8.140}$$

where

$$ds = \sqrt{dx_\mu dx^\mu} = \sqrt{c^2 dt^2 - d\mathbf{r}^2} = c\,dt\sqrt{1 - \frac{v^2}{c^2}} \tag{8.141}$$

is the invariant length increment of Eq. (8.61). Note that for physical systems ds is a timelike vector and thus always real. The invariant quantity

$$\boxed{d\tau = \frac{ds}{c} = dt\sqrt{1 - \beta^2}} \tag{8.142}$$

is called the increment of *proper time*, or *eigentime*. The four-velocity u^μ thus measures the (frame-dependent) change dx^μ against an eigentime increment $d\tau$ that is the same for every observer. Calculating the length of the four-velocity, we find

$$u^2 = u_\mu u^\mu = c^2\frac{dx_\mu}{ds}\frac{dx^\mu}{ds} = c^2\frac{dx_\mu dx^\mu}{ds^2} = c^2, \tag{8.143}$$

i.e., it is not only a timelike vector (as it ought to be), it always has the same fixed length.[160] This means that the components u^μ are not independent

---

[160]The unit vector $\hat{e}_0 = u/c$ always points along the time axis in the particle's rest frame; see Fig. 8.12.

of each other. We may, for example, always write $u^0 = \sqrt{c^2 + \mathbf{u}^2}$ for its time component.[161]

Next, defining the *four-momentum p* of a particle with mass m by

$$\boxed{p^\mu = mu^\mu = m\frac{dx^\mu}{d\tau}}, \tag{8.144}$$

its invariant length is found to be

$$p^2 = p_\mu p^\mu = m^2 c^2. \tag{8.145}$$

We conclude, therefore, that in its rest frame [also called the center-of-momentum (CM) frame], where by definition $\mathbf{u} = 0$, we have

$$p \to P_{\mathrm{CM}} = \begin{pmatrix} mc \\ \mathbf{0} \end{pmatrix}. \tag{8.146}$$

The mass parameter m appearing here in the time component[162] is therefore referred to as the *rest mass* of the particle.[163]

When boosting this four-momentum from the rest frame with a boost vector $\boldsymbol{\beta} = \mathbf{v}/c$, we obtain

$$p \equiv \begin{pmatrix} E/c \\ \mathbf{p} \end{pmatrix} = \Lambda_{\mathrm{B}}(\mathbf{v})P_{\mathrm{CM}} = \gamma m \begin{pmatrix} c \\ \mathbf{v} \end{pmatrix}, \tag{8.147}$$

where, following Einstein, we have identified the zeroth momentum component with the energy, $p^0 = E/c$. Hence, we have

$$E = \gamma mc^2 = \sqrt{m^2 c^4 + \mathbf{p}^2 c^2}, \tag{8.148a}$$

$$\mathbf{p} = \gamma m\mathbf{v} = \frac{m\mathbf{v}}{\sqrt{1 - \beta^2}}, \tag{8.148b}$$

for the energy and the three-momentum, respectively. The rest-frame result for the energy, $E_{\mathrm{CM}} = mc^2$, provides Einstein's famous relationship about the equivalence of mass and energy.

If we write the three-momentum in the form familiar from nonrelativistic physics,

$$\mathbf{p} = m^*\mathbf{v}, \tag{8.149}$$

---

[161]Then again, in some dynamical formulations it will be advantageous to treat all components u^μ as if they were independent and impose the condition $u^2 = c^2$ only at the very last step as a subsidiary condition to the resulting equations of motion.

[162]For momenta, in particular, the time component also is called the 'energy' component.

[163]Note that particles (like the photon, for example) described by a null vector, $p^\mu = (|\mathbf{k}|, \mathbf{k})$, are necessarily massless and they do not have a rest frame.

we must introduce the notion of a velocity-dependent mass,

$$m^* = \gamma m = \frac{m}{\sqrt{1 - \beta^2}}, \tag{8.150}$$

that increases from its rest-mass value m beyond all bounds to infinity as $\beta \to 1$. Since it requires an infinite amount of energy to overcome the inertia of infinitely massive objects, it follows that massive particles cannot be accelerated to reach the speed of light. Conversely, it means that only massless particles can move at the speed of light.

In analogy to the velocity, the acceleration is defined as the rate of change of the velocity per unit eigentime, i.e.,

$$\boxed{a^\mu = \frac{du^\mu}{d\tau} = \frac{d^2 x^\mu}{d\tau^2}}. \tag{8.151}$$

The constant length of the velocity then shows

$$a^\mu u_\mu = \frac{du^\mu}{d\tau} u_\mu = \frac{1}{2} g_{\mu\nu} \frac{d(u^\mu u^\nu)}{d\tau} = \frac{1}{2} \frac{du^2}{d\tau} = 0. \tag{8.152}$$

In other words, in Minkowski spacetime the velocity and the acceleration are always 'perpendicular' to each other.

At this stage, we only know that the relativistic velocities, momenta, and accelerations defined here are proper four-vectors. We do not know, however, whether they are the 'correct' kinematical variables to describe the physics of relativistic systems. Not to worry — they are! This will be shown in Sec. 8.2.8.

8.2.7.3   *Thomas precession*

The product of two Lorentz transformations is itself a Lorentz transformation,

$$\Lambda_3 = \Lambda_2 \Lambda_1. \tag{8.153}$$

If Λ_1 and Λ_2 provide pure boosts along the *same* direction, Λ_3 will also be a pure boost along this very same direction.[164] If, however, the two successive boosts happen along *different* directions, Λ_3 will contain both a boost and a rotation, i.e.,

$$\Lambda_3 = \Lambda_\mathsf{R} \Lambda_\mathsf{B} = \Lambda_\mathsf{B}' \Lambda_\mathsf{R}', \tag{8.154}$$

where the primed differ from unprimed quantities by the order of the corresponding boost and rotation. The pure rotations found here,

$$\Lambda_\mathsf{R} = \Lambda_2 \Lambda_1 \Lambda_\mathsf{B}^{-1} \quad \text{or} \quad \Lambda_\mathsf{R}' = \Lambda_\mathsf{B}'^{-1} \Lambda_2 \Lambda_1, \tag{8.155}$$

give rise to an important effect called the *Thomas precession*.

---

[164]One may show quite easily that boost operations along a given, fixed direction form a subgroup of the restricted group.

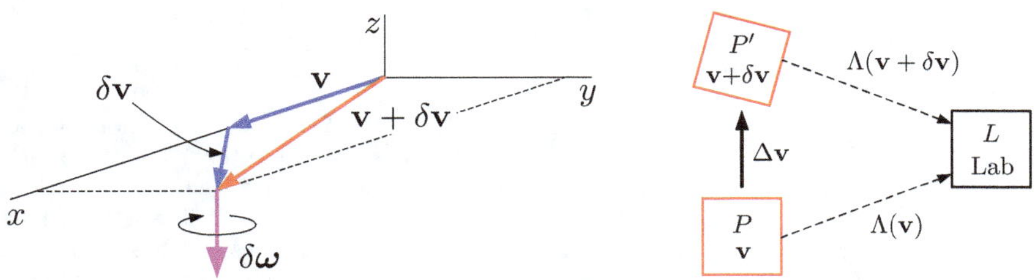

Fig. 8.11   The figure on the left shows the coordinates used to explain the Thomas precession; see text for details. The xyz-frame is the inertial lab frame L, whose relationship to the particle frames P and P' before and after the velocity increment $\delta\mathbf{v}$ is added is shown in the diagram on the right. (For an explanation how the boost increment $\Delta\mathbf{v}$ differs from $\delta\mathbf{v}$, see text.) The Λ's indicate the respective *passive* Lorentz boosts that transmit the information to the lab frame.

To understand this effect in more detail, let us consider a particle that has a velocity $\mathbf{v}$ in the laboratory frame at some time t. An infinitesimal time δt later it has a velocity $\mathbf{v} + \delta\mathbf{v}$, with an infinitesimal increment $\delta\mathbf{v}$. Without lack of generality, we may arrange the spatial coordinate axes in the lab frame such that $\mathbf{v}$ is aligned with the x-axis and that the velocity increment $\delta\mathbf{v}$ lies in the xy-plane, as shown in Fig. 8.11.

As far as the particle is concerned, whatever happens may *always* be viewed as happening in the particle's rest frame. The interpretations of our measurements in the laboratory frame, therefore, involves transformations from the particle frame to the lab frame. The problem is complicated by the fact that *the* particle rest frame does not exist as a unique entity. To better understand this, let us first introduce the concept of an *instantaneous rest frame*. To this end, consider a multitude of inertial frames — infinitely many really — that can be obtained by pure Lorentz boosts from the inertial laboratory frame and that move with constant velocities ranging from 0 to c in all directions with respect to the lab frame. As the particle moves along its world line with varying velocity $u^\mu(\tau)$, it will always be at rest, at least for an instant, with respect to one of these frames, and this frame is called the instantaneous rest frame. Thus, when we make the *first* measurement that involves a boost to the particle's rest frame, this establishes a baseline particle rest frame $\{\hat{e}_0, \hat{e}_1, \hat{e}_2, \hat{e}_3\}$ for all subsequent measurements that may be viewed as being carried with the particle as its eigentime τ evolves, to be compared with later measurements (see Fig. 8.12). Let us call this frame the *Thomas rest frame*. A boost to the particle's rest frame at some later time will provide an instantaneous

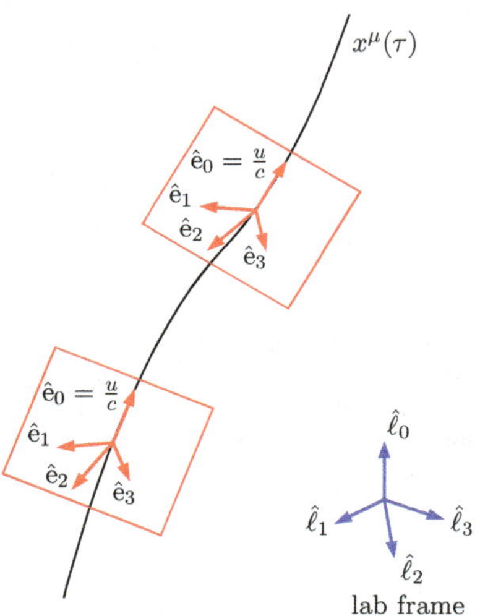

Fig. 8.12 Schematic depiction of the Thomas basis (see text) $\{\hat{e}_0, \hat{e}_1, \hat{e}_2, \hat{e}_3\}$ carried along with the particle as it moves along its world line $x^\mu(\tau)$. The time unit vector $\hat{e}_0$ is always aligned with the four-velocity $u^\mu = dx^\mu/d\tau$, i.e., $\hat{e}_0 = u/c$. A boost from the lab frame $\{\hat{\ell}_0, \hat{\ell}_1, \hat{\ell}_2, \hat{\ell}_3\}$ produces an instantaneous rest frame $\{\hat{e}'_0, \hat{e}'_1, \hat{e}'_2, \hat{e}'_3\}$ where $\hat{e}'_0 = \hat{e}_0 = u/c$, but the spatial basis $\{\hat{e}'_1, \hat{e}'_2, \hat{e}'_3\}$ my differ by a rotation from the basis $\{\hat{e}_1, \hat{e}_2, \hat{e}_3\}$ carried by the particle.

rest-frame basis $\{\hat{e}'_0, \hat{e}'_1, \hat{e}'_2, \hat{e}'_3\}$. The Lorentz boost properties ensure that the time unit vector is the same as the one resulting from the time evolution of the previous measurement, i.e., $\hat{e}'_0 = \hat{e}_0 = u/c$. However, the spatial basis $\{\hat{e}'_1, \hat{e}'_2, \hat{e}'_3\}$ may be different from the (time-evolved) Thomas basis $\{\hat{e}_1, \hat{e}_2, \hat{e}_3\}$ established by the previous measurement. In general, therefore, to determine the correlation with the first measurement, one must take into account the rotation between the frame $\{\hat{e}_1, \hat{e}_2, \hat{e}_3\}$ carried by the particle (as established by a previous measurement) and the spatial frame $\{\hat{e}'_1, \hat{e}'_2, \hat{e}'_3\}$ of the subsequent measurement.[165]

Suppose now we want to measure some particle property — generically denoted by $\mathcal{A}_\mathrm{P}$ — that changes as the particle's velocity is incremented from $\mathbf{v}$ to $\mathbf{v} + \delta\mathbf{v}$. In the particle's instantaneous rest frame as seen from the laboratory, let us call the respective quantities $\mathcal{A}_\mathrm{P}$ and $\mathcal{A}'_\mathrm{P} = \mathcal{A}_\mathrm{P} + \Delta\mathcal{A}_\mathrm{P}$. These quantities are measured in the lab frame L as $\mathcal{A}_\mathrm{L}$ and $\mathcal{A}'_\mathrm{L} = \mathcal{A}_\mathrm{L} + \delta\mathcal{A}_\mathrm{L}$, respectively. They are related to the corresponding particle rest-frame quantities by *passive* Lorentz boosts. Symbolically we can write (cf. the right-hand side diagram in Fig. 8.11)

$$\text{At time } t: \qquad \mathcal{A}_\mathrm{P} = \Lambda_\mathrm{B}(-\mathbf{v})\mathcal{A}_\mathrm{L}, \tag{8.156a}$$

$$\text{At time } t + \delta t: \qquad \mathcal{A}_\mathrm{P} + \Delta\mathcal{A}_\mathrm{P} = \Lambda_\mathrm{B}(-\mathbf{v} - \delta\mathbf{v})\big(\mathcal{A}_\mathrm{L} + \delta\mathcal{A}_\mathrm{L}\big), \tag{8.156b}$$

$$\Delta\mathcal{A}_\mathrm{P} = \Lambda_\mathrm{B}(-\mathbf{v} - \delta\mathbf{v})\delta\mathcal{A}_\mathrm{P} \tag{8.156c}$$

---

[165] In this context note that, in general, the instantaneous rest-frame bases of observers in different inertial frames will also differ by spatial rotations.

We recall here that the explicit boosts introduced in Sec. 8.2.6 were active boosts; hence the minus signs in the arguments of Λ_{B} to obtain the corresponding passive boosts. The last equation describes how the particle-frame increment $\Delta\mathcal{A}_{\mathrm{P}}$ by itself would appear in the lab frame as $\delta\mathcal{A}_{\mathrm{P}}$. Now, inverting the boost in the second equation, substituting $\mathcal{A}_{\mathrm{P}}$ from the first, and then employing the third equation produces

$$\begin{aligned}
\mathcal{A}_{\mathrm{L}} + \delta\mathcal{A}_{\mathrm{L}} &= \Lambda_{\mathrm{B}}(\mathbf{v} + \delta\mathbf{v})\big[\Lambda_{\mathrm{B}}(-\mathbf{v})\mathcal{A}_{\mathrm{L}} + \Delta\mathcal{A}_{\mathrm{P}}\big] \\
&= \Lambda_{\mathrm{B}}(\mathbf{v} + \delta\mathbf{v})\Lambda_{\mathrm{B}}(-\mathbf{v})\mathcal{A}_{\mathrm{L}} + \Lambda_{\mathrm{B}}(\mathbf{v} + \delta\mathbf{v})\Delta\mathcal{A}_{\mathrm{P}} \\
&= \Lambda_{\mathrm{B}}(\mathbf{v} + \delta\mathbf{v})\Lambda_{\mathrm{B}}(-\mathbf{v})\mathcal{A}_{\mathrm{L}} + \delta\mathcal{A}_{\mathrm{P}},
\end{aligned} \qquad (8.157)$$

and finally

$$\delta\mathcal{A}_{\mathrm{L}} = \delta\mathcal{A}_{\mathrm{P}} + \big[\Lambda_{\mathrm{B}}(\mathbf{v} + \delta\mathbf{v})\Lambda_{\mathrm{B}}(-\mathbf{v}) - \mathbb{1}\big]\mathcal{A}_{\mathrm{L}}, \qquad (8.158)$$

This tells us that the full effect of the change $\delta\mathcal{A}_{\mathrm{L}}$ measured in the lab has two contributions: the change $\delta\mathcal{A}_{\mathrm{P}}$ of what happens in the particle frame (suitably translated to the lab frame) and a purely kinematical effect that arises from making successive measurements employing different boosts. Clearly, if there is no change in velocity ($\delta\mathbf{v} = 0$), the latter effect vanishes. This therefore is an effect only present for particles that are being accelerated, and the corresponding particle rest frames are then non-inertial, i.e., they are being *rotated*.

The Lorentz transformation

$$\Lambda_{\mathrm{T}} = \Lambda_{\mathrm{B}}(\mathbf{v} + \delta\mathbf{v})\Lambda_{\mathrm{B}}(-\mathbf{v}) \qquad (8.159)$$

in (8.157) thus describes the kinematical effect that enter the measurement of $\mathcal{A}'_{\mathrm{L}}$. To evaluate this transformation explicitly, we write

$$\delta\mathbf{v} = c\begin{pmatrix} \delta\beta_1 \\ \delta\beta_2 \\ 0 \end{pmatrix} \qquad (8.160)$$

and $\beta = v_x/c$, and we employ the explicit form (8.125) to obtain

$$\Lambda_{\mathrm{B}}(-\mathbf{v}) = \begin{pmatrix} \gamma & -\gamma\beta & 0 & 0 \\ -\gamma\beta & \gamma & 0 & 0 \\ 0 & 0 & 1 & 0 \\ 0 & 0 & 0 & 1 \end{pmatrix}, \qquad (8.161)$$

and

$$\Lambda_{\mathsf{B}}(\mathbf{v} + \delta\mathbf{v}) = \Lambda_{\mathsf{B}}(\mathbf{v}) + \begin{pmatrix} \gamma^3 \beta \, \delta\beta_1 & \gamma^3 \delta\beta_1 & \gamma \, \delta\beta_2 & 0 \\ \gamma^3 \delta\beta_1 & \gamma^3 \beta \, \delta\beta_1 & \alpha\beta \, \delta\beta_2 & 0 \\ \gamma \, \delta\beta_2 & \alpha\beta \, \delta\beta_2 & 0 & 0 \\ 0 & 0 & 0 & 0 \end{pmatrix}, \qquad (8.162)$$

where only contributions linear in the velocity increments are retained. Straightforward matrix multiplication and then employing the generator matrices (8.95) yields

$$\Lambda_{\mathsf{B}}(\mathbf{v} + \delta\mathbf{v})\Lambda_{\mathsf{B}}(-\mathbf{v}) = \mathbb{1} + \begin{pmatrix} 0 & \gamma^2 \delta\beta_1 & \gamma \, \delta\beta_2 & 0 \\ \gamma^2 \delta\beta_1 & 0 & \alpha\beta \, \delta\beta_2 & 0 \\ \gamma \, \delta\beta_2 & -\alpha\beta \, \delta\beta_2 & 0 & 0 \\ 0 & 0 & 0 & 0 \end{pmatrix}$$

$$= \mathbb{1} - i \left(\gamma^2 \delta\beta_1 \mathbb{K}_x + \gamma \, \delta\beta_2 \mathbb{K}_y - \alpha \, \beta \, \delta\beta_2 \mathbb{J}_z \right). \quad (8.163)$$

According to (8.98), this corresponds to two *infinitesimal* Lorentz transformations: a boost in the xy-plane (generated by $\mathbb{K}_x$ and $\mathbb{K}_y$) along the direction $(\gamma \, \delta\beta_1, \delta\beta_2, 0)$ and a rotation about the z-axis (generated by $\mathbb{J}_z$) by an angle $-\alpha\beta \, \delta\beta_2$. Being infinitesimal, they commute, of course. Noting that in our coordinate system (see Fig. 8.11), where $\delta\beta_1 = \delta v_x/c$ is parallel to $\mathbf{v}$ and $\delta\beta_2 = \delta v_y/c$ is perpendicular, we can write this result in a coordinate-independent way as

$$\Lambda_{\mathsf{B}}(\mathbf{v} + \delta\mathbf{v})\Lambda_{\mathsf{B}}(-\mathbf{v}) = \mathbb{1} - i \left(\delta\boldsymbol{\xi} \cdot \mathbb{K} + \delta\boldsymbol{\omega} \cdot \mathbb{J} \right), \qquad (8.164)$$

where the infinitesimal boost and rotation parameters are given vectorially by

$$\delta\boldsymbol{\xi} = \frac{\gamma^2 \, \delta\mathbf{v}_{\|} + \gamma \, \delta\mathbf{v}_{\perp}}{c} \quad \text{and} \quad \delta\boldsymbol{\omega} = \frac{\gamma^2}{\gamma + 1} \frac{\delta\mathbf{v} \times \mathbf{v}}{c^2}, \qquad (8.165)$$

respectively, with $\delta\mathbf{v} = \delta\mathbf{v}_{\|} + \delta\mathbf{v}_{\perp}$ being the decomposition of the velocity increment into contributions parallel and perpendicular to $\mathbf{v}$. We may write

$$\delta\boldsymbol{\xi} = \gamma\frac{\delta\mathbf{v}}{c} + \delta\boldsymbol{\xi}_{\|}, \quad \text{with} \quad \delta\boldsymbol{\xi}_{\|} = \gamma(\gamma - 1)\frac{\delta\mathbf{v}_{\|}}{c}, \qquad (8.166)$$

where the $\delta\mathbf{v}$ contribution corresponds to the infinitesimal boost $\Lambda_{\mathsf{B}}(\delta\mathbf{v})$. Hence, we find

$$\Lambda_{\mathsf{B}}(\mathbf{v} + \delta\mathbf{v})\Lambda_{\mathsf{B}}(-\mathbf{v}) = \mathbb{1} - i \left(\delta\boldsymbol{\xi}_{\|} \cdot \mathbb{K} + \frac{\gamma}{c}\delta\mathbf{v} \cdot \mathbb{K} + \delta\boldsymbol{\omega} \cdot \mathbb{J} \right)$$

$$= \left(\mathbb{1} - i\delta\boldsymbol{\omega} \cdot \mathbb{J} \right)\left(\mathbb{1} - i\delta\boldsymbol{\xi}_{\|} \cdot \mathbb{K} \right)\left(\mathbb{1} - i\frac{\gamma}{c}\delta\mathbf{v} \cdot \mathbb{K} \right)$$

$$= \Lambda_{\mathsf{R}}(\delta\boldsymbol{\omega})\Lambda_{\mathsf{B}}(\delta\boldsymbol{\xi}_{\|})\Lambda_{\mathsf{B}}(\delta\mathbf{v}), \qquad (8.167)$$

where the three infinitesimal Lorentz transformations on the right may appear in any order. Multiplying the entire equation by $\Lambda_\mathsf{B}(\mathbf{v})$ from the right produces

$$\Lambda_\mathsf{B}(\mathbf{v} + \delta\mathbf{v}) = \Lambda_\mathsf{R}(\delta\boldsymbol{\omega})\Lambda_\mathsf{B}(\delta\boldsymbol{\xi}_\parallel)\Lambda_\mathsf{B}(\delta\mathbf{v})\Lambda_\mathsf{B}(\mathbf{v})$$
$$= \Lambda_\mathsf{R}(\delta\boldsymbol{\omega})\Lambda_\mathsf{B}(\Delta\mathbf{v})\Lambda_\mathsf{B}(\mathbf{v}), \tag{8.168}$$

where

$$\Lambda_\mathsf{B}(\Delta\mathbf{v}) = \Lambda_\mathsf{B}(\delta\boldsymbol{\xi}_\parallel)\Lambda_\mathsf{B}(\delta\mathbf{v}), \tag{8.169}$$

with

$$\Delta\mathbf{v} = \delta\mathbf{v} + (\gamma - 1)\delta\mathbf{v}_\parallel = \gamma\delta\mathbf{v}_\parallel + \delta\mathbf{v}_\perp, \tag{8.170}$$

is the incremental boost necessary to ensure that the observed velocity increase is $\delta\mathbf{v}$ since relativistically velocities cannot be simply added [see (8.118)]. The correction only affects the direction parallel to the initial velocity, of course, and — equally of course — it vanishes in the nonrelativistic limit (where $\gamma = 1$).

Equation (8.168) shows most clearly what happens here. The left-hand side provides the *passive* Lorentz boost that transmits the information from the particle coordinates at $t + \delta t$ to the lab frame that is moving away at a relative velocity $-\mathbf{v} - \delta\mathbf{v}$. On the right-hand side we see that the particle with a velocity $\mathbf{v}$ at time t first is boosted, during a time interval δt, by a velocity increment $\Delta\mathbf{v}$ (that ensures that its *observed* increment is $\delta\mathbf{v}$). Then its coordinate system is rotated by an angle $\delta\boldsymbol{\omega}$ — and it is this rotated system that is being observed at $t + \delta t$. This is a genuine effect of relativistic kinematics and — most remarkably — it does not become negligible in the nonrelativistic limit.

The *angular velocity* associated with this rotation relative to the lab frame is then obtained as

$$\boxed{\boldsymbol{\omega}_\mathrm{T} = \lim_{\delta t \to 0} \frac{\delta\boldsymbol{\omega}}{\delta t} = \frac{\gamma^2}{\gamma + 1}\frac{\mathbf{a} \times \mathbf{v}}{c^2} \xrightarrow{\text{NR}} \frac{\mathbf{a} \times \mathbf{v}}{2c^2}}, \tag{8.171}$$

where $\mathbf{a} = d\mathbf{v}/dt$ is the observed acceleration in the lab frame. The rightmost expression is the nonrelativistic limit. This is the *Thomas angular frequency*. It implies that if a physical vector quantity $\mathbf{A}$ is changing in the particle's accelerated rest frame, its effect in the inertial, non-rotating lab frame will manifest itself according to [cf. Eq. (1.253)]

$$\left.\frac{d\mathbf{A}}{dt}\right|_\mathrm{lab} = \left.\frac{d\mathbf{A}}{dt}\right|_\mathrm{rest} + \boldsymbol{\omega}_\mathrm{T} \times \mathbf{A}. \tag{8.172}$$

This effect, known as the *Thomas precession*, has some very important consequences. Historically, one of the most important ones is the fact that it provides an additive term for the spin-orbit interaction for an atomic electron that cancels half of the contribution obtained without it, thus leaving only half of the original contribution. The resulting overall factor of $1/2$ is usually called the *Thomas factor*.

8.2.7.4   *Relativistic collisions: Mandelstam variables*

Many collision processes of interest between relativistic particles are two-body in nature, i.e, there are two particles going into the interaction region and two particles — not necessarily the same — coming out. Typical examples are elastic nucleon–nucleon scattering $(N+N \to N+N)$, Compton scattering off the electron $(\gamma + e \to \gamma + e)$, or photon-induced pion production off the nucleon $(\gamma + N \to \pi + N)$, etc. Such processes can be depicted schematically as in Fig. 8.13. To calculate them requires a detailed description of the dynamics of what is happening inside the interaction region, which usually is very complex. However, as far as the kinematics of the process is concerned, one can find a number of helpful invariants that make the transformation between various frames much easier.

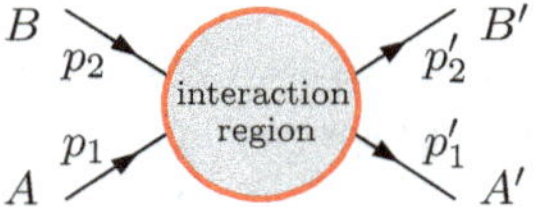

Fig. 8.13   The figure above shows a generic depiction of a two-body collision reaction, where the circle indicates the interaction region, with the two incoming particles having four-momenta p_1 and p_2, and the two outgoing ones p'_1 and p'_2. Four-momentum conservation relates them by $p_1 + p_2 = p'_1 + p'_2$. The three diagrams below show the primary ways of how the momenta can 'flow' through the interaction region to ensure overall four-momentum conservation with individual four-momentum conservation holding true at any of the three-line vertices indicated by the solid dots. The diagrams are classified by the squared four-momenta of the internal momentum-exchange lines that link these vertices as s-, u-, and t-channel diagrams, as indicated by the respective expressions below the diagrams.

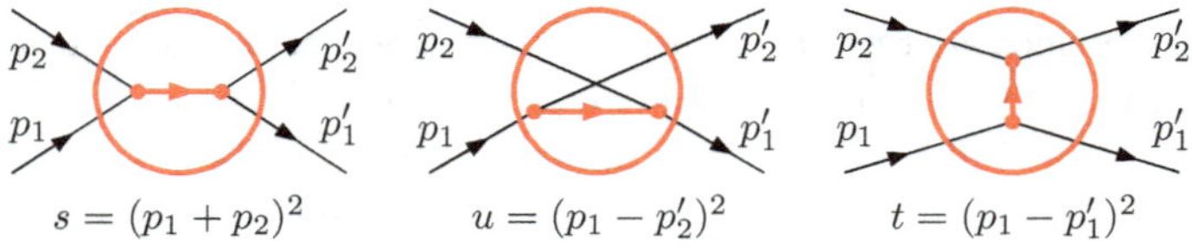

$$s = (p_1 + p_2)^2 \qquad u = (p_1 - p'_2)^2 \qquad t = (p_1 - p'_1)^2$$

Let us consider a generic two-body process

$$A(p_1) + B(p_2) \rightarrow A'(p_1') + B'(p_2') \tag{8.173}$$

for two incoming particles A and B going over into two outgoing particles A' and B', as shown in Fig. 8.13. The arguments here provide the four-momenta of the respective particles. Energy and momentum conservation of the process is very concisely given by *four-momentum conservation*, i.e.,

$$p_1 + p_2 = p_1' + p_2', \tag{8.174}$$

or explicitly

$$\begin{pmatrix} E_1/c \\ \mathbf{p}_1 \end{pmatrix} + \begin{pmatrix} E_2/c \\ \mathbf{p}_2 \end{pmatrix} = \begin{pmatrix} E_1'/c \\ \mathbf{p}_1' \end{pmatrix} + \begin{pmatrix} E_2'/c \\ \mathbf{p}_2' \end{pmatrix}, \tag{8.175}$$

where the time and space components of this equation respectively provide energy and three-momentum conservation. For a physical process, the energy for particle i here is given by [cf. (8.148)]

$$E_i = \sqrt{m_i^2 c^4 + \mathbf{p}_i^2 c^2}, \tag{8.176}$$

where m_i is the particle's rest mass; a similar expression holds for the primed particles.

There are three primary ways one can picture the momenta to 'flow' through the interaction region that is consistent with four-momentum conservations (and non-trivial[166] interaction mechanisms); this is shown in the three diagrams at the bottom of Fig. 8.13. Each of these primary ways involves just one single exchange of momentum between the particles internally and it can therefore be calculated directly in terms of the external momenta. The squares of the corresponding exchanged momenta are the *Mandelstam variables* given by

$$\boxed{s = (p_1 + p_2)^2, \quad u = (p_1 - p_2')^2, \quad \text{and} \quad t = (p_1 - p_1')^2} \,, \tag{8.177}$$

and the associated diagrams are called, respectively, s-, u-, and t-channel diagrams.[167] Clearly, being four-vector squares, the values of s, u, and

---

[166] A trivial 'interaction' mechanism would be to have *no* interaction taking place between the incoming and outgoing particles, in which case one would necessarily have either $p_1 = p_1'$ and $p_2 = p_2'$ or $p_1 = p_2'$ and $p_2 = p_1'$.

[167] We mention in this context that their significance in the reaction dynamics of particle physics is that these diagrams also correspond to the primary interaction mechanisms for the two-body process, with the internal exchanged momentum being carried by an actual particle that mediates the interaction.

t are *invariants* that do not change when transforming between frames. Moreover, one easily shows that

$$\boxed{s + u + t = (m_1^2 + m_2^2 + m_1'^2 + m_2'^2)c^2}\,, \qquad (8.178)$$

i.e., the Mandelstam variables are related to each other by the sum of the squared masses of all reaction partners.

Theorists like to do their calculations in the so-called center-of-momentum frame (CM) where the total three-momentum is zero,

$$\text{CM:} \qquad \mathbf{P}_{\text{CM}} = \mathbf{p}_{1,\text{CM}} + \mathbf{p}_{2,\text{CM}} = \mathbf{p}_{1,\text{CM}}' + \mathbf{p}_{2,\text{CM}}' = 0. \qquad (8.179)$$

Experimentalists have no choice; they must do their measurements in the laboratory (LAB) frame, where usually the target (say, particle B) is at rest,

$$\text{Target } B \text{ at rest in lab:} \qquad \mathbf{p}_{2,\text{LAB}} = 0. \qquad (8.180)$$

Instead of having to perform laborious Lorentz boosts to go back and forth between the LAB and the CM frame, one can employ the invariant Mandelstam variables.

We may now write without lack of generality

$$p_{1,\text{CM}} = \begin{pmatrix} E_1^{\text{CM}}/c \\ \mathbf{p}_{\text{CM}} \end{pmatrix} \quad \text{and} \quad p_{2,\text{CM}} = \begin{pmatrix} E_2^{\text{CM}}/c \\ -\mathbf{p}_{\text{CM}} \end{pmatrix} \qquad (8.181)$$

for the initial four-momenta in the CM frame, with

$$E_1^{\text{CM}} = c\sqrt{m_1^2 c^2 + \mathbf{p}_{\text{CM}}^2} \quad \text{and} \quad E_1^{\text{CM}} = c\sqrt{m_2^2 c^2 + \mathbf{p}_{\text{CM}}^2}, \qquad (8.182)$$

and

$$p_{1,\text{LAB}} = \begin{pmatrix} E_1^{\text{LAB}}/c \\ \mathbf{p}_{\text{LAB}} \end{pmatrix} \quad \text{and} \quad p_{2,\text{LAB}} = \begin{pmatrix} m_2 c \\ 0 \end{pmatrix} \qquad (8.183)$$

for the same momenta in the laboratory frame, with

$$E_1^{\text{LAB}} = c\sqrt{m_1^2 c^2 + \mathbf{p}_{\text{LAB}}^2}, \qquad (8.184)$$

taking the (fixed) target to be particle 2.

In the CM frame, we find[168]

$$s = \frac{(E_1^{\text{CM}} + E_2^{\text{CM}})^2}{c^2} \geq (m_1 + m_2)^2 c^2 \qquad (8.185)$$

which shows that the corresponding total energy is essentially given by s, i.e.,

$$E_{\text{tot}}^{\text{CM}} = E_1^{\text{CM}} + E_2^{\text{CM}} = c\sqrt{s}. \qquad (8.186)$$

---

[168]The lower bound given here is based on kinematics alone. Depending on the particular process, the lower bound may actually be larger.

Moreover,

$$\mathbf{p}_{\text{CM}}^2 = \frac{(s - m_1^2 c^2 - m_2^2 c^2)^2 - 4 m_1^2 m_2^2 c^4}{4s}, \tag{8.187}$$

and thus

$$E_1^{\text{CM}} = c \frac{s + m_1^2 c^2 - m_2^2 c^2}{2\sqrt{s}} \quad \text{and} \quad E_2^{\text{CM}} = c \frac{s + m_2^2 c^2 - m_1^2 c^2}{2\sqrt{s}}. \tag{8.188}$$

For the LAB frame, one has

$$s = \frac{(E_1^{\text{LAB}} + m_1 c^2)^2}{c^2} - \mathbf{p}_{\text{LAB}}^2, \tag{8.189}$$

from which one finds

$$E_1^{\text{LAB}} = \frac{s - m_1^2 c^2 - m_2^2 c^2}{2 m_2} \tag{8.190}$$

and

$$\mathbf{p}_{\text{LAB}}^2 = \frac{(s - m_1^2 c^2 - m_2^2 c^2)^2 - 4 m_1^2 m_2^2 c^4}{4 m_2^2 c^2}. \tag{8.191}$$

Comparing the last equation with (8.187) provides

$$\frac{\mathbf{p}_{\text{LAB}}^2}{s} = \frac{\mathbf{p}_{\text{CM}}^2}{m_2^2 c^2}, \tag{8.192}$$

which is a very simple relationship between the three-momenta in the CM and LAB frames.

All energy and momentum transformations between the two frames are thus seen to be facilitated by the invariant variable s. In a similar manner we may also express the relations between the collision angles in the two frames by utilizing the variable t. The details are left as an exercise.

8.2.8 *Relativistic point-particle mechanics*

Let us consider a free particle of rest mass m moving through space with a velocity $v = |\dot{\mathbf{r}}|$. We expect that the motion is such that the invariant action integral

$$S = -mc \int ds = -mc^2 \int d\tau = -mc^2 \int dt \sqrt{1 - \frac{v^2}{c^2}} \tag{8.193}$$

is minimized, with the particle moving along a Minkowski geodesic defined by the variation $\delta \int ds = 0$.[169] We have used here

$$ds = \sqrt{c^2 dt^2 - d\mathbf{r}^2} = c\, dt \sqrt{1 - \frac{1}{c^2} \left(\frac{d\mathbf{r}}{dt} \right)^2}. \tag{8.194}$$

---

[169]Because of the peculiarities of the Minkowski metric, $\int ds$ has its *maximum* value for a straight line connecting two events, i.e., any other curvilinear path connecting the two events has a lower value. This is the reason for the minus sign in (8.193); it ensures that the physical paths for a free particle — straight lines — correspond to a minimum of S.

Equating S with $\int dt\, L$, the corresponding Lagrange function is

$$L = -mc^2 \sqrt{1 - \frac{v^2}{c^2}}. \tag{8.195}$$

The nonrelativistic limit of this expression is

$$L_{\mathrm{NR}} = -mc^2 \left(1 - \frac{1}{2}\frac{v^2}{c^2}\right) = -mc^2 + \frac{m}{2}\dot{\mathbf{r}}^2, \tag{8.196}$$

which, apart from an inconsequential constant term,[170] is just the kinetic energy of the free nonrelativistic particle, as it ought to be.

The relativistic free-particle Lagrangian provides now the canonical three-momentum

$$\mathbf{p} = \boldsymbol{\nabla}_{\dot{\mathbf{r}}} L = \frac{m}{\sqrt{1 - \frac{v^2}{c^2}}} \frac{d\mathbf{r}}{dt} = mc\frac{d\mathbf{r}}{ds} = m\frac{d\mathbf{r}}{d\tau} \tag{8.197}$$

in terms of the three-velocity

$$\mathbf{u} = \frac{d\mathbf{r}}{d\tau}. \tag{8.198}$$

Squaring the three-momentum relation, we find

$$\frac{v^2}{c^2} = \frac{\mathbf{p}^2}{\mathbf{p}^2 + m^2 c^2} \tag{8.199}$$

which we need for eliminating the velocities when calculating the Hamilton function,

$$H = \mathbf{p}\cdot\dot{\mathbf{r}} - L = \left(\frac{\mathbf{p}^2}{m} + mc^2\right)\sqrt{\frac{m^2 c^2}{\mathbf{p}^2 + m^2 c^2}}$$

$$= \sqrt{\mathbf{p}^2 c^2 + m^2 c^4}. \tag{8.200}$$

Since this describes a free particle, this is the conserved energy E of the particle. We may thus write

$$\frac{E}{c} = \sqrt{\mathbf{p}^2 + m^2 c^2} = \frac{mc}{\sqrt{1 - \frac{v^2}{c^2}}}$$

$$= \frac{mc}{\sqrt{1 - \frac{v^2}{c^2}}}\frac{c\,dt}{c\,dt}$$

---

[170]The extra term mc^2 arises from the rest-mass energy that acts like a constant 'potential' contribution in the nonrelativistic limit. It can be removed in the limit, i.e., $L + mc^2 \to L_{\mathrm{NR}} = mv^2/2$, just as one can remove it from the nonrelativistic limit of the energy, $E - mc^2 \to E_{\mathrm{NR}} = mv^2/2$ since nonrelativistically the energy is measured against the (constant) rest-mass energy mc^2 as a reference level.

$$= mc\frac{d(ct)}{ds}$$

$$= m\frac{d(ct)}{d\tau}. \tag{8.201}$$

Combined with (8.197), this implies that

$$p \equiv \begin{pmatrix} E/c \\ \mathbf{p} \end{pmatrix} = m\frac{d}{d\tau}\begin{pmatrix} ct \\ \mathbf{r} \end{pmatrix} \tag{8.202}$$

properly transforms as a Lorentz four-vector since $(ct, \mathbf{r})$ is a four-vector and the eigentime element $d\tau$ is an invariant. In other words, the momentum four-vector introduced by fiat in Sec. 8.2.7.2 is indeed the one that is consistent with Hamilton's principle.

8.2.8.1 *Relativizing Newton's second law*

The problem of finding equations of motion for mechanical systems of relativistic particles subject to some interaction is not at all a clear-cut one. Given what we have learned so far, we may suspect that the generic structure will by like a relativized version of Newton's second law,

$$\frac{d(mu^\mu)}{d\tau} = K^\mu, \tag{8.203}$$

where the four-vector on the right is referred to as the four-force or *Minkowski force*. To relate this to Newton's equation of motion, we write out the space components as

$$\gamma\frac{d(mu^i)}{dt} = K^i. \tag{8.204}$$

The problem is now that in most cases the nonrelativistic expressions $\mathbf{F} = (F^1, F^2, F^3)$ of extant forces provide only insufficient guidance as to how to choose the components K^μ of the corresponding Minkowski force since simply requiring the nonrelativistic limit $K^i \to \gamma F^i$ as $\beta \to 0$ to ensure $d(m\dot{x}^i)/dt = F^i$ does not sufficiently constrain the choices for K^μ.

There is one problem, however, where this is not the case, and this is the problem that started it all. One of the motivating factors for developing special relativity was to accommodate the findings of Maxwell's electromagnetic theory. We expect, therefore, that we can find clean answers for this case.

8.2.8.2   *Particle in electromagnetic field: Lorentz force*

We had treated the case of a particle of mass m and charge e in an electromagnetic field in Sec. 2.2.2.5 and found its (nonrelativistic) Lagrangian as

$$L_{\mathrm{NR}} = \frac{m}{2}\dot{\mathbf{r}}^2 - U, \tag{8.205}$$

where we had introduced the generalized potential

$$U = \frac{e}{c}\left[c\,\Phi(\mathbf{r},t) - \dot{\mathbf{r}}\cdot\mathbf{A}(\mathbf{r},t)\right]. \tag{8.206}$$

We recall that Φ and $\mathbf{A}$ here are the electromagnetic scalar and vector potentials, respectively, that provide the electric and magnetic fields,

$$\mathbf{E}(\mathbf{r},t) = -\boldsymbol{\nabla}\Phi(\mathbf{r},t) - \frac{1}{c}\frac{\partial\mathbf{A}(\mathbf{r},t)}{\partial t}, \tag{8.207a}$$

$$\mathbf{B}(\mathbf{r},t) = \boldsymbol{\nabla}\times\mathbf{A}(\mathbf{r},t), \tag{8.207b}$$

respectively. Both fields together produce the *Lorentz force*,

$$\mathbf{F}_{\mathrm{L}} = e\mathbf{E} + \frac{e}{c}\dot{\mathbf{r}}\times\mathbf{B}, \tag{8.208}$$

as had been shown in Eq. (2.152).

Similar to the procedure for the free particle, the task is now to find a manifestly form-invariant action $S = \int d\tau\,\mathscr{L}$ that allows us to deduce how to modify the Lagrange function (8.205) to account for special relativity. And then, we seek to find the corresponding manifestly form-invariant equations of motion, which in this case means finding a Minkowski force whose space part provides the Lorentz force, at least in the nonrelativistic limit.

From the free-particle case of the previous section, we know already that we should make the replacement

$$\frac{m}{2}\dot{\mathbf{r}}^2 \;\longrightarrow\; -mc\sqrt{c^2 - \dot{\mathbf{r}}^2} \tag{8.209}$$

for the kinetic energy in L_{NR}. What about the potential term? First, note that the four-velocity can be written as

$$u^\mu = \frac{dx^\mu}{d\tau} = \gamma\frac{dx^\mu}{dt} = \gamma\begin{pmatrix}c\\\dot{\mathbf{r}}\end{pmatrix} \tag{8.210}$$

and that combining the scalar and the vector potentials into a four-vector according to

$$\boxed{A^\mu = \begin{pmatrix}\Phi\\\mathbf{A}\end{pmatrix}}, \tag{8.211}$$

their scalar product produces

$$u_\mu A^\mu = \gamma\left(c\,\Phi - \dot{\mathbf{r}}\cdot\mathbf{A}\right) = \gamma\frac{c}{e}U, \qquad (8.212)$$

which apart from an overall factor is exactly U. In doing so, let us just accept for the moment that A^μ — called the *four-potential* — is indeed a proper four-vector (and not just some arbitrary collection of elements put together to give the appearance of a four-vector, but without the correct transformation properties); we will get back to this point later, in connection with Eq. (8.237) below. Next, let us look at

$$\frac{e}{c}\int d\tau\, u_\mu A^\mu = \frac{e}{c}\int \frac{dt}{\gamma}\gamma\frac{c}{e}U = \int dt\, U, \qquad (8.213)$$

and we find that the form-invariant integral on the left is exactly the same as the potential contribution for the nonrelativistic action integral on the right.

We may therefore write the action integral as

$$S = \int d\tau\,\mathscr{L} = \int dt\, L, \qquad (8.214)$$

where the first integral is manifestly covariant, with a form-invariant Lagrange density

$$\mathscr{L} = -mc^2 - \frac{e}{c}u_\mu A^\mu, \qquad (8.215)$$

and the second one employs the Lagrangian

$$L = -\frac{mc^2}{\gamma} - U = -mc\sqrt{c^2 - \dot{\mathbf{r}}^2} - \frac{e}{c}\left(cA^0 - \dot{\mathbf{r}}\cdot\mathbf{A}\right) \qquad (8.216)$$

which differs from (8.205) only in the kinetic part — the interaction U stays the same.

The canonical three-momentum of the Lagrangian L is given by

$$\mathbf{p} = \nabla_{\dot{\mathbf{r}}}L = m\gamma\dot{\mathbf{r}} + \frac{e}{c}\mathbf{A}, \qquad (8.217)$$

where we find that the previous three-momentum part for the free particle, $m\gamma\dot{\mathbf{r}}$, has now been changed by the additive vector-potential term. The corresponding Euler–Lagrange equation,

$$\frac{d\mathbf{p}}{dt} - \nabla_{\mathbf{r}}L = 0 \quad \Rightarrow \quad \frac{d}{dt}\left(m\gamma\dot{\mathbf{r}} + \frac{e}{c}\mathbf{A}\right) + \nabla_{\mathbf{r}}U = 0, \qquad (8.218)$$

may be rearranged and written out in space components as

$$\frac{d}{dt}(mu^i) = \partial^i U - \frac{e}{c}\frac{d}{dt}A^i$$

$$= \frac{e}{\gamma c}(\partial^i A^\nu)u_\nu - \frac{e}{c}\frac{d}{dt}A^i, \tag{8.219}$$

for $i = 1, 2, 3$, where the contravariant partial derivative notation was employed (see Sec. 8.2.3.1). Use was also made of the fact that velocities are functions of time only. Multiplying the entire equation by γ, we then obtain

$$\frac{d}{d\tau}(mu^i) = \frac{e}{c}\left[(\partial^i A^\nu)u_\nu - \frac{dA^i}{d\tau}\right], \tag{8.220}$$

which has the structure of (8.204) and thus suggests to consider the right-hand side here as the space part K^i of the Minkowski force for the electromagnetic case. Indeed, one easily shows by direct evaluation that the right-hand side here is $\gamma F_{\rm L}^i$, where $F_{\rm L}^i$ is a component of the Lorentz force (8.208). Derivation of this result is deferred to Eq. (8.225) below.

Electromagnetic field tensor

Using the identity

$$\frac{dA^i}{d\tau} = \frac{\partial A^i}{\partial x_\nu}\frac{dx_\nu}{d\tau} = (\partial^\nu A^i)u_\nu, \tag{8.221}$$

we can rewrite (8.220) equivalently as

$$\frac{d}{d\tau}(mu^i) = \frac{e}{c}\left(\partial^i A^\nu - \partial^\nu A^i\right)u_\nu, \tag{8.222}$$

which shows conclusively that for the right-hand side here to provide the space components of the Minkowski force K^μ, the four-vector K^μ *must* have the covariant structure of a second-rank Lorentz tensor $F^{\mu\nu}$ contracted with the four-velocity u_ν, i.e., $K^\mu = (e/c)F^{\mu\nu}u_\nu$. Equation (8.222) directly provides 12 of the 16 elements of $F^{\mu\nu}$ according to

$$F^{i0} = \partial^i A^0 - \partial^0 A^i = E^i, \tag{8.223a}$$

$$F^{ij} = \partial^i A^j - \partial^j A^i = -\varepsilon_{ijk}B^k, \tag{8.223b}$$

which relates the elements to electric and magnetic field components as given in (8.207). Extending the anti-symmetry of $F^{ji} = -F^{ij}$ to $F^{0i} = -F^{i0}$ (see also footnote 171), one then obtains

$$F^{\mu\nu} = \partial^\mu A^\nu - \partial^\nu A^\mu = \begin{pmatrix} 0 & -E_x & -E_y & -E_z \\ E_x & 0 & -B_z & B_y \\ E_y & B_z & 0 & -B_x \\ E_z & -B_y & B_x & 0 \end{pmatrix}. \tag{8.224}$$

This *anti-symmetric* matrix is called the contravariant *electromagnetic field tensor* (also known by *field-strength* tensor or *Faraday* tensor, etc.). Contracting this tensor with the four-velocity, we obtain by straightforward matrix multiplication

$$F^{\mu\nu}u_\nu = \gamma \begin{pmatrix} 0 & -E_x & -E_y & -E_z \\ E_x & 0 & -B_z & B_y \\ E_y & B_z & 0 & -B_x \\ E_z & -B_y & B_x & 0 \end{pmatrix} \begin{pmatrix} c \\ -\dot{x} \\ -\dot{y} \\ -\dot{z} \end{pmatrix}$$

$$= \gamma \begin{pmatrix} \mathbf{E} \cdot \dot{\mathbf{r}} \\ c\mathbf{E} + \dot{\mathbf{r}} \times \mathbf{B} \end{pmatrix}. \tag{8.225}$$

Apart from some factors, the space component of the right-hand side is just the Lorentz force of (8.208), as alluded to above. Thus, one finds that

$$\boxed{\frac{d(mu^\mu)}{d\tau} = \frac{e}{c}F^{\mu\nu}u_\nu} \tag{8.226}$$

is indeed the correct form-invariant generalization of Newton's second law for this case, with the Minkowski force given by the contracted field tensor on the right that may be written as

$$\boxed{K^\mu = \frac{e}{c}F^{\mu\nu}u_\nu = \frac{e}{c}\left[(\partial^\mu A^\nu)u_\nu - \frac{dA^\mu}{d\tau}\right]}. \tag{8.227}$$

This is the covariant extension of the space components of Eq. (8.220) obtained via the Lagrange formalism. As shown by the explicit expression (8.225), the time component here describes the power transferred to the charged particle by the electric field.[171]

One must realize now that accepting this result, it follows necessarily that the electric and magnetic fields do *not* transform as the space parts of proper four-vectors. Instead, *the* $\mathbf{E}$ *and* $\mathbf{B}$ *fields transform as the elements of the field tensor* F. Indeed, subjecting the force equation to a Lorentz transformation Λ, we have

$$\Lambda^\mu{}_\sigma \frac{d(mu^\sigma)}{d\tau} = \frac{d}{d\tau}\Lambda^\mu{}_\sigma(mu^\sigma) = \frac{d(mu'^\mu)}{d\tau} \tag{8.228}$$

---

[171]The time component K^0 does not contain any information independent of its space part $\mathbf{K}$, i.e., with the space part given, one can find the corresponding time-component equation. This is true for *any* four-force $K^\mu = (K^0, \mathbf{K})$ with a Newtonian equation of motion, $dp^\mu/d\tau = K^\mu$. The proof is left as an exercise. — Note that this also determines the four $F^{0\nu}$ elements in the first row of the field-strength tensor, in addition to the twelve space elements $F^{i\nu}$ given by (8.223).

for the left-hand side of (8.226), where the prime indicates transformed entities. Using the same transformation on the right-hand side, we obtain

$$\Lambda^{\mu}{}_{\sigma}\,(Fu)^{\sigma} = \underbrace{\Lambda^{\mu}{}_{\sigma}F^{\sigma\rho}\Lambda^{\nu}{}_{\rho}}_{=F'^{\mu\nu}}\,\underbrace{\Lambda_{\nu}{}^{\tau}u_{\tau}}_{=u'_{\nu}} = F'^{\mu\nu}u'_{\nu}, \qquad (8.229)$$

with

$$\Lambda^{\nu}{}_{\rho}\Lambda_{\nu}{}^{\tau} = (\Lambda^{-1})_{\rho}{}^{\nu}\Lambda_{\nu}{}^{\tau} = \delta_{\rho}{}^{\tau}. \qquad (8.230)$$

As expected, F thus indeed transforms like a rank-two Lorentz tensor,

$$\boxed{F'^{\mu\nu} = \Lambda^{\mu}{}_{\sigma}\Lambda^{\nu}{}_{\rho}F^{\sigma\rho} = (\Lambda F \Lambda^{\mathrm{T}})^{\mu\nu}}. \qquad (8.231)$$

Moreover, identifying the element entries of the transformed tensor with the corresponding electric and magnetic field elements in the primed frame according to (8.224) will tell us how the $\mathbf{E}$ and $\mathbf{B}$ fields transform.

Viewing the electric and magnetic fields of an (unprimed) reference system at rest from a (primed) system moving with uniform relative velocity $\mathbf{u}$, one easily finds by applying the passive Lorentz boost $\Lambda_{\mathsf{B}}(-\mathbf{u})$ that the field components parallel ($\parallel$) and perpendicular ($\perp$) to $\mathbf{u}$ transform like

$$\mathbf{E}'_{\parallel} = \mathbf{E}_{\parallel}, \qquad\qquad \mathbf{E}'_{\perp} = \gamma\,(\mathbf{E}_{\perp} + \boldsymbol{\beta} \times \mathbf{B}_{\perp})\,, \qquad (8.232\mathrm{a})$$

$$\mathbf{B}'_{\parallel} = \mathbf{B}_{\parallel}, \qquad\qquad \mathbf{B}'_{\perp} = \gamma\,(\mathbf{B}_{\perp} - \boldsymbol{\beta} \times \mathbf{E}_{\perp})\,, \qquad (8.232\mathrm{b})$$

where $\boldsymbol{\beta} = \mathbf{u}/c$ is the boost vector and γ the associated Lorentz factor; the sums of parallel and perpendicular components uniquely determine the respective fields (e.g., $\mathbf{E} = \mathbf{E}_{\parallel} + \mathbf{E}_{\perp}$, with $\mathbf{E}_{\parallel} = (\mathbf{E}\cdot\boldsymbol{\beta})\boldsymbol{\beta}/\beta^{2}$, etc.). The details of this derivation are left as an exercise. Experimentally, one finds that the fields determined in this manner indeed describe the physics correctly.

The fact that electric and magnetic fields thus are inextricably linked by their joint transformation properties in terms of the field tensor suggests that it is more appropriate to speak of a unifying *electromagnetic tensor field* $F^{\mu\nu}$ rather than of individual, separate $\mathbf{E}$ and $\mathbf{B}$ fields since the latter manifestations are frame-dependent, as shown by the transformations (8.232).

8.2.8.3   *Minimal substitution*

Let us now return to the canonical three-momentum $\mathbf{p}$ of Eq. (8.217). The momentum

$$\boldsymbol{\pi} \equiv m\gamma\dot{\mathbf{r}} = \mathbf{p} - \frac{e}{c}\mathbf{A} \qquad (8.233)$$

is the particle's *kinematical,* or *mechanical, momentum.* With the previous result (8.200) for the free particle, replacing $\mathbf{p}$ by $\boldsymbol{\pi}$, we immediately determine the Hamilton function associated with L of (8.216) as

$$H = \mathbf{p} \cdot \dot{\mathbf{r}} - L = c\sqrt{\left(\mathbf{p} - \frac{e}{c}\mathbf{A}\right)^2 + m^2c^2} + eA^0. \tag{8.234}$$

Subtracting eA^0 from both sides and squaring the resulting expression then produces

$$\left(\frac{E}{c} - \frac{e}{c}A^0\right)^2 - \left(\mathbf{p} - \frac{e}{c}\mathbf{A}\right)^2 = m^2c^2, \tag{8.235}$$

where we have also used the fact that the Hamilton function is equal to the energy, $E = H$. This looks exactly like the squared four-momentum length (8.145), but with the replacement

$$\boxed{p \;\longrightarrow\; p - \frac{e}{c}A = \begin{pmatrix} E/c \\ \mathbf{p} \end{pmatrix} - \frac{e}{c}\begin{pmatrix} A^0 \\ \mathbf{A} \end{pmatrix}}. \tag{8.236}$$

This replacement is called *minimal substitution* and it provides the general prescription for how to introduce the electromagnetic interaction for a free (structureless) particle of four-momentum p within a Hamiltonian formalism.

Note here that the invariance of (8.235), i.e.,

$$\left(p - \frac{e}{c}A\right)^2 = m^2c^2, \tag{8.237}$$

is proof that $A^\mu = (A^0, \mathbf{A}) = (\Phi, \mathbf{A})$ is indeed a proper four-vector since we know that p^μ is one.

Appendices

A   Coordinates, Vector Operations, etc.

For reference purposes, this Appendix provides a listing of various relations in Cartesian, cylindrical, spherical, and arbitrary curvilinear coordinates, without derivations and without detailed explanations. More details can be found in textbooks on mathematical physics (for example, [8, 9]).

A.1   *Cartesian coordinates x, y, z*

Unit vectors:

$$\mathbf{e}_x = \begin{pmatrix} 1 \\ 0 \\ 0 \end{pmatrix}, \qquad \mathbf{e}_y = \begin{pmatrix} 0 \\ 1 \\ 0 \end{pmatrix}, \qquad \mathbf{e}_z = \begin{pmatrix} 0 \\ 0 \\ 1 \end{pmatrix}. \tag{A.1}$$

Position vector:

$$\mathbf{r} = x\mathbf{e}_x + y\mathbf{e}_y + z\mathbf{e}_z. \tag{A.2}$$

Position-vector increment:

$$d\mathbf{r} = \mathbf{e}_x dx + \mathbf{e}_y dy + \mathbf{e}_z dz. \tag{A.3}$$

Line element:

$$ds = |d\mathbf{r}| = \sqrt{dx^2 + dy^2 + dz^2}. \tag{A.4}$$

Volume element:

$$d^3 r = dx\, dy\, dz. \tag{A.5}$$

Gradient:

$$\boldsymbol{\nabla} f = \mathbf{e}_x \frac{\partial}{\partial x} f + \mathbf{e}_y \frac{\partial}{\partial y} f + \mathbf{e}_z \frac{\partial}{\partial z} f. \tag{A.6}$$

Divergence:

$$\boldsymbol{\nabla} \cdot \mathbf{F} = \frac{\partial}{\partial x} F_x + \frac{\partial}{\partial y} F_y + \frac{\partial}{\partial z} F_z. \tag{A.7}$$

Curl:

$$\boldsymbol{\nabla} \times \mathbf{F} = \mathbf{e}_x \left(\frac{\partial}{\partial y} F_z - \frac{\partial}{\partial z} F_y \right) + \mathbf{e}_y \left(\frac{\partial}{\partial z} F_x - \frac{\partial}{\partial x} F_z \right)$$
$$+ \mathbf{e}_z \left(\frac{\partial}{\partial x} F_y - \frac{\partial}{\partial y} F_x \right). \tag{A.8}$$

Laplacian:

$$\boldsymbol{\nabla}^2 \Phi = \frac{\partial^2}{\partial x^2} \Phi + \frac{\partial^2}{\partial y^2} \Phi + \frac{\partial^2}{\partial z^2} \Phi. \tag{A.9}$$

A.2 *Cylindrical coordinates ρ, ϕ, z*

Coordinate transformations (see Fig. A.1):

$$\rho = \sqrt{x^2 + y^2}, \qquad \phi = \arctan \frac{y}{x}, \qquad z = z. \tag{A.10}$$

Unit vectors:

$$\mathbf{e}_\rho = \begin{pmatrix} \cos\phi \\ \sin\phi \\ 0 \end{pmatrix}, \qquad \mathbf{e}_\phi = \begin{pmatrix} -\sin\phi \\ \cos\phi \\ 0 \end{pmatrix}, \qquad \mathbf{e}_z = \begin{pmatrix} 0 \\ 0 \\ 1 \end{pmatrix}. \tag{A.11}$$

Position vector:

$$\mathbf{r} = \rho\mathbf{e}_\rho + z\mathbf{e}_z. \tag{A.12}$$

Position-vector increment:

$$d\mathbf{r} = \mathbf{e}_\rho d\rho + \mathbf{e}_\phi \rho\, d\phi + \mathbf{e}_z dz. \tag{A.13}$$

Line element:

$$ds = |d\mathbf{r}| = \sqrt{d\rho^2 + \rho^2\, d\phi^2 + dz^2}. \tag{A.14}$$

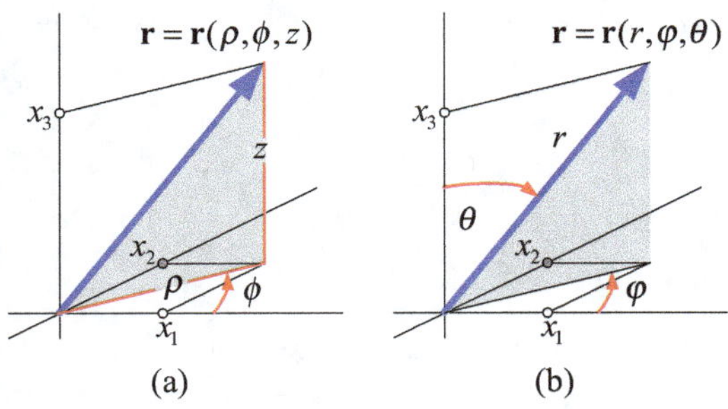

Fig. A.1 Relationship of (a) cylindrical coordinates (ρ, ϕ, z) and (b) spherical coordinates (r, φ, θ) to Cartesian coordinates $(x_1, x_2, x_3) \equiv (x, y, z)$.

Volume element:

$$d^3r = d\rho(\rho\,d\phi)dz = \rho\,d\rho\,d\phi\,dz. \tag{A.15}$$

Gradient:

$$\boldsymbol{\nabla} f = \mathbf{e}_\rho \frac{\partial}{\partial\rho} f + \mathbf{e}_\phi \frac{1}{\rho}\frac{\partial}{\partial\phi} f + \mathbf{e}_z \frac{\partial}{\partial z} f. \tag{A.16}$$

Divergence:

$$\boldsymbol{\nabla}\cdot\mathbf{F} = \frac{1}{\rho}\frac{\partial}{\partial\rho}(\rho F_\rho) + \frac{1}{\rho}\frac{\partial}{\partial\phi} F_\phi + \frac{\partial}{\partial z} F_z. \tag{A.17}$$

Curl:

$$\boldsymbol{\nabla}\times\mathbf{F} = \mathbf{e}_\rho \left[\frac{1}{\rho}\frac{\partial}{\partial\phi} F_z - \frac{\partial}{\partial z} F_\phi \right] + \mathbf{e}_\phi \left[\frac{\partial}{\partial z} F_\rho - \frac{\partial}{\partial\rho} F_z \right]$$

$$+ \mathbf{e}_z \frac{1}{\rho} \left[\frac{\partial}{\partial\rho}(\rho F_\phi) - \frac{\partial}{\partial\phi} F_\rho \right]. \tag{A.18}$$

Laplacian:

$$\boldsymbol{\nabla}^2\Phi = \frac{1}{\rho}\frac{\partial}{\partial\rho}\left(\rho\frac{\partial}{\partial\rho}\Phi\right) + \frac{1}{\rho^2}\frac{\partial^2}{\partial\phi^2}\Phi + \frac{\partial^2}{\partial z^2}\Phi. \tag{A.19}$$

A.3   *Spherical coordinates r, φ, θ*

Coordinate transformations (see Fig. A.1):

$$r = \sqrt{x^2 + y^2 + z^2}, \quad \theta = \arctan\frac{\sqrt{x^2+y^2}}{z}, \quad \varphi = \arctan\frac{y}{x}. \tag{A.20}$$

Unit vectors:

$$\mathbf{e}_r = \begin{pmatrix} \sin\theta\cos\varphi \\ \sin\theta\sin\varphi \\ \cos\theta \end{pmatrix}, \quad \mathbf{e}_\theta = \begin{pmatrix} \cos\theta\cos\varphi \\ \cos\theta\sin\varphi \\ -\sin\theta \end{pmatrix}, \quad \mathbf{e}_\varphi = \begin{pmatrix} -\sin\varphi \\ \cos\varphi \\ 0 \end{pmatrix}. \tag{A.21}$$

Position vector:

$$\mathbf{r} = r\mathbf{e}_r. \tag{A.22}$$

Position-vector increment:

$$d\mathbf{r} = \mathbf{e}_r dr + \mathbf{e}_\theta r\,d\theta + \mathbf{e}_\varphi r\sin\theta\,d\varphi. \tag{A.23}$$

Line element:

$$ds = |d\mathbf{r}| = \sqrt{dr^2 + r^2\,d\theta^2 + r^2\sin^2\theta\,d\varphi^2}. \tag{A.24}$$

Volume element:

$$d^3r = dr(r\,d\theta)(r\sin\theta\,d\varphi) = r^2\sin\theta\,dr\,d\theta\,d\varphi. \tag{A.25}$$

Gradient:

$$\boldsymbol{\nabla} f = \mathbf{e}_r\frac{\partial}{\partial r}f + \mathbf{e}_\theta\frac{1}{r}\frac{\partial}{\partial\theta}f + \mathbf{e}_\varphi\frac{1}{r\sin\theta}\frac{\partial}{\partial\varphi}f. \tag{A.26}$$

Divergence:

$$\boldsymbol{\nabla}\cdot\mathbf{F} = \frac{1}{r^2}\frac{\partial}{\partial r}(r^2 F_r) + \frac{1}{r\sin\theta}\frac{\partial}{\partial\theta}(\sin\theta F_\theta) + \frac{1}{r\sin\theta}\frac{\partial}{\partial\varphi}F_\varphi. \tag{A.27}$$

Curl:

$$\begin{aligned}
\boldsymbol{\nabla}\times\mathbf{F} =\ & \mathbf{e}_r\frac{1}{r\sin\theta}\left[\frac{\partial}{\partial\theta}(\sin\theta F_\varphi) - \frac{\partial}{\partial\varphi}F_\theta\right] \\[1ex]
& + \mathbf{e}_\theta\frac{1}{r}\left[\frac{1}{\sin\theta}\frac{\partial}{\partial\varphi}F_r - \frac{\partial}{\partial r}(rF_\varphi)\right] \\[1ex]
& + \mathbf{e}_\varphi\frac{1}{r}\left[\frac{\partial}{\partial r}(rF_\theta) - \frac{\partial}{\partial\theta}F_r\right].
\end{aligned} \tag{A.28}$$

Laplacian:

$$\begin{aligned}
\boldsymbol{\nabla}^2\Phi =\ & \frac{1}{r^2}\frac{\partial}{\partial r}\left(r^2\frac{\partial}{\partial r}\Phi\right) + \frac{1}{r^2\sin\theta}\frac{\partial}{\partial\theta}\left(\sin\theta\frac{\partial}{\partial\theta}\Phi\right) \\[1ex]
& + \frac{1}{r^2\sin^2\theta}\frac{\partial^2}{\partial\varphi^2}\Phi.
\end{aligned} \tag{A.29}$$

A.4 *Arbitrary curvilinear coordinates q_1, q_2, q_3*

Coordinate transformations:

$$q_1 = f_1(x,y,z), \qquad q_2 = f_2(x,y,z), \qquad q_3 = f_3(x,y,z). \tag{A.30}$$

Unit vectors:

$$\mathbf{u}_i = \frac{1}{h_i}\frac{\partial\mathbf{r}}{\partial q_i}, \quad \text{for}\quad i = 1,2,3 \quad [\text{defines } h_i = h_i(q_1,q_2,q_3)]\,. \tag{A.31}$$

The unit vectors $\mathbf{u}_1$, $\mathbf{u}_2$, $\mathbf{u}_3$ here are expressed in terms of the Cartesian unit vectors $\mathbf{e}_x$, $\mathbf{e}_y$, $\mathbf{e}_z$ by inverting (A.30) and replacing x, y, z in (A.2) by expressions in q_1, q_2, q_3 accordingly. The partial derivatives $\partial\mathbf{r}/\partial q_i$ then follow as

$$h_i\mathbf{u}_i = \frac{\partial\mathbf{r}}{\partial q_i} = \frac{\partial x}{\partial q_i}\mathbf{e}_x + \frac{\partial y}{\partial q_i}\mathbf{e}_y + \frac{\partial z}{\partial q_i}\mathbf{e}_z, \quad \text{for}\quad i = 1,2,3, \tag{A.32}$$

where

$$h_i = h_i(q_1, q_2, q_3) = \sqrt{\left(\frac{\partial x}{\partial q_i}\right)^2 + \left(\frac{\partial y}{\partial q_i}\right)^2 + \left(\frac{\partial z}{\partial q_i}\right)^2} \qquad \text{(A.33)}$$

normalizes the $\mathbf{u}_i$ to unity.
Position-vector increment:

$$d\mathbf{r} = \mathbf{u}_1 h_1\, dq_1 + \mathbf{u}_2 h_2\, dq_2 + \mathbf{u}_3 h_3\, dq_3. \qquad \text{(A.34)}$$

Line element:

$$ds = |d\mathbf{r}| = \sqrt{h_1^2\, dq_1^2 + h_2^2\, dq_2^2 + h_3^2\, dq_3^2}. \qquad \text{(A.35)}$$

Volume element:

$$d^3 r = (h_1\, dq_1)(h_2\, dq_2)(h_3\, dq_3) = h_1 h_2 h_3\, dq_1\, dq_2\, dq_3. \qquad \text{(A.36)}$$

Gradient:

$$\boldsymbol{\nabla} f = \mathbf{u}_1 \frac{1}{h_1} \frac{\partial}{\partial q_1} f + \mathbf{u}_2 \frac{1}{h_2} \frac{\partial}{\partial q_2} f + \mathbf{u}_3 \frac{1}{h_3} \frac{\partial}{\partial q_3} f. \qquad \text{(A.37)}$$

Divergence:

$$\boldsymbol{\nabla} \cdot \mathbf{F} = \frac{1}{h_1 h_2 h_3} \left[\frac{\partial}{\partial q_1}(h_2 h_3\, F_1) + \frac{\partial}{\partial q_2}(h_3 h_1\, F_2) + \frac{\partial}{\partial q_3}(h_1 h_2\, F_3) \right]. \qquad \text{(A.38)}$$

Curl:

$$\boldsymbol{\nabla} \times \mathbf{F} = \sum_{k=1}^{3} (\boldsymbol{\nabla} \times \mathbf{F})_k\, \mathbf{u}_k = \sum_{i,j,k=1}^{3} \frac{1}{h_i h_j} \frac{\partial}{\partial q_i}(h_j\, F_j)\varepsilon_{ijk}\mathbf{u}_k. \qquad \text{(A.39)}$$

Laplacian:

$$\nabla^2 \Phi = \frac{1}{h_1 h_2 h_3} \left[\frac{\partial}{\partial q_1}\left(\frac{h_2 h_3}{h_1}\frac{\partial}{\partial q_1}\Phi\right) + \frac{\partial}{\partial q_2}\left(\frac{h_3 h_1}{h_2}\frac{\partial}{\partial q_2}\Phi\right) \right.$$
$$\left. + \frac{\partial}{\partial q_3}\left(\frac{h_1 h_2}{h_3}\frac{\partial}{\partial q_3}\Phi\right) \right]. \qquad \text{(A.40)}$$

The special examples of Cartesian, cylindrical, and spherical coordinates
are given in Table 8.3.

Table 8.3 Curvilinear-coordinate parameters for the special cases of Cartesian, cylindrical, and spherical coordinates.

Curvilinear	Cartesian	Cylindrical	Spherical
q_1	x	ρ	r
q_2	y	ϕ	θ
q_3	z	z	φ
h_1	1	1	1
h_2	1	ρ	r
h_3	1	1	$r\sin\theta$

A.5 *Vector-operator identities*

$$\boldsymbol{\nabla}\cdot(\boldsymbol{\nabla}\phi) = \boldsymbol{\nabla}^2\phi, \tag{A.41}$$

$$\boldsymbol{\nabla}\cdot(\boldsymbol{\nabla}\times\mathbf{F}) = 0, \tag{A.42}$$

$$\boldsymbol{\nabla}\times(\boldsymbol{\nabla}\phi) = 0, \tag{A.43}$$

$$\boldsymbol{\nabla}\times(\boldsymbol{\nabla}\times\mathbf{F}) = \boldsymbol{\nabla}(\boldsymbol{\nabla}\cdot\mathbf{F}) - \boldsymbol{\nabla}^2\mathbf{F} \quad \text{(see note below)}, \tag{A.44}$$

$$\boldsymbol{\nabla}(\phi\psi) = (\boldsymbol{\nabla}\phi)\psi + \phi(\boldsymbol{\nabla}\psi), \tag{A.45}$$

$$\boldsymbol{\nabla}(\mathbf{F}\cdot\mathbf{G}) = (\mathbf{F}\cdot\boldsymbol{\nabla})\,\mathbf{G} + \mathbf{F}\times(\boldsymbol{\nabla}\times\mathbf{G})$$
$$+ (\mathbf{G}\cdot\boldsymbol{\nabla})\,\mathbf{F} + \mathbf{G}\times(\boldsymbol{\nabla}\times\mathbf{F}), \tag{A.46}$$

$$\boldsymbol{\nabla}\cdot(\phi\mathbf{F}) = (\boldsymbol{\nabla}\phi)\cdot\mathbf{F} + \phi(\boldsymbol{\nabla}\cdot\mathbf{F}), \tag{A.47}$$

$$\boldsymbol{\nabla}\cdot(\mathbf{F}\times\mathbf{G}) = (\boldsymbol{\nabla}\times\mathbf{F})\cdot\mathbf{G} - (\boldsymbol{\nabla}\times\mathbf{G})\cdot\mathbf{F}, \tag{A.48}$$

$$\boldsymbol{\nabla}\times(\phi\mathbf{F}) = (\boldsymbol{\nabla}\phi)\times\mathbf{F} + \phi(\boldsymbol{\nabla}\times\mathbf{F}), \tag{A.49}$$

$$\boldsymbol{\nabla}\times(\mathbf{F}\times\mathbf{G}) = (\boldsymbol{\nabla}\cdot\mathbf{G})\,\mathbf{F} - (\boldsymbol{\nabla}\cdot\mathbf{F})\,\mathbf{G}$$
$$+ (\mathbf{G}\cdot\boldsymbol{\nabla})\,\mathbf{F} - (\mathbf{F}\cdot\boldsymbol{\nabla})\,\mathbf{G}. \tag{A.50}$$

Note: The Laplacian of a vector is the vector whose Cartesian components are obtained by taking the Laplacians of the respective Cartesian vector components. In arbitrary coordinates, the Laplacian of a vector is *defined* by (A.44).

A.6 *Integral theorems*

Gauss: Simply connected volume V, bounded by closed surface S, with area elements $d\mathbf{A}$ oriented towards the outside of the volume.

$$\int_V \boldsymbol{\nabla}\cdot\mathbf{F}\,d^3r = \oint_S \mathbf{F}\cdot d\mathbf{A}. \tag{A.51}$$

Green's theorem follows be choosing $\mathbf{F} = \psi\boldsymbol{\nabla}\phi - \phi\boldsymbol{\nabla}\psi$:

$$\int_V \left(\psi\boldsymbol{\nabla}^2\phi - \phi\boldsymbol{\nabla}^2\psi\right) d^3r = \oint_S \left(\psi\boldsymbol{\nabla}\phi - \phi\boldsymbol{\nabla}\psi\right)\cdot d\mathbf{A}. \tag{A.52}$$

Stokes: Simply connected surface S, with area elements $d\mathbf{A}$, bounded by closed curve C, with path elements $d\mathbf{s}$. The directions of $d\mathbf{A}$ and $d\mathbf{s}$ are linked by the right-hand rule: if the fingers of the right hand curl around the direction of $d\mathbf{s}$, the thumb of the right hand points along $d\mathbf{A}$.

$$\oint_C \mathbf{F}\cdot d\mathbf{s} = \int_S (\boldsymbol{\nabla}\times\mathbf{F})\cdot d\mathbf{A}. \tag{A.53}$$

Other integral theorems:

$$\int_S (d\mathbf{A}\times\boldsymbol{\nabla}\phi) = \oint_C \phi\,d\mathbf{s}, \tag{A.54}$$

$$\int_V \boldsymbol{\nabla}\phi\,d^3r = \oint_S \phi\,d\mathbf{A}, \tag{A.55}$$

$$\int_V (\boldsymbol{\nabla}\times\mathbf{F})\,d^3r = \oint_S (d\mathbf{A}\times\mathbf{F}), \tag{A.56}$$

$$\int_V (\boldsymbol{\nabla}\cdot\mathbf{G} + \mathbf{G}\cdot\boldsymbol{\nabla})\,\mathbf{F}\,d^3r = \oint_S \mathbf{F}\,(\mathbf{G}\cdot d\mathbf{A}). \tag{A.57}$$

B   Dirac δ Distribution

This Appendix will give a brief introduction to the Dirac delta (δ) distribution. The presentation is only intended to provide some working knowledge of the basic properties and it is by no means rigorous. For mathematically rigorous treatments, see the mathematical literature (see, for example, Refs. [8] or [9] and references found therein).

While also referred to as the delta function, it is not a function in the usual sense; it is a *generalized function*, or a *distribution*, defined by

$$\int_{-\infty}^{+\infty} dx'\,\delta(x-x')f(x') = f(x) \tag{B.1}$$

for any (sufficiently well-behaved) function $f(x)$ that does not diverge at infinity.

We emphasize that *all* integrals involving the δ distribution are to be understood as *Lebesgue integrals*. The main distinction, from a practical point of view, between Riemann and Lebesgue integrations is the fact that integrands that are equal except for a set of *measure zero* produce the same result (see the mathematical literature on measure theory). Functions that agree except for on a measure-zero set are said to be equal 'almost everywhere'. For example, Lebesgue integrations are well-defined even if the integrand contains infinitely many discontinuities, as long as they form a set of measure zero, as is the case with the function defined over the interval $[0, 1]$ that is unity for all irrational x, but zero for all rational x, which is discontinuous *everywhere*. Its integral is not defined in the Riemann sense, but its Lebesgue integral over $[0, 1]$ is unity since the rational numbers $\mathbb{Q}$ form a set of measure zero (all countable sets have measure zero). In addition, the Lebesgue integral formalizes if and when the integration can be interchanged with certain limits. This makes it possible, for example, to introduce integrations involving the Dirac delta function via limits of integrations over certain continuous functions; see, e.g., Eq. (B.7).

Some properties of the δ function are

$$\delta(x) = \delta(-x), \tag{B.2a}$$

$$f(x)\delta(x) = f(0)\delta(x), \quad \text{in particular,} \quad x\delta(x) = 0, \tag{B.2b}$$

$$f(x')\delta(x - x') = f(x)\delta(x - x'), \tag{B.2c}$$

$$\delta(ax) = \frac{1}{|a|}\delta(x), \tag{B.2d}$$

$$\delta\big(f(x)\big) = \sum_i \frac{\delta(x_i)}{|f'(x_i)|}, \tag{B.2e}$$

where the sum extends over all simple zeros x_i of $f(x)$ for which the first derivatives, $f'(x_i)$, do not vanish.

There are many representations of the δ distribution. A particularly useful one is

$$\delta(x - x') = \frac{d}{dx}\theta(x - x'), \tag{B.3}$$

where

$$\theta(x - x') = \begin{cases} 1 & \text{for } x > x', \\ 0 & \text{for } x < x' \end{cases} \tag{B.4}$$

is the *step function* (also called *Heaviside function*). The understanding in the definition (B.3) is that it is to be evaluated in terms of an integration by parts,

$$\int\limits_{-\infty}^{+\infty} dx'\, f(x') \frac{d\theta(x'-x)}{dx'} = f(x')\theta(x'-x)\Big|_{-\infty}^{+\infty} - \int\limits_{-\infty}^{+\infty} dx'\, \frac{df(x')}{dx'}\theta(x'-x)$$

$$= f(+\infty) - \int\limits_{x}^{+\infty} dx'\, \frac{df(x')}{dx'}$$

$$= f(+\infty) - f(+\infty) + f(x)$$

$$= f(x). \tag{B.5}$$

Other representations use continuous, integrable functions $h(x)$, with $h(0) \neq 0$, normalized such that

$$\int\limits_{-\infty}^{+\infty} h(x)\, dx = 1, \tag{B.6}$$

which form the bases of various representations of the δ distribution via

$$\delta(x) = \lim_{\varepsilon \to 0} \delta_\varepsilon(x), \quad \text{where} \quad \delta_\varepsilon(x) = \frac{1}{\varepsilon}\, h\!\left(\frac{x}{\varepsilon}\right). \tag{B.7}$$

For $\varepsilon \to 0$, the functional form of $\delta_\varepsilon(x)$ is seen to become more and more concentrated around $x = 0$ — since only $|x| \lesssim \varepsilon$ will provide sufficiently small arguments x/ε to obtain nonvanishing values of $h(x/\varepsilon)$ — while getting larger and larger (see Fig. B.1), such that the integral of $\delta_\varepsilon(x)$ remains normalized to unity, as in (B.6). The rule for evaluating the limit is that it is to be taken after the integration over x has been performed, i.e.,

$$\int\limits_{-\infty}^{+\infty} dx'\, \delta(x-x')f(x') \equiv \lim_{\varepsilon \to 0} \int\limits_{-\infty}^{+\infty} dx'\, \delta_\varepsilon(x-x')f(x'). \tag{B.8}$$

The following list gives some of these functions:

$$\delta_\varepsilon(x) = \frac{e^{-x^2/\varepsilon^2}}{\varepsilon\sqrt{\pi}} \qquad \text{(see Fig. B.1)}, \tag{B.9a}$$

$$\delta_\varepsilon(x) = \frac{1}{2\pi} \int\limits_{-\infty}^{+\infty} e^{ikx-\varepsilon|k|}\, dk = \frac{1}{\pi} \frac{\varepsilon}{x^2 + \varepsilon^2}, \tag{B.9b}$$

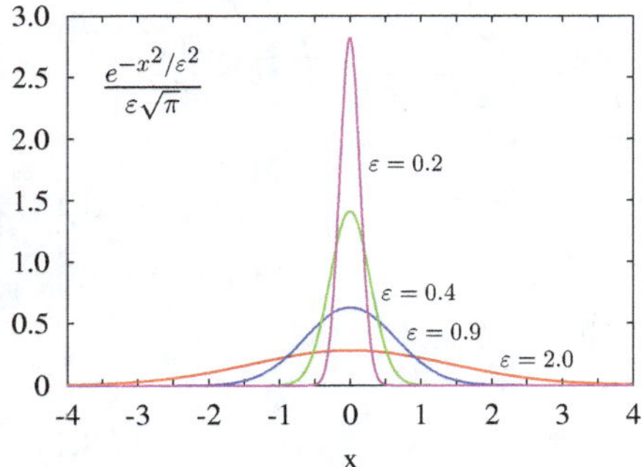

Fig. B.1 Plot of the Gaussian $\delta_\varepsilon(x) = e^{-x^2/\varepsilon^2}/\varepsilon\sqrt{\pi}$ for various values of ε. The peak becomes narrower and narrower as $\varepsilon \to 0$. The areas under the respective curves, however, remain the same.

$$\delta_\varepsilon(x) = \frac{1}{2\pi}\int_{-1/\varepsilon}^{+1/\varepsilon} e^{ikx}\,dk = \frac{1}{\pi}\frac{\sin(x/\varepsilon)}{x}, \tag{B.9c}$$

$$\delta_\varepsilon(x) = \frac{1}{2\varepsilon}\,e^{-|x|/\varepsilon}. \tag{B.9d}$$

Let us take (B.9b) to illustrate how these representations are to be utilized. The understanding here is that the limit is to be exchanged with the integration, i.e.,

$$\lim_{\varepsilon\to 0}\frac{\varepsilon}{\pi}\int_{-\infty}^{+\infty} dx' \frac{f(x')}{(x-x')^2 + \varepsilon^2} = \lim_{\varepsilon\to 0} f(x')\frac{1}{\pi}\arctan\frac{x'-x}{\varepsilon}\Big|_{-\infty}^{+\infty}$$

$$- \lim_{\varepsilon\to 0}\frac{1}{\pi}\int_{-\infty}^{+\infty} dx' \frac{df(x')}{dx'}\arctan\frac{x'-x}{\varepsilon}$$

$$= \frac{1}{2}\Big[f(+\infty) + f(-\infty)\Big]$$

$$- \lim_{\varepsilon\to 0}\frac{1}{\pi}\Bigg[\int_{-\infty}^{x} dx' \frac{df(x')}{dx'}\arctan\frac{x'-x}{\varepsilon}$$

$$+ \int_{x}^{+\infty} dx' \frac{df(x')}{dx'}\arctan\frac{x'-x}{\varepsilon}\Bigg]$$

$$= \frac{1}{2}\Big[f(+\infty) + f(-\infty)\Big]$$

$$+ \frac{1}{2} \int_{-\infty}^{x} dx' \, \frac{df(x')}{dx'} - \frac{1}{2} \int_{x}^{+\infty} dx' \, \frac{df(x')}{dx'}$$

$$= f(x), \tag{B.10}$$

where $d \arctan x / dx = 1/(1 + x^2)$ and $\arctan(\pm\infty) = \pm\pi/2$ was used.

For Eq. (B.9c), the δ distribution written as

$$\delta(x) = \frac{1}{2\pi} \int_{-\infty}^{+\infty} e^{ikx} \, dk, \tag{B.11}$$

referred to as the *Fourier representation*, *formally* equates the δ function to the Fourier transform of unity. This equation, in particular, together with the basic property (B.1), allows one to utilize the δ distribution *formally* in practical calculations as if it were a normal function, without the need for explicitly carrying out the implied limits of the underlying definitions.

The δ function can easily be generalized to the three-dimensional case as a product of one-dimensional δ functions. For example,

$$\delta(\mathbf{r} - \mathbf{r}') = \begin{cases} \delta(x - x') \, \delta(y - y') \, \delta(z - z'), \\[2mm] \dfrac{\delta(r - r')}{r^2} \, \delta(\phi - \phi') \, \delta(\cos\theta - \cos\theta'), \\[2mm] \dfrac{\delta(\rho - \rho')}{\rho} \, \delta(\varphi - \varphi') \, \delta(z - z'), \end{cases} \tag{B.12}$$

for Cartesian, spherical, and cylindrical coordinates, respectively. In the literature, one often finds the alternative notation $\delta^{(3)}(\mathbf{r} - \mathbf{r}')$ to emphasize that this distribution acts in three dimensions.

C   Green's Function Method: An Example

As an illustration of the Green's function method introduced in Sec. 1.3.6, let us consider the second-order differential equation

$$\left(\frac{d^2}{dt^2} + \omega_0^2 \right) x(t) = f(t) \tag{C.1}$$

for an externally driven harmonic oscillator. This case will exemplify the following important generally valid result: To construct a Green's function $g(t, t')$, and thus the general solution of the *inhomogeneous* equation according to (1.122), it is sufficient to know the general solution of the *homogeneous* equation.

Assuming that the force function $f(t)$ is piecewise continuous, we can approximate it by a series of constant rectangular forces

$$f(t) \approx \sum_i f(t_i)\delta_i(t), \tag{C.2}$$

centered around points t_i, with $t_{i+1} = t_i + \Delta t$, where Δt is a fixed positive time increment and

$$\delta_i(t) = \begin{cases} 0 & \text{for } t < t_i - \Delta t/2, \\ 1 & \text{for } |t - t_i| < \Delta t/2, \\ 0 & \text{for } t > t_i + \Delta t/2. \end{cases} \tag{C.3}$$

As $\Delta t \to 0$, this approximation for $f(t)$ can be made arbitrarily accurate.

In view of the linearity of the differential equation, we can now build up a particular solution $x_\mathrm{p}(t)$ as a superposition of solutions for each of the impulse-type forces $f(t_i)$ acting only for a period of time around $t = t_i$ by letting $\Delta t \to 0$, i.e.,

$$x_\mathrm{p}(t) = \lim_{\Delta t \to 0} \sum_i g_{\Delta t}(t, t_i)\, f(t_i), \tag{C.4}$$

where $g_{\Delta t}$ is a solution of

$$\frac{d^2}{dt^2} g_{\Delta t}(t, t_i) + \omega_0^2\, g_{\Delta t}(t, t_i) = \delta_i(t). \tag{C.5}$$

For each of these solution, we must match the functions and their first derivatives at the endpoints $t_i \pm \Delta t/2$ of each interval. This can easily be achieved by using the initial conditions

$$g_{\Delta t}(t, t_i) = 0, \quad \text{and} \quad \frac{d}{dt} g_{\Delta t}(t, t_i) = 0 \quad \text{at } t = t_i - \frac{\Delta t}{2}, \tag{C.6}$$

thus choosing the trivial solution

$$g_{\Delta t}(t, t_i) = 0 \tag{C.7}$$

for all $t < t_i - \Delta t/2$.

To solve (C.5) for the other regions note that for a constant force f_0, the solutions of $\ddot{y}(t) + \omega_0^2\, y(t) = f_0$ are directly found from the solutions of the homogeneous equation $\ddot{z}(t) + \omega_0^2\, z(t) = 0$ via $y(t) = f_0/\omega_0^2 + z(t)$. The effect of the constant force, therefore, is simply a change of the equilibrium point by the constant distance f_0/ω_0^2, but otherwise the general solution of simple harmonic motion applies.

Following these considerations, the remaining pieces of the solution are

$$g_{\Delta t}(t, t_i) = \frac{1}{\omega_0^2} + A_i \sin \omega_0 t + B_i \cos \omega_0 t \qquad (C.8)$$

for $t_i - \Delta t/2 < t < t_i + \Delta t/2$ and

$$g_{\Delta t}(t, t_i) = a_i \sin \omega_0 t + b_i \cos \omega_0 t \qquad (C.9)$$

for $t > t_i + \Delta t/2$. Matching the functions and their derivatives at the interval boundaries determines the coefficients a_i, b_i, A_i, and B_i, i.e.,

$$g_{\Delta t}(t, t_i) = \frac{1}{\omega_0^2} \begin{cases} 0 & \text{for } t < t_i - \Delta t/2, \\[2ex] 1 - \cos\left[\omega_0\left(t - t_i + \frac{\Delta t}{2}\right)\right] & \text{for } |t - t_i| < \Delta t/2, \\[2ex] 2\sin\frac{\omega_0 \Delta t}{2} \sin\left[\omega_0(t - t_i)\right] & \text{for } t > t_i + \Delta t/2. \end{cases}$$
$$(C.10)$$

If we let now $\Delta t \to 0$, so that terms of order Δt^2, and higher, are negligible, we can write

$$g_{\Delta t}(t, t_i) = g(t, t_i)\, \Delta t, \qquad (C.11)$$

where

$$g(t, t') = \frac{1}{\omega_0} \sin\left[\omega_0(t - t')\right] \theta(t - t'), \qquad (C.12)$$

with the step function $\theta(t - t')$ defined in (B.4). We indeed find now

$$x_{\mathrm{p}}(t) = \lim_{\Delta t \to 0} \sum_i g_{\Delta t}(t, t_i)\, f(t_i) = \lim_{\Delta t \to 0} \sum_i g(t, t_i)\, f(t_i)\, \Delta t$$

$$= \int_{-\infty}^{+\infty} g(t, t')\, f(t')\, dt', \qquad (C.13)$$

i.e., $g(t, t')$ is the desired Green's function. Note that it is obvious from the derivation that the force function $f(t)$ need not be continuous.

The physical meaning of this result is that $g(t, t_i)f(t_i)\Delta t$ is the solution of an oscillator that is initially at rest and then experiences a brief (impulse-type) force, like the strike of a hammer. The effect of a series of such 'hammer strikes' is given by a summation of the individual solutions (see Fig. C.1), according to the superposition principle.

It is straightforward to verify that the Green's function satisfies Eq. (1.119). Using the properties of the δ function given in Appendix B, the derivatives

$$\frac{d}{dt} g(t, t') = \cos\left[\omega_0(t - t')\right] \theta(t - t') + \underbrace{\frac{1}{\omega_0} \sin\left[\omega_0(t - t')\right] \delta(t - t')}_{=0}, \qquad (C.14a)$$

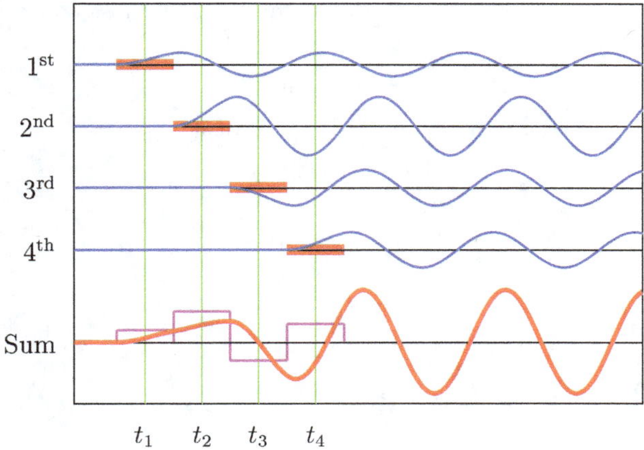

Fig. C.1   Illustration of how the particular solution is built up by the individual solutions of Eq. (C.10) for four successive times t_i. The lowest curve shows the sum of the upper four curves. The solid horizontal bars centered around t_i for each of the four solutions shows the duration Δt of the corresponding forces $f(t_i)$. The curves are generated by the forces indicated by the box columns, with values $f(t_1) = 0.2$, $f(t_2) = 0.5$, $f(t_3) = -0.3$, and $f(t_4) = 0.3$, in arbitrary units; the time increment is $\Delta t = 0.4T$, where T is the period.

$$\frac{d^2}{dt^2} g(t, t') = -\omega_0 \sin\left[\omega_0(t - t')\right] \theta(t - t') + \underbrace{\cos\left[\omega_0(t - t')\right] \delta(t - t')}_{=\delta(t-t')},$$

$$\text{(C.14b)}$$

indeed produce

$$\left(\frac{d^2}{dt^2} + \omega_0^2\right) g(t, t') = \delta(t - t'), \tag{C.15}$$

as stipulated.

Adding now the general homogeneous solution to (C.13), we obtain the general solution for Eq. (C.1),

$$x(t) = a \sin \omega_0 t + b \cos \omega_0 t + \frac{1}{\omega_0} \int_{-\infty}^{t} \sin\left[\omega_0(t - t')\right] f(t') \, dt', \tag{C.16}$$

with two integration constants a and b to be determined from the initial conditions.

To illustrate this, let us consider an example with initial conditions $x(0) = 0$ and $\dot{x}(0) = 0$, which corresponds to an oscillator at rest that is subject to a force for $t > 0$. Implementation of these initial conditions produces

$$x(0) = 0: \qquad b = \frac{1}{\omega_0} \int_{-\infty}^{0} \sin \omega_0 t' f(t')\, dt', \qquad \text{(C.17a)}$$

$$\dot{x}(0) = 0: \qquad a = - \int_{-\infty}^{0} \cos \omega_0 t' f(t')\, dt', \qquad \text{(C.17b)}$$

which provides

$$\left. \begin{array}{l} x(0) = 0 \\ \dot{x}(0) = 0 \end{array} \right\} \qquad x(t) = \frac{1}{\omega_0} \int_{0}^{t} \sin \left[\omega_0 (t - t') \right] f(t')\, dt' \qquad \text{(C.18)}$$

as the solution for this special case.

Finally, it should be emphasized that the Green's function derived here is not unique since it depends on the particular boundary conditions chosen in (C.6). Other boundary conditions will lead to other Green's functions. For example, the function

$$g(t, t') = \frac{1}{\omega_0} \left[\sin \omega_0 t \, \cos \omega_0 t' \, \theta(t - t') + \sin \omega_0 t' \, \cos \omega_0 t \, \theta(t' - t) \right] \qquad \text{(C.19)}$$

also satisfies Eq. (1.119).

Index

abstract vector, 3
acceleration
– centripetal, 9
– four-acceleration, 326
– tangential, 9
– vector, 8
action
– function (Hamilton), 189
– functional, 157
– integral (relativistic), 335, 339
angle
– Euler, 200
– scattering, 82
– solid, 82
angular
– frequency, 20
– momentum, 27
– conservation, 27, 137
– rigid body, 215
– total, 36
– velocity, 207
Ansatz, 21
area law, 29
attractor, 282, 289
– strange, 291
Atwood's machine, 125

barometric equation, 273
basis vector
– contravariant, 304
– covariant, 304
– spacetime, 304
Bernoulli's
– equation, 276
– principle, 276
bifurcation, 285
– cascade, 294
body cone, 224

boost
– Lorentz transformation, 315, 320
– along x-axis, 318
– electromagnetic field, 342
– velocity, 320
brachistochrone problem, 142

canonical
– equations, *see* Hamilton's
    equations
– invariance, 180
– momentum, 118
– transformation, 177
canonical systems, 171
Cartesian coordinates, 3, 344
center of mass, 35, 60
center of momentum, 325, 334
central force, 28, 59
– problem, 194
centrifugal force, 55
centripetal acceleration, 9
chaos, 285
characteristic equation, *see* eigenvalue
    equation
closed system, 36
collinearity constraint, 14
collision
– relativistic, 332
– two-body, 80
commutator
– Lie Algebra
    – Lorentz generators, 316
    – rotation, 206
– Poisson bracket, 174
component
– contravariant, 305
– covariant, 305
– vector, 4

compressibility, 270

condition
  – integrability, 111
  – orthogonality, 198

cone
  – body, 224
  – light, 301
  – precession, 226

conic section, 66

conservation
  – angular momentum, 27, 137
  – energy, 29, 32, 34, 135, 166
  – mass, 268
  – momentum, 27, 35, 136, 166

conservative force, 31

constraint
  – classification, 99
  – determination of force, 124
  – force, 95
  – friction, 99
  – geometric, 95
  – holonomic, 96, 149, 154
  – inequality, 133
  – isoperimetric, 149
  – nonholonomic, 96, 111, 131
  – rheonomous, 96
  – scleronomous, 96

contact transformation, 176

continuity equation, 184, 268
  – fluid dynamics, 271

contraction (length), 323

contravariant
  – basis vector, 304
  – component, 305

coordinates
  – Cartesian, 3, 344
  – body and space systems (rigid body), 197
  – curvilinear, 347
  – cylindrical, 6, 345
  – generalized, 109
  – ignorable, 118
  – spherical, 6, 346

Coriolis force, 55

covariance
  – form invariance, 116, 306

covariant
  – basis vector, 304
  – component, 305

cross section
  – Rutherford, 88
  – scattering, 83
  – total, 83

current
  – density
    – scattering, 82
  – density (fluid), 267
  – matter, 267

curvilinear coordinates, 347

cyclic coordinates, *see* ignorable coordinates

cylindrical coordinates, 6, 345

d'Alembert operator, 311

degrees of freedom
  – rigid body, 196

delta
  – Feigenbaum, 286
  – Kronecker, 3
  – function, 26, 350, 354
    – Fourier representation, 354

density
  – Lagrange
    – relativistic, 339
  – Lagrange (for sagging beam), 263
  – Lagrange (for vibrating string), 256
  – current (fluid), 267
  – current (scattering), 82
  – function, 184
  – mass, 266

differential
  – exact, 33
  – perfect, 33, 99
  – scattering cross section, 83
  – total, 33

dilation (time), 324

Dirac delta function (distribution), *see* delta function

direction cosine, 44, 197

displacement
  – real, 98

– virtual (instantaneous), 97
dissipative
  – force, 31
    – Rayleigh's function, 123
distance
  – lightlike, 301
  – spacelike, 301
  – timelike, 301
distribution
  – generalized function, *see* delta
    function
drawbridge with balancing
  counterweight, 128
dyad
  – unit, 4
dyadic product, 4

eccentricity (numerical), 66, 72
eccentricity vector, 71
eigenmodes, 249, 250, 260
eigentime, 324
eigenvalue equation
  – for principal-axis transformation,
    218
  – for small oscillations, 241
  – for triatomic molecule, 252
eigenvectors
  – for normal-mode transformation,
    243
  – for principal-axis transformation,
    218
  – for triatomic molecule, 252
electric field, 338
electromagnetic
  – field
    – Hamilton function, 343
    – boost transformation, 342
  – field tensor, 340
ellipse
  – Kepler, 66
energy
  – conservation, 29, 32, 34, 135, 166
  – kinetic
    – rigid body, 210, 211
    – rotational, 211
  – potential, 17, 31, 32

– total, 31
epsilon symbol, 3
equation of state
  – ideal gas, 270
  – polytropic, 270
equilibrium
  – statistical, 187
equivalence
  – Hamilton–Jacobi and Hamilton's
    principle, 190
equivalent one-body problem, 59
escape velocity, 74
Euclidian space, 2
Euler
  – angle, 200
  – equation
    – fluid dynamics, 270, 271
    – variational calculus, 144
    – variational calculus
      (generalizations), 147
  – equations of motion (rigid body),
    220
  – theorem (rigid body), 219
  – torque-free equations of motion,
    221
exact differential, 33
example
  – Earth as rotating top, 230–236
  – Atwood's machine, 125
  – Galilei invariance (Noether), 167
  – Hamilton's equations, 171
  – bead on rotating rod, 106
  – blade motion on plane, 131
  – brachistochrone problem, 145
  – central force (Hamilton–Jacobi),
    194
  – chain in field of gravity, 152
  – constraining force for planar
    pendulum, 125
  – constructing Green's function for
    driven harmonic oscillator,
    354
  – drawbridge with balancing
    counterweight, 128
  – free fall (Hamilton–Jacobi), 193

– free-fall motion in accelerated
     frame, 117
– geodesic, 155
– harmonic motion, 19
     – with external force, 23
     – with damping, 21
– liquid at rest in gravitational
     field, 272
– point mass in spherical
     coordinates, 118
– rolling disk on plane, 132
– rotating liquid in gravitational
     field, 273
– rotation about z-axis, 200
– sagging beam in external force
     field, 261
– shortest distance, 145
– sliding mass on sphere, 134
– small oscillations in two
     dimensions, 247
– sound propagation, 277
– stability of uniform rotation, 222
– time independence with
     Noether's theorem, 165
– two coupled pendula, 249
– variable mass
     – rocket launch, 24
– vibrating string, 254
– vibrating triatomic molecule, 251
external force, 34, 246

Faraday tensor, *see* electromagnetic
     field tensor
Feigenbaum δ, 286, 295
field
– electric, 338
– electromagnetic
     – boost transformation, 342
– magnetic, 338
– pressure, 267, 274
– tensor (electromagnetic), 340
– velocity, 266
figure axis, 223
flow
– phase space, 182
– potential, 276

fluid, 266
– dynamics, 266
     – Euler equation, 270
force
– Coriolis, 55
– Lorentz, 123, 338
– Minkowski, 337, 341
– central, 28, 59
– centrifugal, 55
– conservative, 31
– constraint, 95
     – generalization, 112
– dissipative, 31
     – Rayleigh's function, 123
– external, 34, 246
– generalized, 112
– gravitational, 65
– internal, 34
– lost (d'Alembert), 103
– recoil, 15
– two-body, 34
forced oscillation, 23, 247, 281
form invariance
– covariance, 116, 306
Foucault pendulum, 58
four-acceleration, 326
four-dimensional spacetime, 303
four-divergence, 311
four-force, 337, 341
four-gradient, 310
four-momentum, 325
four-potential, 338
four-vector, 303, 304
four-velocity, 324
Fourier
– representation (delta function),
     354
– series expansion, 259
fractal, 291, 295
frame of reference
– center of mass, 61
– center of momentum, 325, 334
– heliocentric, 69
– inertial, 11, 42
– accelerated, 52
frequency

– angular, 20
– natural, 24, 281
friction, 21, 270, 272
– constraint, 99
functional
– action, 157
– generic, 143
– minimal time, 142
– path length, 141
– potential, 148
– small variation, 143
fundamental lemma
– variational calculus, 144

Galilei
– invariance, 167
gauge
– transformation, 159, 167
– electrodynamics, 160
Gauss's theorem, 349
Gedankenexperiment, 73
generalized
– coordinates, 109
– force, 112
– force of constraint, 112
– momentum, 118, 169
– potential, 338
– definition, 122
– electromagnetic forces, 122
– rotating frame, 120
– velocity, 109
generator
– rotation, 204
– transformation
– Lorentz, 314
– canonical, 177–180
geodesic, 149, 155–157
geoid, 230
geometric constraint, 95
gradient, 32
gravitational
– force, 65
– mass, 12
Green's theorem, 350
group
– $SU(2)$, 206

– Lorentz
– homogeneous $O(1,3)$, 308, 311
– proper $SO(1,3)$, 313
– Poincaré $P(1,3)$, 309
– definition, 199
– proper rotation $SO(3)$, 200
– rotation $O(3)$, 199

Hamilton
– action or principal function, 189
– function, 170
– for electromagnetic field, 343
– relativistic (free particle), 336
Hamilton's equations, 171
– with dissipative forces, 171
Hamilton's principle
– advantage of, 158
Hamilton–Jacobi equation, 189
– time-independent, 191
– variables completely separable, 191
Hamiltonian systems, 171
hard sphere scattering, 85
harmonic oscillator
– potential, 20
– relationship to inverse-square force, 70
– with damping and external force, 23, 280
harmonics, 283
Heaviside function, 352
heliocentric frame of reference, 69
holonomic constraint, 96, 149, 154
homogeneous
– Lorentz group $O(1,3)$, 311
– Lorentz transformation, 308
homogeneous space, 2
Hooke's law, 19
hurricane, 58
hydrodynamics, *see* fluid dynamics
hypersurface, 98, 301

ideal fluid
– definition, 267

– equations, 271
ideal gas, *see* ideal fluid
– equation of state, 270
ignorable coordinates, 118
impact parameter, 84
inertia
– moment of, 216
– tensor, 211
inertial
– frame of reference, 11, 42
– mass, 12
infinitesimal
– rotation, 203, 205
– transformation, 163
– Lorentz, 313
inhomogeneous
– Lorentz (Poincaré)
transformation, 308
integrability condition, 33, 111
internal force, 34
invariance
– Galilei, 167
– canonical, 180
– of squared spacetime distance,
301
– rotational, 137
– translational, 2, 136
irrotational, 276
isoperimetric constraint, 149
iterated map, *see* Poincaré map

Jacobi identity, 175
Jacobian, 183

Kepler
– ellipse, 66
– laws, 67
– problem, 65
kinematical momentum, 342
kinetic
– energy
– total, 38
– translational and rotational,
211
Kronecker delta, 3

laboratory system
– rotating Earth, 56
– scattering, 89
Lagrange
– density
– for sagging beam, 263
– for vibrating string, 256
– relativistic, 339
– equations (form invariance), 116
– equations (summary), 138
– equations and Newton's 2nd law,
120
– equations in vector form, 120
– equations of first kind, 100
– equations of mixed type, 113
– equations of second kind
(conservative), 115
– equations of second kind
(conservative, holonomic),
115
– function
– electromagnetic field, 123
– for coupled oscillators, 241
– heavy top, 226
– relativistic (free particle),
336
– function (rotational invariance),
137
– function (translational
invariance), 136
– multiplier, 100
laminar, 276
Laplace's equation, 276
Lebesgue integral, 351
length
– contraction, 323
– path, 141
– proper, 322
– vector, 199
Lie algebra
– commutator (Lorentz
generators), 316
– commutator (rotation), 206
– restricted Lorentz group, 316
– rotation group SO(3), 204–207
light cone, 301

lightlike distance, 301
linear transformation, 5, 307
Liouville
  – equation, 185
  – theorem, 182
logistic map, 293–296
longitudinal wave, 278
Lorentz
  – factor γ, 318
  – force, 123, 338
  – group
    – homogeneous $O(1,3)$, 311
    – proper $SO(1,3)$, 313
  – scalar, 310–311
  – transformation, 307
    – boost, 315, 320
    – exponential representation, 315
    – generators, 314, 315
    – homogeneous, 308
    – infinitesimal, 313
    – inhomogeneous (Poincaré), 308
    – inverse, 309
    – non-orthochronous, 312
    – orthochronous, 312
    – rotation, 315, 317
lost force (d'Alembert), 103
Lyapunov exponent, 287

magnetic field, 338
Mandelstam variables, 333
mass, 1
  – center of, 35
  – conservation, 268
  – density, 266
  – gravitational, 12
  – inertial, 12
  – point, 1
  – reduced, 60
  – total, 35, 61
  – variable, 12
matrix
  – representation, 5
  – rotation, 198
matter current, 267

mechanical momentum, 342
metric
  – space, 303
  – tensor, 303
    – inverse, 304
minimal substitution, 343
Minkowski
  – force, 337, 341
  – space, 303
mode
  – normal, 244
moment of inertia, 216
momentum, 11
  – angular, 27
    – rigid body, 215
    – total, 36
  – canonical, 118
  – center of, 325, 334
  – conservation, 27, 35, 136, 166
  – four-momentum, 325
  – generalized, 118, 169
  – kinematical, 342
  – mechanical, 342
  – total, 35
multiplier
  – Lagrange, 100

natural frequency, 24, 281
Navier-Stokes equation, 272
Newton's
  – axioms, 11–14
  – second law (relativistic), 337, 341
Noether's theorem
  – conservation
    – energy, 166
    – momentum, 166
non-orthochronous Lorentz transformation, 312
nonholonomic constraint, 96, 111, 131
nonlinear pendulum, 281
  – with damping and external force, 281
normal mode, 244
null vector, *see* lightlike vector
nutation, 229

Occam's razor, 10
operator
  – parity, 200
  – vector, 205, 207
orbital symmetry, 64
orthochronous Lorentz
  transformation, 312
orthogonal transformation, 199, 213
orthogonality condition, 198
oscillation
  – forced, 23, 247, 281

paraboloid
  – rotation, 274
parallel axis theorem, 214
parameter
  – impact, 84
parity, 312
  – operator, 200
path
  – length, 141
path-independent work, 33
pendulum
  – Foucault, 58
  – constraining force, 125
  – nonlinear, 281
  – nonlinear, with damping and
     external force, 281
  – planar, 95, 104
perfect differential, 33, 99
period doubling, 284, 294
  – cascade, 285
  – universality, 285
Pfaffian forms, 99
phase space
  – diagrams, 289
  – flow, 182
  – state space, 182
  – trajectory, 289
planar pendulum, 95, 104
plane wave, 278
Poincaré
  – map, 290
    – orbit, 290
  – transformation, 307, 308
point

  – mass, 1
  – transformation, 116
Poisson bracket, 174
  – properties, 175
  – quantum-mechanical
     commutator, 175
polytropic equation of state, 270
position vector, 3
potential
  – *vs.* potential energy, 18
  – energy, 17
  – flow, 276
  – four-potential, 338
  – functional, 148
  – generalized, 338
     – definition, 122
     – electromagnetic forces, 122
  – generalized, in rotating frame,
     120
  – harmonic oscillator, 20
  – power-law, 65
  – velocity, 276
potential energy, 31, 32
  – *vs.* potential, 18
power, 30
power-law potential, 65
precession, 226
  – Thomas, *see* Thomas precession
  – cone, 226
  – velocity (Earth), 235
  – velocity (Moon), 235
presssure field, 267
pressure field, 274
principal
  – axis transformation, 216–219
  – function (Hamilton), 189
principle
  – integral *vs.* differential, 158
principle of relativity
  – nonrelativistic (Galilei), 42
  – relativistic (Einstein), 299
problem
  – Kepler, 65
  – brachistochrone, 142
  – central force, 194
  – equivalent one-body, 59

– reduction to one-dimensional, 62
projectile
 – scattering, 80
proper length, 322
pseudoforce
 – Coriolis, 55
 – centrifugal, 55

quadratic map, *see* logistic map

rapidity, 319
Rayleigh's dissipative-force function, 123
real displacement, 98
recoil force
 – variable mass, 15
reduced mass, 60
reduction to one-dimensional problem, 62
relativistic
 – Hamilton function (free particle), 336
 – Lagrange density, 339
 – Lagrange function (free particle), 336
 – Newton's second law, 337, 341
 – action integral, 335, 339
 – collision, 332
 – spacetime, 303
representation
 – matrix, 5
 – unitary, of rotation, 205
 – vector, 3, 4
resonance, 24
rheonomous constraint, 96
rigid body
 – body and space coordinates, 197
 – Euler equations of motion, 220
 – degrees of freedom, 196
 – energy
   – kinetic, 210, 211
 – momentum
   – angular, 215
 – torque-free Euler equations of motion, 221
rotating Earth

– laboratory system, 56
rotation
 – Lorentz transformation, 315, 317
 – generator, 204
 – group $O(3)$, 199
 – group, proper $SO(3)$, 200
 – infinitesimal, 203, 205
 – matrix, 198
 – paraboloid, 274
 – torque-free (rigid body), 221
rotational
 – energy
   – rigid body, 211
 – invariance, 137
Runge–Lenz vector, 71
Rutherford
 – cross section, 88
 – scattering, 87–93

scattering
 – Rutherford cross section, 88
 – angle, 82
 – cross section, 83
 – current density, 82
 – elastic, 91
 – hard sphere, 85
 – projectile, 80
 – target, 80
 – trajectory, 86
scleronomous constraint, 96
self-similarity, 291, 295
separation ansatz, 258
similarity transformation, 213
simultaneity, 301
small variation
 – functional, 143
solid angle, 82
sound
 – speed, 278
 – wave equation, 277
space
 – Euclidian, 2
 – four-dimensional (Minkowski), 303
 – homogeneous, 2
 – metric, 303

– three-dimensional (Euclidian), 2
space-inversion, *see* parity
spacelike distance, 301
spacetime, 300
   – basis vector, 304
   – four-dimensional, 303
   – relativistic, 303
   – scalar product, 304
speed of light, 2, 299
spherical
   – coordinates, 6, 346
   – wave, 278
state space
   – phase space, 182
statistical equilibrium, 187
step function, 352
Stokes's theorem, 350
strange attractor, 291
streamline, 275
string tension, 256, 260
subharmonics, 284
superposition principle, 11
symmetric top, 223
symmetry
   – orbital, 64
system
   – body and space coordinates
       (rigid body), 197
   – closed, 36

tangential acceleration, 9
target
   – scattering, 80
tensor
   – field
       – electromagnetic, 340
   – inertia, 211
   – metric, 303
theorem
   – Euler (rigid body), 219
   – Gauss's, 349
   – Green's, 350
   – Liouville, 182
   – Stokes's, 350
   – parallel axis, 214
Thomas precession, 326–332

time, 1
   – dilation, 324
   – independence
       – Noether's theorem, 165
   – proper, 324
   – reversal, 312
timelike distance, 301
top
   – heavy, 226–230
       – Lagrange function, 226
   – symmetric, 223
   – torque-free symmetric, 223–226
torque, 27, 37, 232
total
   – cross section, 83
   – differential, 33
   – energy, 31
       – kinetic, 38
   – mass, 35, 61
   – momentum, 35
       – angular, 36
trajectory
   – phase space, 289
   – scattering, 86
transformation
   – Lorentz, 307
       – boost, 315, 320
       – exponential representation,
        315
       – generators, 314, 315
       – homogeneous, 308
       – infinitesimal, 313
       – inhomogeneous, 308
       – non-orthochronous, 312
       – orthochronous, 312
       – rotation, 315, 317
   – Poincaré, 307, 308
   – canonical, 177
       – generator, 177–180
   – contact, 176
   – electromagnetic field (boost), 342
   – gauge, 159, 167
       – electrodynamics, 160
   – infinitesimal, 163
   – linear, 5, 307
   – orthogonal, 199, 213

– point, 116
– principal axis, 216–219
– similarity, 213
transient, 24, 282, 287
translational
  – energy
    – rigid body, 211
  – invariance, 2, 136
tropical cyclone, 58
tropics of cancer and capricorn, 234
two-body
  – collision, 80
  – force, 34

unit
  – dyad, 4
  – vector, 3
unitary representation
  – rotation, 205
universality, 294, 296
  – Feigenbaum δ, 286
  – period doubling, 285

variable mass, 12
  – recoil force, 15
variational calculus
  – fundamental lemma, 144
vector
  – Runge–Lenz, 71
  – abstract, 3
  – acceleration, 8
  – component, 4
  – eccentricity, 71
  – length, 4, 199

– operator, 205, 207
– position, 3
– representation, 3, 4
– unit, 3
– velocity, 7
– wave, 278
vector-operator identities, 349
velocity
  – angular, 207
  – boost, 320
  – escape, 74
  – field, 266
  – four-velocity, 324
  – generalized, 109
  – potential, 276
  – precession (Earth), 235
  – precession (Moon), 235
  – vector, 7
  – wave, 257
virtual (instantaneous) displacement, 97
viscosity, 270, 272

wave
  – equation
    – force-free string, 257
    – sound, 277
  – longitudinal, 278
  – plane, 278
  – spherical, 278
  – vector, 278
  – velocity, 257
work, 30, 31
  – path-independent, 33

www.ingramcontent.com/pod-product-compliance
Lightning Source LLC
Chambersburg PA
CBHW071727150726
47998CB00005B/1534